Valves Revisited

by
Bengt Grahn, SM0YZI

Radio Society of Great Britain

Published by the Radio Society of Great Britain, 3 Abbey Court, Fraser Road, Priory Business Park, Bedford, MK44 3WH.

First published 2011

© Radio Society of Great Britain 2011. All rights reserved. No part of this publication may be reproduced, stored in a retrieval system, or transmitted in any form, or by any means, electronic, mechanical, photocopying, recording or otherwise, without the prior written agreement of the Radio Society of Great Britain.

ISBN 9781 9050 8670 9

Publisher's note
The opinions expressed in this book are those of the author and not necessarily those of the RSGB. While the information presented is believed to be correct, the author, publisher and their agents cannot accept responsibility for consequences arising from any inaccuracies or omissions.

Cover design: Kim Meyern
Typography, editing and design: Mike Dennison, G3XDV of Emdee Publishing
Production: Mark Allgar, M1MPA

Printed in Great Britain by Nuffield Press Ltd of Abingdon

Contents

1. Evolution of the Valve 1
2. How does a Valve Work? 7
3. Characteristics of Valves 25
4. Connecting Stages Together 41
5. Tuned Circuits 53
6. Amplifiers 69
7. Modulation 95
8. Receivers 103
9. The Superhet in Detail 113
10. Designing a Receiver 141
11. Hi-fi Amplifiers 147
12. Construction with Valves 175
13. The Power Supply 181
14. Oscillators 193
15. A Signal Generator 205
16. Measurements 223
17. Fault Finding 239
18. Transmitters 249
19. Further Information on the Web 257
    Appendix: European Valve Designations 263

Foreword

LITTLE DID I KNOW when in the 1950s, at the age of ten, I stepped into one of the local radio shops to ask for scrap items, how important radio would come to be to me. I began reading about radios, building radios (starting, of course, with a crystal set). Many of my friends think that I am crazy to have collected valve radios from the tender age of 50.

However, there is something irresistibly fascinating about valves and valve technology. The best evidence of that is undoubtedly the fact that so many people around the world are, somehow, involved in valve technology, plus the fantastic prices some people are willing to pay for a valve amplifier.

Lately, thermionic valves seem to be going through a renaissance. More and more companies specialise in valves and valve accessories. Undoubtedly, therefore, there is a need for facts about these fascinating components.

However, trying to find modern literature on the subject is not easy. Believe me, I have tried, not least at car boot sales in England and Wales. So, in desperation, I decided to fill the gap by writing a book myself.

It is my sincere hope that it will be of use to its readers.

Bengt Grahn
Bergshamra, Sweden, 13 April 2011

Editor's note:
The terms 'aerial' and 'antenna' have been used interchangeably in this book. Although 'aerial' is now considered archaic, it will still be encountered in the labelling and documentation of domestic valve radios. Similarly, the terms 'overtone' and 'harmonic' are interchangeable in the context used here. In the schematic drawings the old 'zig-zag line' symbol is used for resistors, rather than the more modern 'rectangular box' symbol. Again this was felt to be in keeping with the documentation associated with the valve era.

1

Evolution of the Valve

THE VALVE WAS BORN out of an ordinary lamp. In fact, for a while, a valve was called "lamp" or "bulb". Sir John Ambrose Fleming (1849 - 1945) found that lamps (which at the time had a carbon filament) deposited a dark layer on the inside of the bulb. He wanted to find out how and why. He inserted a metallic electrode inside the bulb. What he found was that electrons could be captured by the new electrode if it was connected to a positive voltage in relation to the filament. He found a current flow through the lamp. One of Fleming's valves is shown in **Fig 1**.

The filament was given the name 'cathode' and the new electrode 'anode'. These are names that have survived even until today. The diode was used already before Fleming's patent as a detector of radio waves in experiments carried out in Fleming's lab.

Valves can be directly or indirectly heated or of the cold cathode type, depending on how/if they are heated. In a directly heated valve, the filament doubles as the cathode (like in Fleming's lamp and in more recent battery powered valves). The filament is coated with an agent that easily emits electrons when heated. The drawback is that directly heated valves must be powered by a DC source to avoid hum entering the signal path. So, the indirectly heated valve was developed, where the cathode is a cylinder that surrounds the filament which heats it. The principle reduces filament hum considerably. These days most valves are indirectly heated.

The American Lee de Forest (1873 - 1961) added in 1906 yet another electrode to Fleming's valve - a grid shaped structure, which he placed between the anode and the cathode. This became what now is known as a triode. De Forest called it the 'audion'. It was patented in 1907. A gadget was born which could rectify, amplify and oscillate.

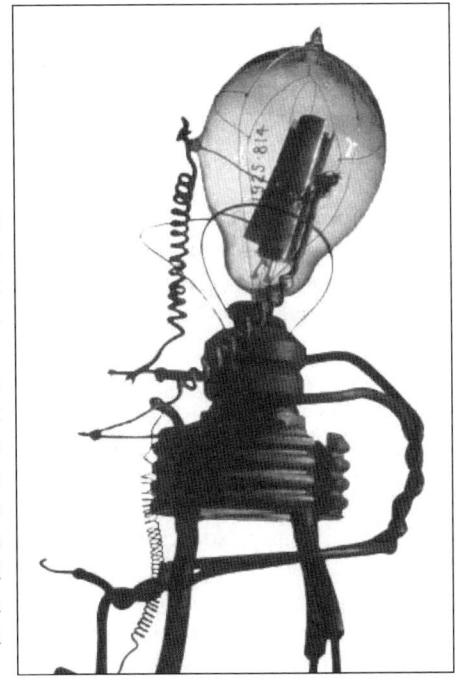

Fig 1: One of Fleming's valves

1

Fig 2: Restored 1941 Canadian General Electric KL-500 4-band radio [source: Wikipedia]

Radio Receivers

All valves were directly heated in the very beginning. The shortcomings of the triode (see below) caused the receivers of those days to be insensitive and their selectivity was bad. Various tricks were invented to overcome those shortcomings and to make radio receivers more efficient. Positive feedback was implemented to increase selectivity and sensitivity. However, a stage with positive feedback increases the noise level. Positive feedback (reaction) played an extremely important role in radio reception in the 1920s and early 1930s. It was the only way in which a receiver could be built with sufficient selectivity and sensitivity. All domestic radios in those days were fitted with some sort of reaction. The degree of reaction could be set with a control on the front panel. In advanced radios, more than one stage had a reaction control. In addition, the tuned circuits, often more than one, had to be tuned to the frequency of interest. Then there were volume controls, band switches, and what have you. Several aerial (antenna) inputs allowed you to choose the best one for reception. The set needed to be properly grounded too. Listening to the radio in those days was a task for the technically-minded.

Receivers designed during the 'golden age' of valve technology, the 1960s, use completely different techniques.

The so called reflex coupling was invented to save the number of (then expensive) valves in a receiver. The principle is that one valve could be used for both RF and AF amplification, so, after detection, AF was channelled back to the beginning of the chain of valves and then separated at the end. However, making the receiver stable could be a problem, because of residual RF.

One or more tuned circuits were added to improve selectivity. This meant in the early days of radio that the front panel could contain a host of controls and the receiver therefore became more difficult to operate. Most of the controls had to do with the tuning of the receiver. Few, if any, ganged variable capacitors were available, so each filter had to be provided with a separate control.

Another control worth mentioning was the volume control. It was a rheostat - a wire wound potentiometer for high currents - which was connected into the heater chain so that the heater current could be controlled, and thence the temperature of the cathode, the cathode emission, and therefore the volume. These days this method of volume control is not recommended since modern valves are designed to work within a fairly narrow band of heater voltages and temperatures and may be damaged if made to work outside their working range.

In the beginning, all receivers were battery powered, and in most cases an accumulator provided the filament current. This meant that the receiver was big and heavy. The loudspeaker was frequently a separate unit, and sometimes there was also sometimes a frame aerial incorporated into the box.

A receiver from the 1920s was, however, a beautiful creation - a black or wood coloured box. The front panel had big knobs, graded 1 - 100. Sometimes the valves were mounted on the panel too, so you could see them glow. There was also frequently a lid or a 'service door' to allow for servicing. Some radios were built such that the chassis could slide in and out of the box.

Valves for battery power are directly heated. They start immediately to function, just like the transistor radio. 45V or 90V anode battery was standard, but other anode voltages were also common. As a matter of fact, a car radio was made where the valves were powered with 12V anode voltage. In some cases, bias voltages were taken from taps in the array of battery cells that constituted the high voltage source.

It is not surprising to learn that people build valve-based electronics. It is more surprising to learn that people are actually making home-made functioning valves even today.

Valve Evolution

The invention of the thermionic valve is ascribed to Fleming. However, Thomas Alva Edison (1847 - 1931) was the real inventor. Other scientists, too, studied the phenomenon of electron emission in a vacuum, such as Eugen Goldstein, Nikola Tesla and Johann Wilhelm Hittorf. Edison couldn't find any applications for the valve, so he didn't bother to patent it. Fleming invented the Kenotron diode in 1904. The actual design of the original valve is better shown in **Fig 3** which comes from Fleming's patent application.

The valve has since then been developed and refined, and development is still going on.

Despite popular belief, the valve is still alive and kicking, even though it has gone through some major changes. Today it is perhaps most often used in audio amplifiers in expensive and exclusive equipment (according to some people, semiconductors have no right to exist in audio amplifiers - not even in the power supply). This is true for RF valves too. A valve amplifier gives its owner high status, and, in certain circles, a valve amplifier is the only possible option.

However, many people are also interested in valves for radio reception and measurement instruments, and this is mainly what this book is all about.

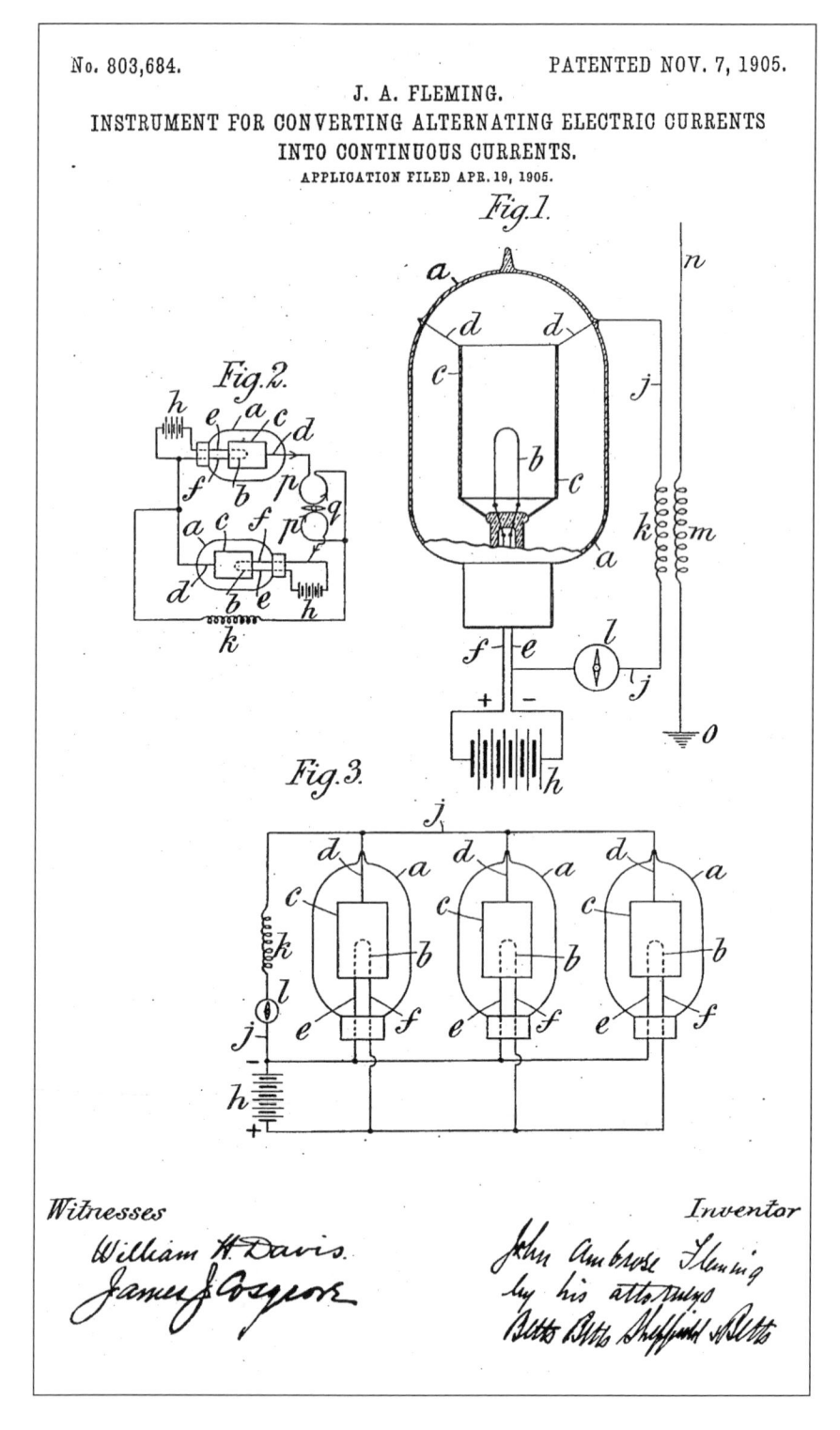

Fig 3: Extract from Fleming's patent application

CHAPTER 1: INTRODUCTION TO VALVES

Fig 4: sub-miniature valve

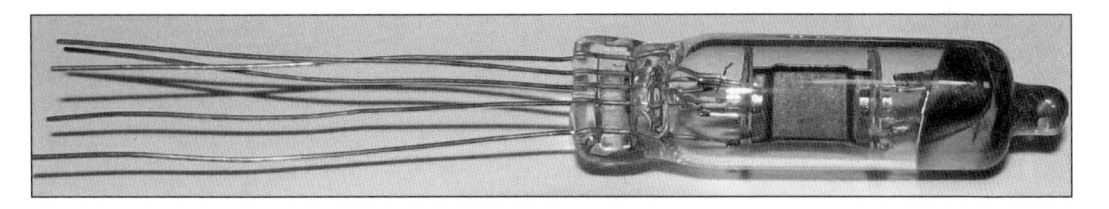

Miniaturising

Already in the 1940s, the valve designers began to feel the need to make smaller valves. They designed miniature and sub-miniature valves. One of the battery powered sub-miniature triodes that were made is the DC70, and a variety of pentodes were also made. Additionally, power pentodes were made that could be used as power amplifiers in transmitters, in receivers and in hearing aids. Their small size made it possible to reduce the inner capacitances of the valve, thereby increasing their usefulness in RF applications. Low anode voltages could be used. They made excellent service in hearing aids and this is perhaps the best known application, even though commercial radio receivers were built using sub-miniature valves (see **Fig 4**). The valve is pen shaped, about 4cm long and 10mm in diameter. They are soldered in place, and the leads are 4cm long.

Yet another miniature valve was launched in 1959: the nuvistor. This is an even smaller creation, with a metal body on top of a ceramic base, as opposed to the glass bottle that housed sub-miniature valves. They were more difficult to manufacture (they were built by robots in vacuum) but are capable of even better performance than the sub-miniature valve. Even lower anode voltage and internal capacitances make them an excellent choice for wideband amplifiers and were included into Tektronix oscilloscopes. They were manufactured as triodes and tetrodes.

A nuvistor is also tube shaped and measures about 2cm by 1cm diameter. They are mounted in special holders as shown in **Fig 5**. The valve itself is shown in **Fig 6**.

Both nuvistors, nuvistor holders and sub-miniature valves can still be purchased at a decent price.

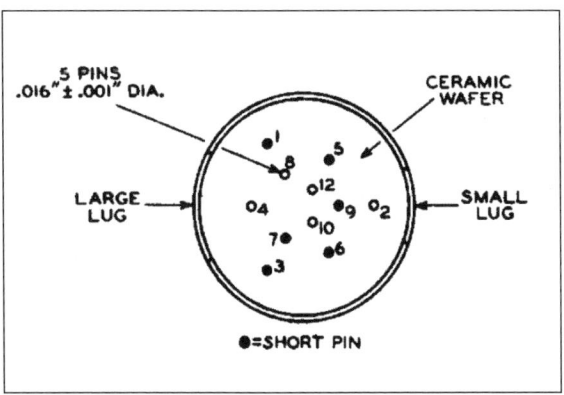

Fig 5: The 6CW4 socket

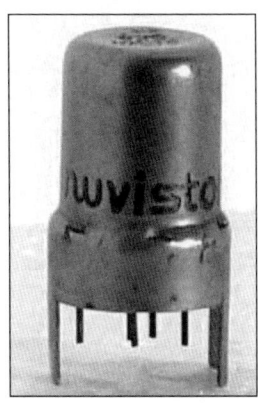

Fig 6: Nuvistor 6CW4

VALVES REVISITED

Fig 7: An acorn valve shown alongside a transistor [source: Wikipedia]

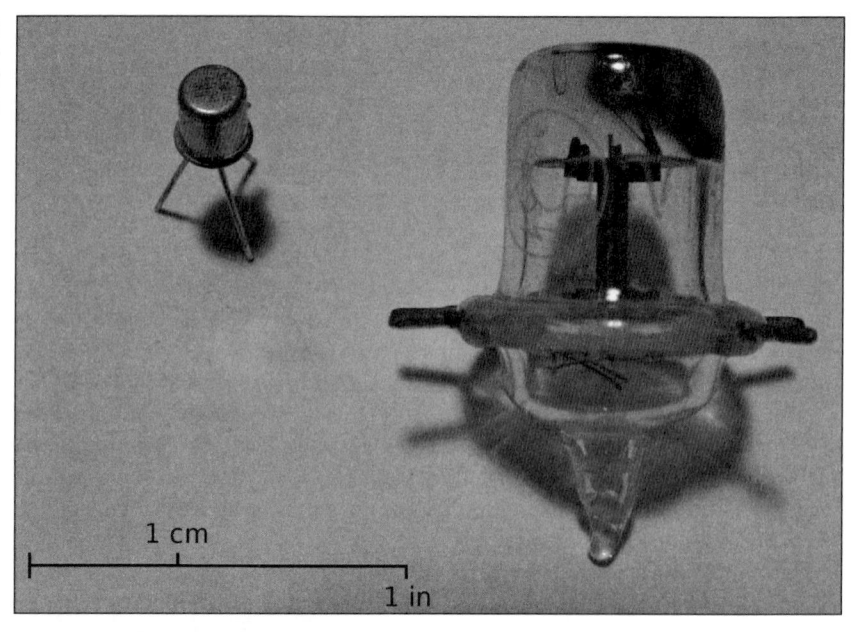

Fig 8: A Nixie valve

The need for higher frequencies and better performance necessitated development of valves with a somewhat different set of connections. **Fig 7** shows an acorn valve. Note that the electrodes are pulled out the side of the valve for lower inter-electrode capacitances.

The acorn valve was developed during the mid 1930s as an answer to the need for higher frequencies. They could handle frequencies up to 400MHz. It doesn't sound much today, but in those days, the number was mind-boggling.

They were made as triodes and pentodes, and there is actually nothing particular about the way they were used, except for the high frequency capability and the requirements that followed.

The list of special valves could be made a lot longer. Higher power handling capabilities was also included in the wish list. Air and water cooled transmitter valves were developed, valves that could display digits ('Nixie valves', **Fig 8**) for instruments such as frequency counters etc, tuning indicators, television screens, and a wide range of other valves came on the market.

2

So, How Does a Valve Work?

LET US BEGIN BY LOOKING at a directly heated diode, which is the simplest valve. At the centre of a thermionic diode is a filament which is heated when a voltage is applied to it. Surrounding the filament is a cylinder. When the filament is heated, it begins to emit electrons which gather as a cloud around the filament. They have nowhere to go, though. As you know, equal charges repel each other and different charges attract each other. Without anode voltage, though, there is no charge to attract the electrons, as shown in **Fig 9**. This cloud is called 'space charge'. **Fig 10** shows what happens when a voltage is connected between anode and filament/cathode.

Indirectly heated valves still require heat, so at the centre of those we will still find a filament. Around the filament, however, but inside the anode, there is another cylinder. This is the cathode. The surface of the cathode has a thin layer of material that easily emits electrons when heated. In an indirectly heated valve, the filament is usually powered by AC. If you connect an AC voltage to

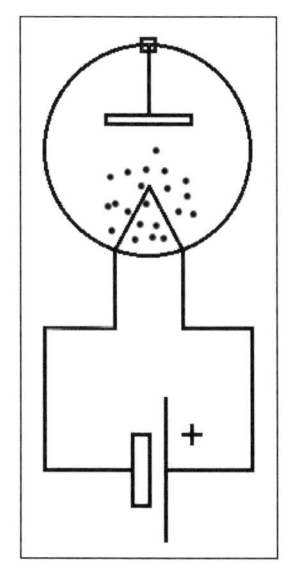

Fig 9: Directly heated diode
Fig 10: The anode is connected and a current flows.

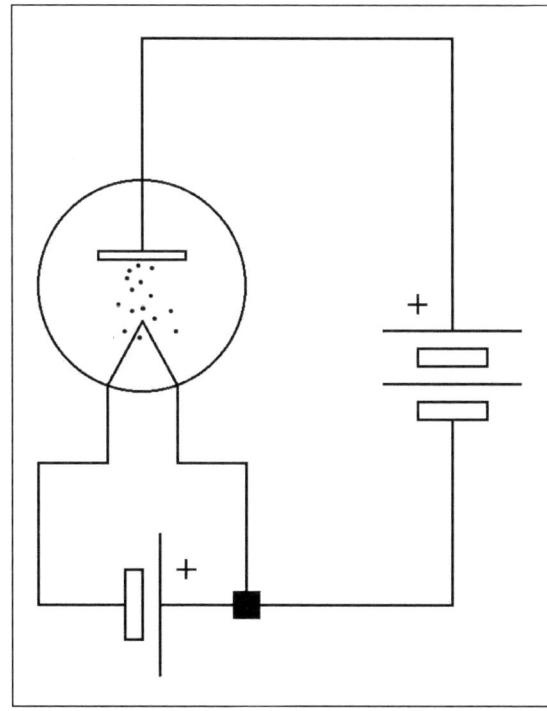

the filament of a directly heated valve, hum from the filament is added to the current through the valve. The advantage of an indirectly heated valve is that it reduces the hum from the filament to a minimum.

Then we connect a positive voltage to the anode and the negative side to the filament. The electrons then begin to move. They feel the positive field from the anode and move towards it. We get a current through the diode. According to conventions, the current flows from the anode to the cathode. This is because of a misunderstanding during the childhood of electronics which has remained until today. In reality, the electrons move from the filament (or cathode) to the anode.

The actual voltage at the anode doesn't matter - there is still a current flowing. However, the polarity does matter. If you connect the positive pole of the battery to the filament and the negative pole to the anode, the current flow stops. So, we have a rectifier.

The Diode

Fig 11 shows the symbolic representation of a semiconductor diode (Note that 'di' = two), and an actual component is illustrated in **Fig 12**.

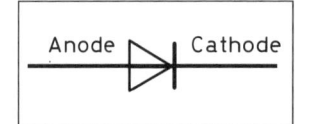

Fig 11: Symbol for a semiconductor diode

Fig 12: Actual implementation of a semiconductor diode

The valve symbol is shown in **Fig 13**, and looks like that shown in **Fig 14** in real life. The valve diode is bigger and has more connection pins. That makes it more awkward to connect. The EAA91 has two independent diodes in the same glass envelope and a seven-pin base.

The screen between the halves, and the fact that they have separate cathodes, enables the diodes to be used in quite different places in a radio without affecting each other's performance. See **Table 1** for a summary.

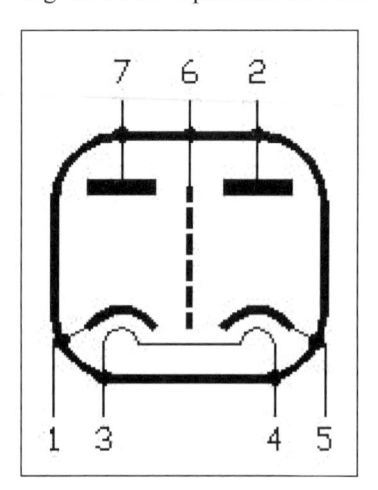

(left) Fig 13: Symbol for two diodes within the same valve (EB91)

(right) Fig 14: The EAA91, a practical implementation of a dual diode. The fingers holding the valve illustrate its size

> **Fact Sheet - Diode**
> **Applications:** Detection, rectification
> **Internal capacitances:** Varies

Table 1: Diode fact sheet

If we begin by looking at the valve symbol (most people these days probably recognise the semiconductor symbol), we find that the anodes are drawn at the top (pins 2 and 7). Pin 6 is connected to the screen between the two diodes to reduce the influence between them as much as possible. Pins 1 and 5 are the cathodes, and 3 and 4 are connected to each end of the filament.

The semiconductor diode is undoubtedly the most common diode these days. However, let us take a closer look at the differences between the semiconductor diode and the valve diode. I will use the 1N4148 and EAA91 for comparison, since both are small signal diodes and are frequently used as signal detectors.

Looking at the curves for the two, **Fig 15** in the valve case and **Fig 16** for semiconductor diodes, we find a number of significant differences:

1. The EAA91 curve passes through zero. This means that when the forward voltage across the diode is zero, the current is zero. This is not the case of the semiconductor diode, where the current remains zero until the voltage across the diode is about 0.6V. This means in turn that the semiconductor diode is not very well suited for small signals, which is the case with the valve diode.
2. The semiconductor diode is strongly temperature dependent, as opposed to the valve diode.

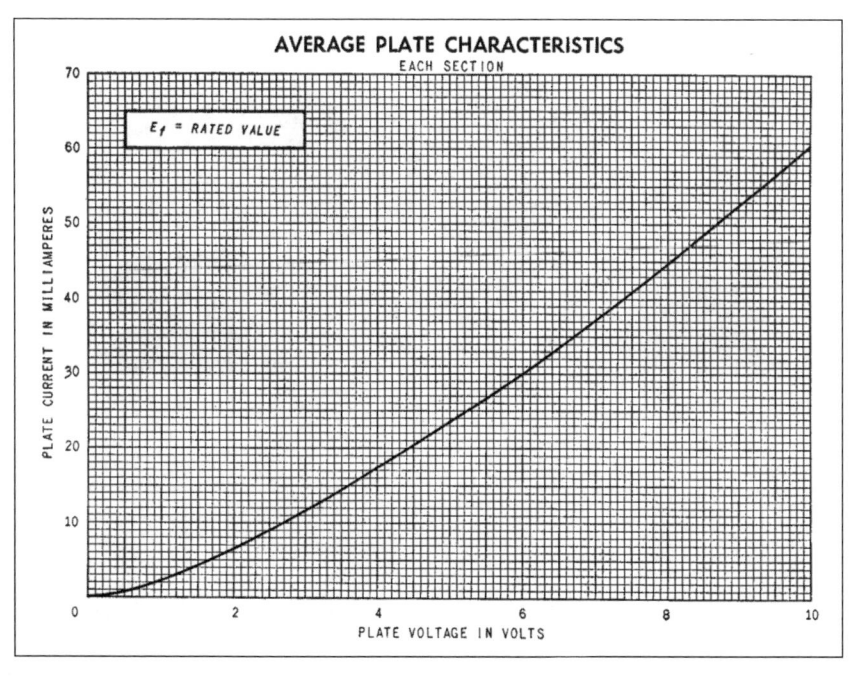

Fig 15: Transfer function for the EAA91 valve

3. The lower part of the semiconductor curve has a bend, which causes distortion. Even though the valve diode has a bend too, it is not as prominent as that of the semiconductor.
4. The semiconductor can handle higher current.
5. The valve diode can handle higher voltages.

Fig 16: Transfer function of the 1N4148 semiconductor

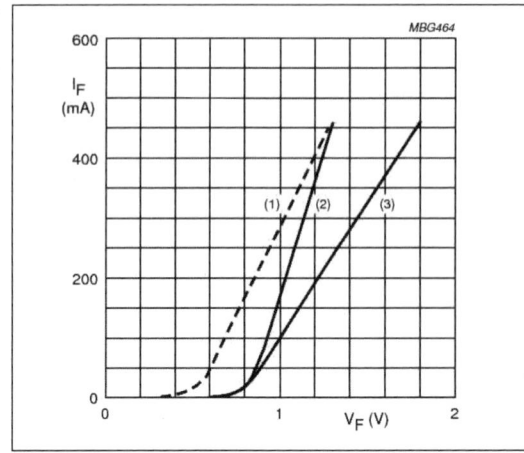

The anode voltage (Va) is shown along the X-axis, and the anode current along the Y-axis. To determine the anode current for a specific anode voltage, first find the anode voltage along the X-axis.

Then follow the grid lines upwards until you meet the curve. A horizontal line to the left from that point gives you the anode current.

The Triode

The triode (tri = three) was invented by the American Lee de Forest (1873-1961), who called it the 'audion'. It is shown in **Fig 17**. A modification of the audion became what we now know as the triode.

What happened when another electrode was placed inside the valve between the cathode and the anode? Well, it turned out that varying the voltage (in relation to the cathode) of the third electrode made it capable of varying the current through the valve, as long as the third electrode was negative in relation to the

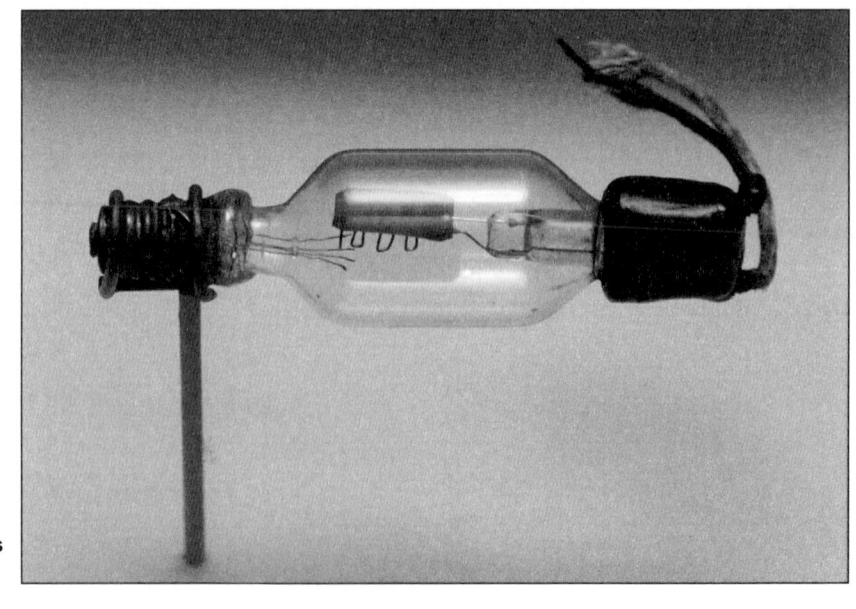

Fig 17: de Forest's audion, 1903

cathode (or filament). The third electrode didn't draw any current. It turned out that the most efficient shape for the electrode was a grid. so it was named the 'grid'. Since this electrode controls the current through the valve, it was called the 'control grid' (or G1). A symbolic representation of a dual triode is shown in **Fig 18**.

Eventually, more and more grids were added, for different reasons. We shall take a look at most of them later.

A triode has the following qualities:

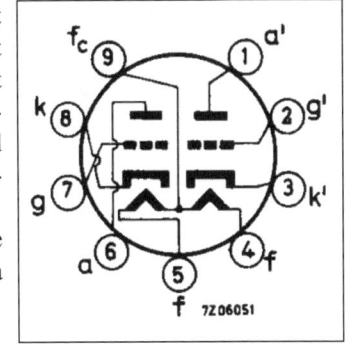

Fig 18: Symbol and connections for a dual triode (ECC82)

- Low internally generated noise.
- Relatively low amplification. The amplification factor depends on the anode voltage.
- Relatively low anode impedance.
- Relatively high capacitances between its electrodes. UHF amplifiers often need to be neutralised.

The fact is that a triode works against itself. The anode current increases as the voltage at the grid becomes less negative. This current needs to be converted to a voltage in preparation for the input of the next stage. A common way of doing this is to connect a resistor between the anode and the power supply. The increased current causes the voltage drop across the anode resistor to increase. The voltage drop subtracts from the power supply voltage, causing the anode voltage to decrease. The amplification of the stage decreases as the anode voltage decreases. This effect limits the possible amplification you can get from a triode.

Fig 19: Ia / Vg-curve for a triode (ECC83)

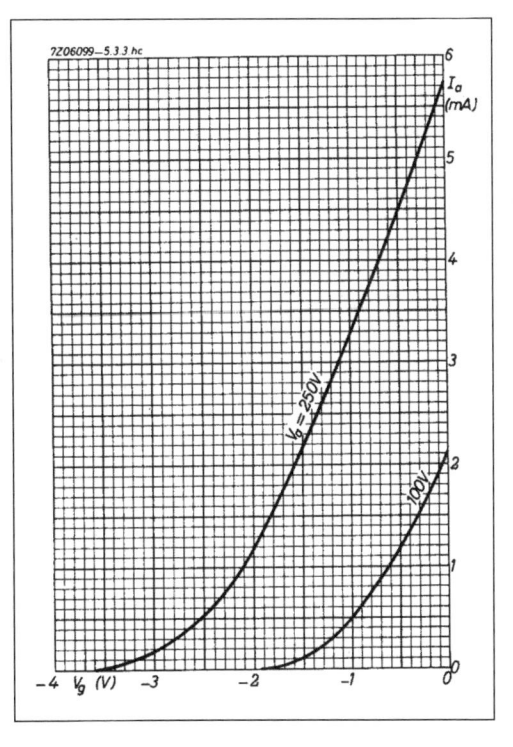

There are methods of getting around the drawbacks (the three last points above), and we will take a closer look at them. It is the first quality - the low noise - that makes triodes so useful in various applications. Noise in various valves is dealt with later.

Under certain circumstances, the inter-electrode capacitances of a triode result in a phase relationship and a coupling between input and output which can cause the stage to oscillate. This can be corrected by neutralisation.

Neutralisation is done by 'compensating away' the internal capacitances with an external capacitor or an inductance. By varying the capacitance or inductance, the phase relationship is also varied so that the stage stops oscillating. This made it a bit more complicated to build stages with single grounded cathode triodes for higher frequencies. Other solutions, as we shall see, addressed the problem in different ways (see the sections about RF and IF amplifiers).

> **Fact Sheet - Triode**
> **Applications:** Detection, amplification
> **Internal capacitances:** High
> **Amplification factor:** Medium
> **Anode resistance:** Medium
> **Benefits:** Low noise
> **Drawbacks:** Amplification dependent on anode voltage.

Table 2: Triode fact sheet

The dependence between amplification and anode voltage can be seen in the curves of **Fig 19**. Here you can clearly see that the slope of the curve varies with varying anode voltage. The higher the anode voltage, the steeper the slope. In other words, the higher the voltage, the higher the amplification. The same thing is shown in the table of typical values in the data sheet. At 100V anode voltage, the transconductance (amplification factor) is 1.25mA/V, and at 250V it is 1.6mA/V. **Table 2** gives a summary of triodes.

One interesting triode that was designed when car radios became more common is the ECC86. It is specified for anode voltages up to 25V, and is designed to run directly off a car battery, 12V or 6V.

Valve Capacitances

A capacitor consists of two isolated metallic objects, separated by a dielectric. A valve contains metallic objects, the cathode, the grid and the anode. Due to their metallic nature, they form capacitances. In a triode, the most important capacitances are the ones between the anode and the grid, and between grid and cathode. Additionally, there is one between the anode and the cathode. The most important one of these is the grid-anode capacitance. The inter-electrode capacitances in a triode are all in the order of one or two picofarads.

Miller Effect

The Miller Effect (discovered and described by the American John Milton Miller in 1920) causes the input capacitance to increase by the amplification of the stage. The capacitances can reach such magnitudes as to affect the bandwidth of the stage even in AF applications. So, a countermeasure would be to decrease the amplification of the stage. Also the stray capacitances (illustrated in **Fig 20**) of the stage are subject to the Miller effect.

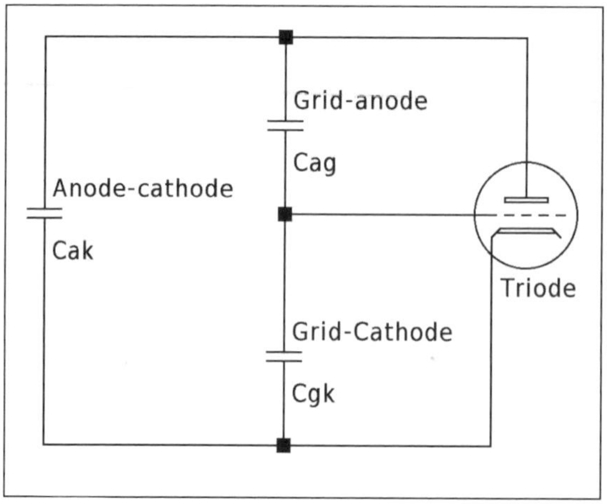

Fig 20: The internal capacitances of a triode

CHAPTER 2: SO, HOW DOES A VALVE WORK?

Other countermeasures would be to:

- decrease the output impedance of the previous stage
- connect a couple of triodes in a cascode or cathode coupled configuration

A wider consequence is that valves for RF applications have to be built with smallest possible input capacitances. An RF stage also needs the smallest possible stray capacitances.

The increase in input capacitance is given by:

$C_M = C * (1 - AV)$

where AV is the voltage gain of the amplifier and C is the feedback capacitance.

The Tetrode

In order to reduce the capacitance between the grid and the anode, a fourth electrode was inserted (tetra = four), as shown in **Fig 21**, by Walter Hermann Schottky (1886 - 1976).

This new electrode is another grid, called the 'screen grid'. As we all know, connecting two capacitors in series yields a total capacitance smaller than the smallest one of the two.

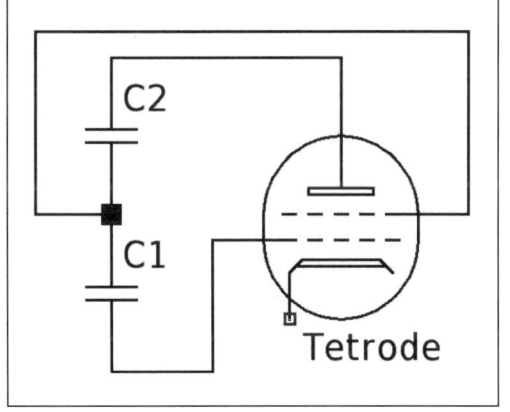

Fig 21: Grid-anode capacitances of a tetrode

In this case, the grid-anode capacitance is made up of the two capacitances C1 and C2.

Secondary Emission

Secondary emission is a phenomenon which occurs when an electron from the cathode strikes the anode. The electrons travel through a valve at high speed, and it is inevitable that each electron hits an atom and knocks one or more electrons from the atom.

Those 'secondary' electrons form a cloud near the anode. In a triode, the secondary emission is not important, because the anode is the electrode with the highest positive voltage, much higher than the cathode and control grid. All secondary electrons are pulled back to the anode.

The situation is different in a valve with a screen grid. Because the voltage of the screen grid is slightly below or equal to that of the anode, not all of the secondary electrons will be pulled back to the anode.

Some of them travel to the screen grid instead. As a result, the screen grid current increases, and the anode current decreases. This will cause distortion of the signal.

A tetrode has a higher output impedance than a triode. Additionally, the tetrode has a higher amplification and the anode current (and thereby the amplification) dependency on the anode voltage is nearly eliminated.

Fig 22: Ia / Va curve for a tetrode (RENS1204)

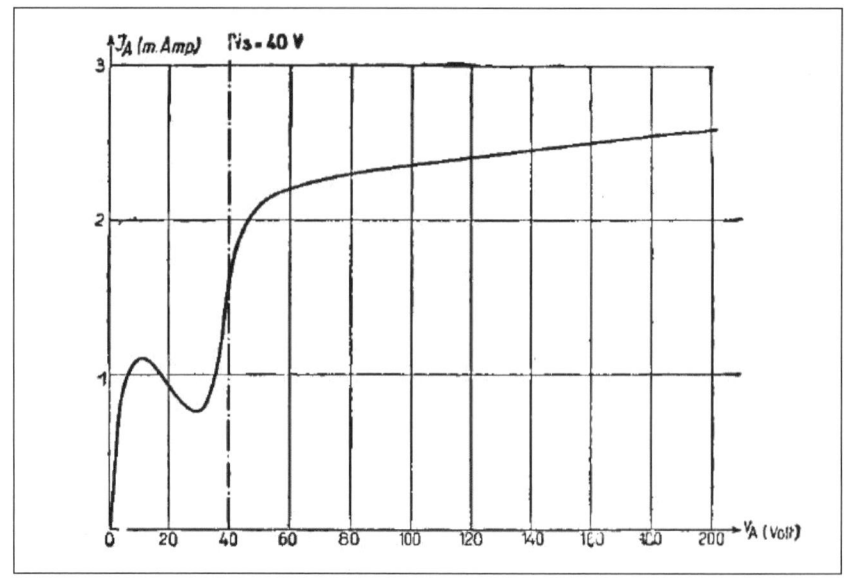

However, a tetrode does have shortcomings. At low anode voltages, the tetrode becomes unstable. This can be deduced from the curve in **Fig 22**.

The curve shows the relation between the anode voltage and anode current at a fixed control grid voltage. Compare this curve to **Fig 23**, the corresponding curves for a triode. The curve is nearly horizontal at anode voltages above about 40V for the tetrode. In other words, the tetrode's anode current is nearly independent on the anode voltage. In the triode, however, the anode current depends heavily on the anode voltage.

Also, some of the electrons from the cathode end up being absorbed by the screen grid, which causes a grid current to flow. The cathode current is consequently equal to the sum of the anode current and the screen grid current.

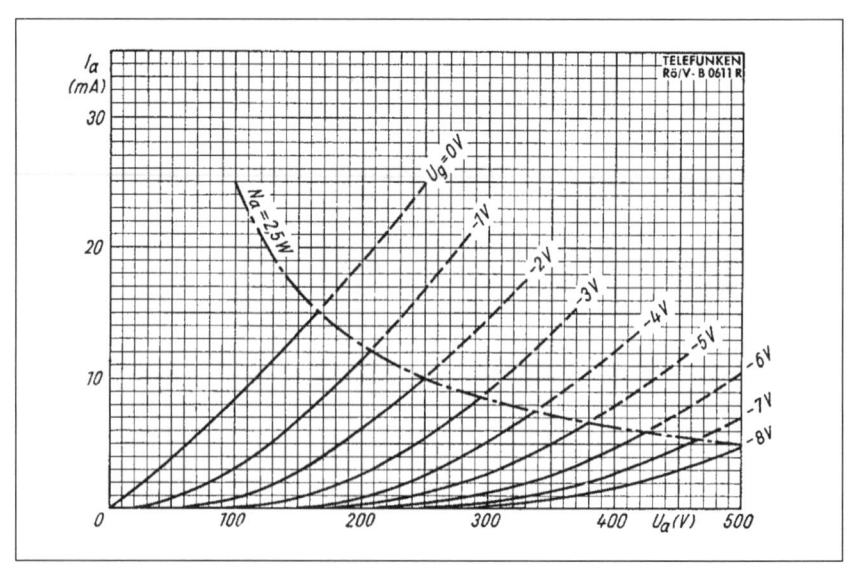

Fig 23: Ia / Va curves for a triode (ECC81)

Another consequence of this is that the internal noise of a tetrode is higher than that of a triode.

Since the stability of a tetrode depends on the stability of the screen grid voltage, the screen grid has to be properly decoupled.

The best known tetrode among radio amateurs would be the legendary 807, shown in **Fig 24**. It was mainly used as the power amplifier in transmitters. These days many people design an 807 into their power amplifiers in audio equipment. Its base connections are shown in **Fig 25**.

As you can see from the tetrode curve, the anode current decreases between 10V and 30V anode voltage. This constitutes a negative resistance which causes instability. Tetrodes other than the one shown in the curve have this area at higher anode voltages, eg between 50V and 150V.

The fact sheet shown in **Table 3** summarises tetrodes.

One method of coping with the problem (without really solving it completely) was to insert a 'box' of metal between the screen grid and the anode.

The box had two slits through which the electrons could pass on to the anode. The box was internally connected to the cathode. However, one of the best solutions turned out to be inserting yet another grid in place of the box. The Pentode was born.

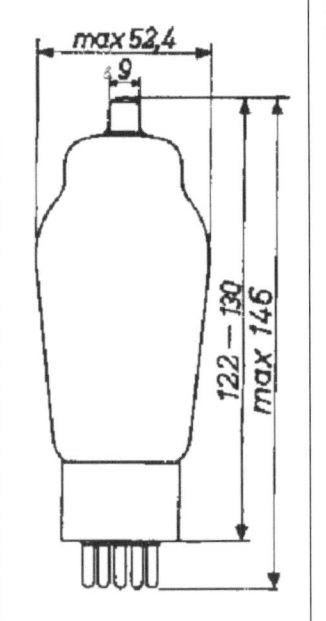

Fig 24: Power tetrode 807

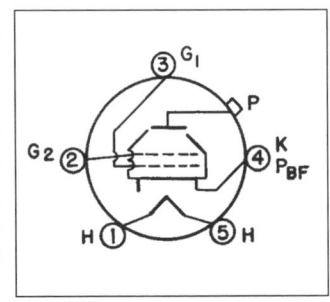

Fig 25: The tetrode symbol and its pin connections

Fact Sheet - Tetrode
Applications: Generally power amplification
Internal capacitances: Medium
Amplification factor: High
Anode resistance: High
Benefits: Amplification independent of anode voltage
Drawbacks: Tendency to instability, 'kink' in curve

Table 3: Tetrode fact sheet

The Pentode

The pentode (penta = five) was invented by the Dutchman Bernard D H Tellegen (1900 - 1990). **Fig 26** show its curves. The pentode has the advantages of the tetrode, but lacks the negative resistance area. Therefore, the pentode is more stable than the tetrode. The inserted grid is called 'suppressor grid'. In some

VALVES REVISITED

Fig 26: Curves for RF pentode EF80

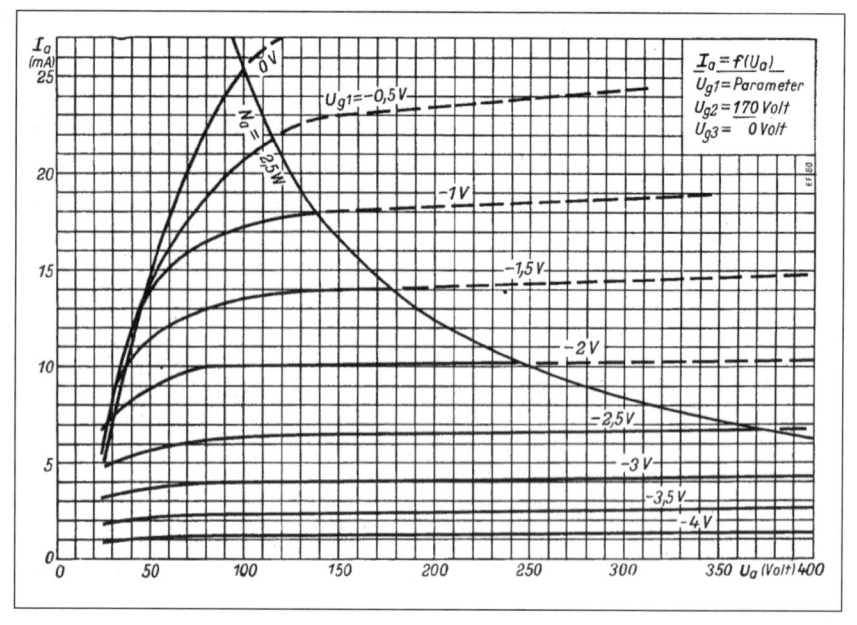

pentodes, this grid is internally connected to the cathode. In others, it is connected to a separate pin in the socket. There are advantages with having access to the suppressor grid. In certain mixer designs, for instance, one signal is connected to the control grid, and the other, usually the oscillator signal, is connected to the suppressor grid.

The sensitivity of the suppressor grid is far lower than that of the control grid, since it is further away from the cathode. **Table 4** summarises pentodes in general.

The low sensitivity of anode current to anode voltage makes the pentode ideal for voltage stabilisation and similar applications. Due to its high output impedance and its low current dependence of anode voltage, it is ideal for generating sawtooth signals and was frequently used in oscilloscopes.

Note that power pentodes, such as those in final stages of an audio amplifier, tend to have a low output impedance instead. Of course that simplifies interfacing to the low impedance speaker. As an example, ECL82, a triode-pentode, has an impedance (Z_{out}) of 5.6kΩ. The EL34 (which is popular in audio applications) has Z_{out} at the lowest 2kΩ. A pentode symbol is shown in **Fig 27**.

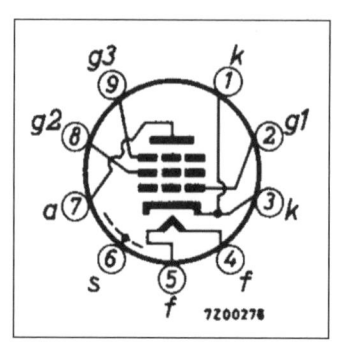

The high noise level makes the pentode less suitable for mixers or RF amplifiers in a receiver or input stages in AF amplifiers.

Pentodes were, however, frequently used in BC (broadcast) receivers (domestic radios) because of their high amplification. They gave more amplification per penny. Performance came second.

The situation is different in communications receivers and amateur radio receivers where performance is more important. For instance

Fig 27: Pentode symbol (EF80)

> ### *Fact Sheet - Pentode*
> **Applications:** Amplification, RF, IF and AF, current generation
> **Internal capacitances:** Low
> **Amplification factor:** High
> **Anode resistance:** High
> **Benefits:** High amplification factor
> **Drawbacks:** Noisy

Table 4: Pentode fact sheet

the Collins professional communications receiver 51-S1 has two triode mixers (first and second) and one pentode mixer (third).

When the desired signal has been sufficiently amplified, the additional noise added by a multi-grid valve has little or no influence on the signal quality. It is therefore possible to benefit from the high amplification and high output impedance of pentodes late in the amplifier chain. High impedance means high Q of the tuned circuits, which in turn means better selectivity.

A power pentode (**Table 5**) is different. It has the ability of handling higher power levels and has a lower anode resistance.

Pentode connections

Various experiments have been carried out to find out how different connections of the grids affect performance. Since the pentode has three grids and two additional electrodes (anode and cathode), it can be connected in five different ways.

- All grids to the anode (in which case it functions as a diode) (see **Fig 28**)
- G2 and G3 to the anode - signal to G1, in which case it functions as a triode
- Using G1 and G2 as inputs and G3 connected to the anode, in which case it functions as a high μ triode
- Connecting G2 and G3 together, in which case it functions as a tetrode
- Using G2 as the input, connecting G1 and G3 to positive voltages, in which case it functions as a space-charge-grid valve

> ### *Fact Sheet - Power Pentode*
> **Applications:** Power amplification, RF and AF, voltage stabilisation
> **Internal capacitances:** Low
> **Amplification factor:** High
> **Anode resistance:** Low
> **Benefits:** Easy to interface to a loudspeaker
> **Drawbacks:** Noisy

Table 5: Power pentode fact sheet

VALVES REVISITED

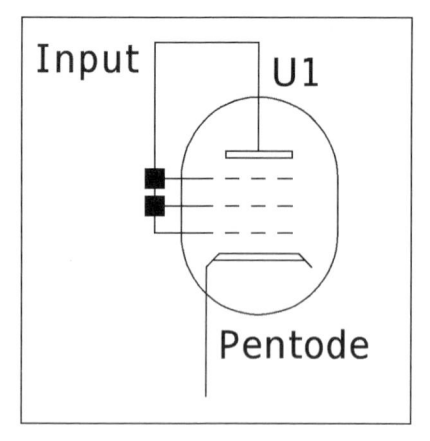

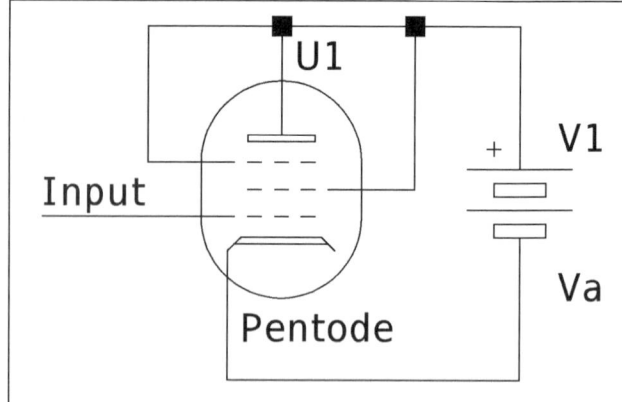

Fig 28: Pentode in diode connection

Fig 29: Pentode in triode connection

With the pentode valve connected as in Fig 28, it acts as a diode. This configuration is not very common. Connecting the grids to the cathode instead is not a good idea, since the grids don't emit electrons. In the triode configuration in **Fig 29** they assist the anode in capturing the electrons emitted from the cathode.

It is fairly common among audio amplifier enthusiasts to connect a pentode as a triode in the output stage. Some valve manufacturers quote valve data for their pentodes in triode connection. G3 could also be connected to the cathode, as is the normal case for a pentode. **Fig 30** shows an extract from the Philips data sheet for the EL34. These data are valid for G3 still connected to the cathode.

```
Operating conditions in triode connection
(g2 connected to anode)
Caractéristiques d'utilisation en connexion triode
(g2 relié à l'anode)
Betriebsdaten in Triodenschaltung
(g2 verbunden mit Anode)
```

		Class A Classe A Klasse A	Class AB Classe AB Klasse AB		
V_b	=	375	400		V
V_{g3}	=	0	0		V
R_k	=	370	220		Ω
$R_a\sim$	=	3	–		kΩ
$R_{aa}\sim$	=	–	5		kΩ
V_i	=	18,9	0	22	V_{eff}
I_a	=	70	2x65	2x71	mA
W_o	=	6	0	16,5	W
d	=	8	–	3	%
$V_i(W_o=50mW)$	=	1,7			V_{eff}

Fig 30: Extract from EL34 data sheet

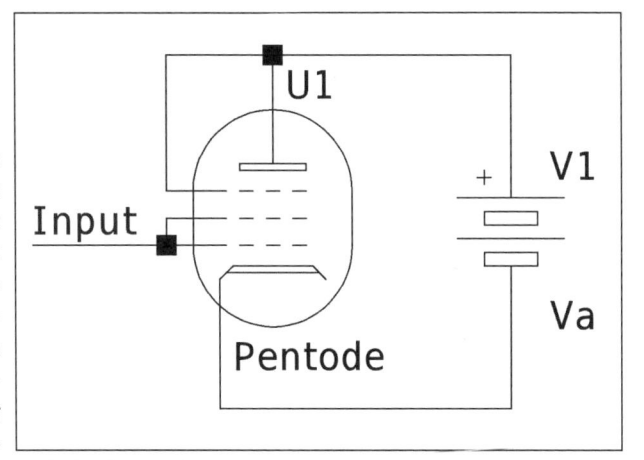

Fig 31: Pentode in hi-mu triode configuration

The connection in **Fig 31** yields a lower distortion and lower output power than a normal pentode connection, although the triode connection allows for higher input voltages (operating Class A in both cases)

This is the high-mu (μ) triode connection of a pentode. Both G1 and G2 participate in controlling the electron current through the valve. G3 and the anode both catch the electrons.

A pentode can also be connected as a tetrode, as in **Fig 32**. This is the tetrode connection of a pentode. The two grids G2 and G3 both act as one, the G2 of a tetrode. AF output valves are sometimes connected like this.

The space-charge-grid valve connection requires some additional explanation. The space charge in a valve is the cloud of electrons that forms around the cathode. It has an important influence on the properties of any valve and is the cause of the bend at the bottom of the Ia/Va curve. In other words, the electron cloud forms an obstacle between the cathode and anode at low anode voltages.

If a (strong) positive voltage is connected to G1 (the control grid), the grid closest to the cathode, the next grid in order (the screen grid) can be negatively biased instead and used as the control grid. This 'removes' the space charge from the cathode, and the cloud is instead formed between the positive G1 and negative G2. This connection yields a high mutual conductance, but, on the other hand, the characteristics of the stage becomes highly non-linear.

A valve designed for this configuration is named 'space-charge grid valve'.

Only a few special space-charge-grid valves were made, such as FP265, FP285, FP400, and a couple more. The valves were made by General Electric and developed for very special purposes, such as RF power amplifiers and highly specialised measurement and research purposes.

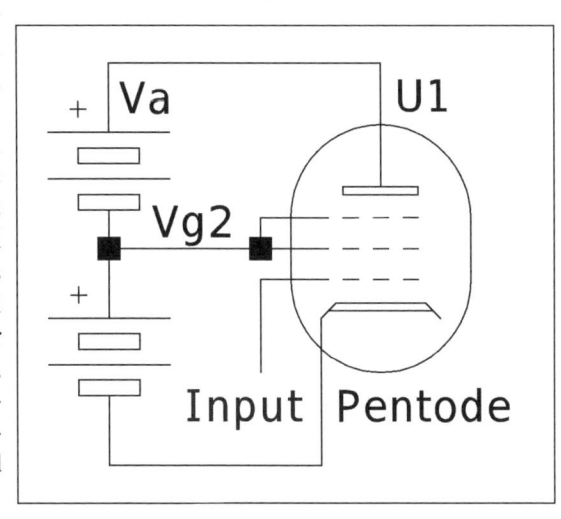

Fig 32: Tetrode connection of a pentode

VALVES REVISITED

Fig 33: Space charge valve

However, the non-linear characteristics and high amplification of a space-charge-grid valve would make this connection (shown in **Fig 32**) suitable for mixer and modulator purposes.

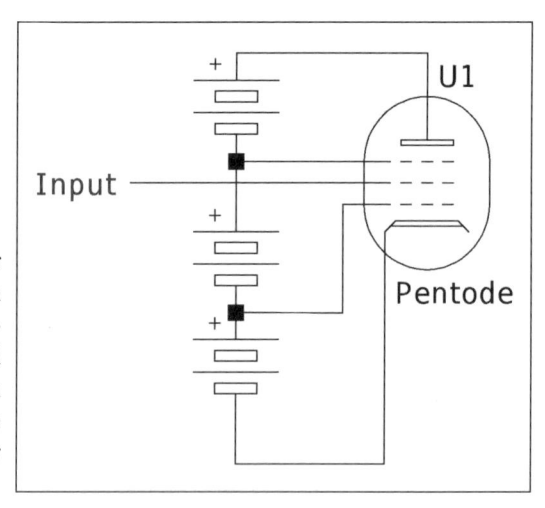

More Grids

Heptode

As valves evolved, more and more specialised valves were developed. Heptodes, valves with 5 grids, (hepta = seven) were developed mainly for mixer applications. They could also be used as IF or RF amplifiers with 'automatic gain control' (AGC), also known as 'automatic volume control' (AVC) or 'remote cut-off'.

Fig 34: Heptode symbol

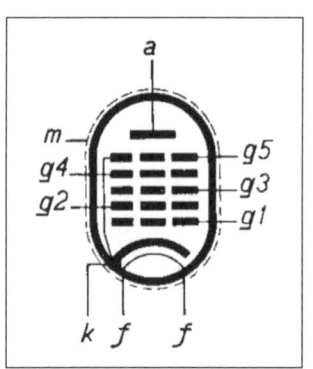

Grids number 2 and 4 were frequently internally connected. Below is a set of curves for a heptode, showing the relationship between control grid voltage (X-axis) and anode current (Y-axis). **Fig 34** shows the heptode symbol and the valve is summarised in **Table 6**.

Heptode curves are shown in **Fig 35**. Note the smooth transition between the right hand portion (high amplification) and the left hand portion (low amplification). Normally, the control grid is biased at the middle of the straight portion of the curve, in this case at -2V. However, if AVC

Fact Sheet - Heptode
Applications: Oscillator/Mixer
Internal capacitances: High
Amplification factor: Low
Anode resistance: High
Benefits: Multiple functions in one envelope
Drawbacks: Noisy

Table 6: Heptode fact sheet

CHAPTER 2: SO, HOW DOES A VALVE WORK?

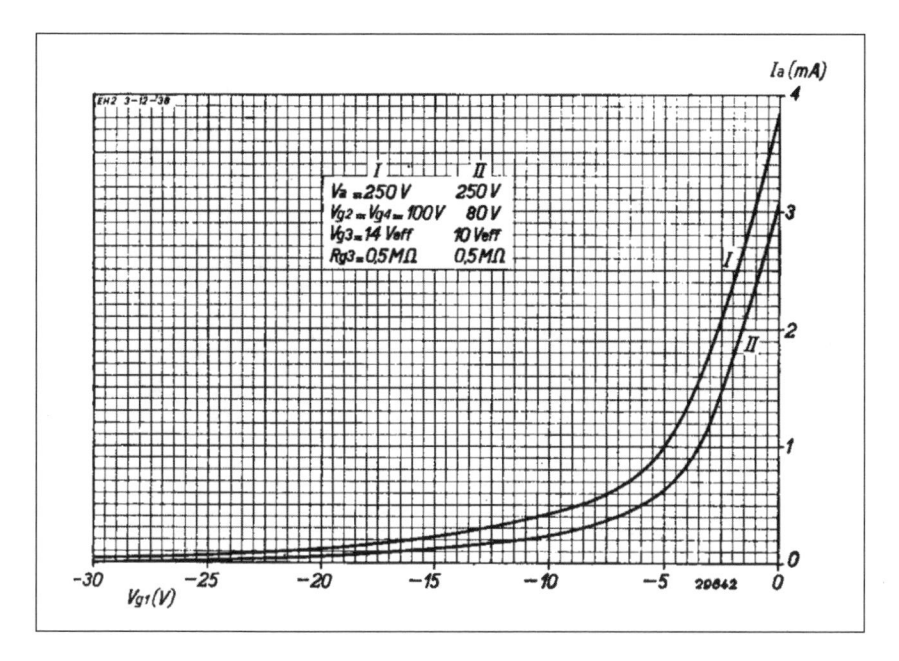

Fig 35: Vg1 / Ia diagram for heptode

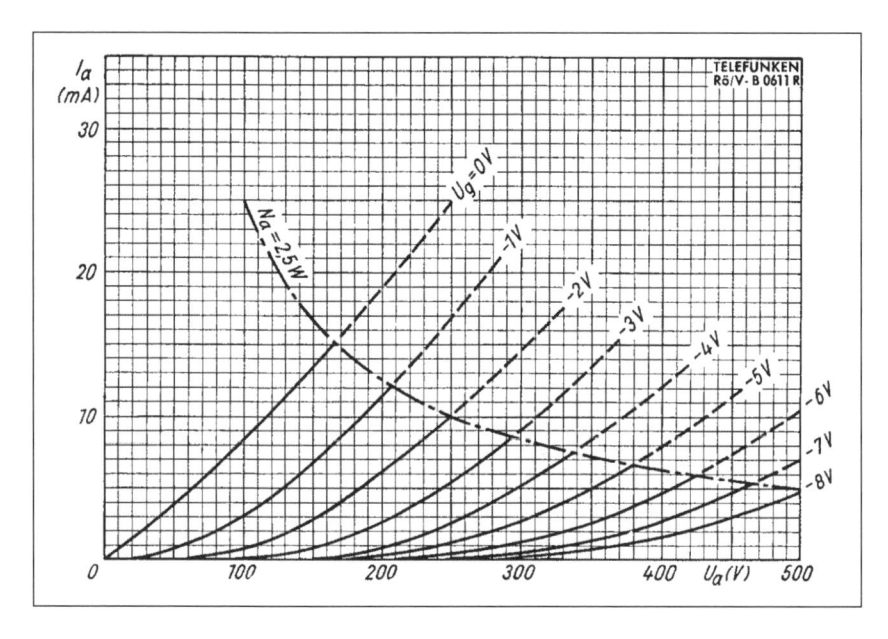

Fig 36: Vg1 / Ia diagram for triode

action is desired, the control grid is made more negative. Compare these curves with the corresponding curves for a triode in **Fig 36**.

The triode curves are more straight. As can be seen, though, pentodes are made with qualities similar to those of heptodes. We shall take a closer look at AVC later on in this book.

Hexode

After the heptode came the hexode, a valve with four grids (hexa = six). Grids 2 and 4 were still internally connected together. The valve was used in mixer applications. The oscillator was in most cases a separate triode, but it didn't take long before it was decided to include the triode into the same glass envelope, forming a triode-hexode. ECH35 was such a combination. The control grid of the triode was internally connected to grid number 3, as can be seen in **Fig 37**. Hexodes do not appear to have been stand-alone valves. Triode-hexodes are obsolete and didn't occur very often in radios.

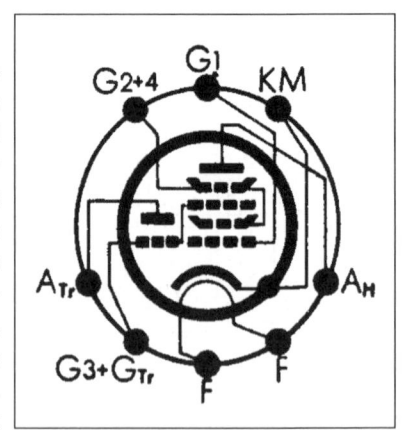

Fig 37: Triode-hexode for mixer applications

Octode

An octode (octa = eight) could be used instead of a triode-hexode. It has yet another grid inserted inside the anode. This new grid was intended to catch the electrons that emanated from secondary emission. It was usually connected internally to the cathode. **Fig 38** shows one example of its usage. The anode voltage is +200V. Let us analyse the diagram.

The oscillator is on the right hand side. It consists of the cathode, R2, C6 and C5 with associated coils. The oscillator occupies the cathode, G1 and G2, where G2 constitutes the anode in a triode. These three electrodes together form a virtual cathode for the rest of the valve. G3 and G5 are connected together and G6 is internally connected to the cathode. That leaves G4 as the control grid for the upper half of the valve. Mixing can occur because the same electron current is

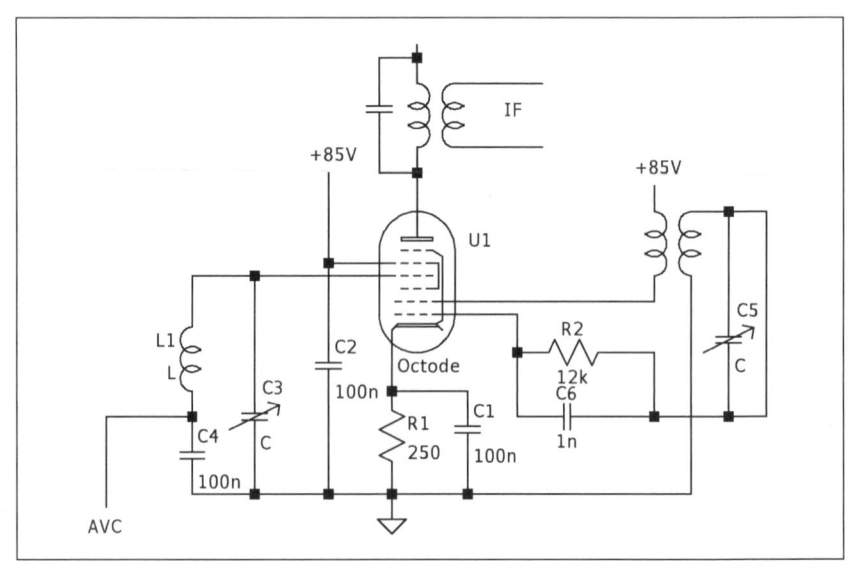

Fig 38: Oscillator / Mixer with octode

CHAPTER 2: SO, HOW DOES A VALVE WORK?

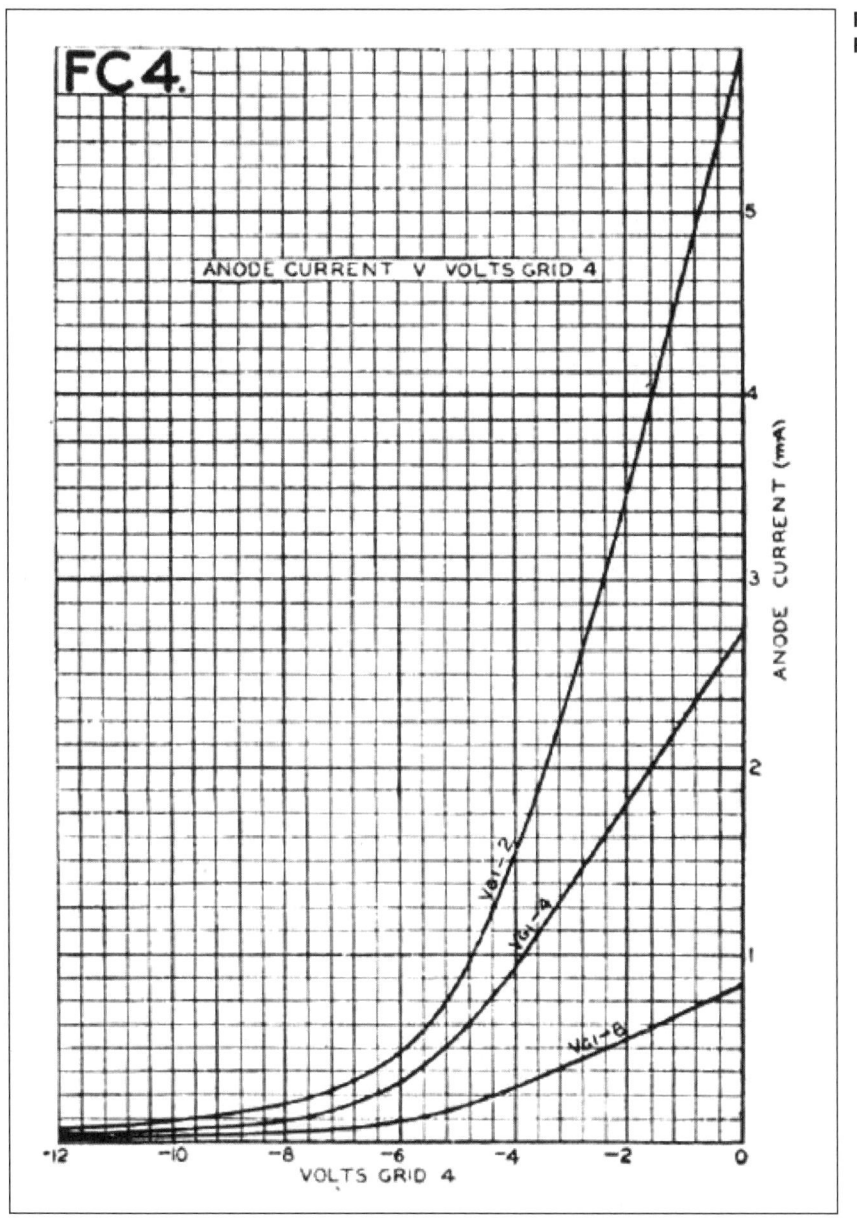

Fig 39: Curves for FC4 octode

controlled by both the oscillator part and the signal part of the valve. A similar method is practised in some oscillators, and is then named an 'electron coupled oscillator' (ECO).

C3 and C5 are, of course, ganged together, ie they are mechanically connected to the same axis. There is an AVC-line coming in from the detector. A glance at the curves in **Fig 39** shows us why. Again we see that the curves are bent towards the negative end of the grid voltage, in this case G4.

Tetrodes and valves with five or more electrodes are included for completeness only. They are rare these days.

Composite valves

In order to reduce manufacturing costs for valves, attempts were made to include as many valve functions as possible into one envelope. We have already seen an example above, the triode-hexode.

Yet others were, however, available. Here are some additional ones:

- Triode-double-diode
- Triode-triple-diode
- Triode-pentode
- Diode-pentode
- Double-diode-pentode
- Double-diode
- Triple-diode
- Double-triode
- Tetrode-pentode
- Double-pentode
- Pentode-indicator

3

Characteristics of Valves

THIS CHAPTER deals with the analysis of valve performance, using several standard measurements. This helps to determine which valve to use for what purpose. Lastly, the causes of valve noise are described.

Valve Characteristics

Amplification factor

The voltage of the control grid controls the anode current. This rule applies to all valves that have a control grid - triodes, tetrodes, pentodes, and all the other types.

The current variations are caught by a load in the anode circuit and are converted to a voltage across the load, which is connected to the next stage. One of the valve characteristics is its amplification factor, or 'µ', pronounced 'mu'. It gives the relationship between the change in anode voltage for a change in control grid voltage and the relationships are shown in Formula 1 below.

$$\mu = \frac{(Anode\ voltage\ change)}{(Control\ grid\ change)} \qquad \textit{Formula 1}$$

Transconductance

The transconductance is measured in milliamps per volt (mA/V) and is given as the relationship between a change in anode current as a result of a change in control grid voltage. Or, as Formula 2 states:

$$gm = \frac{(Anode\ current\ change)}{(Control\ grid\ voltage\ change)} \qquad \textit{Formula 2}$$

Anode resistance

A valve passes only a limited amount of current. The anode resistance is the relationship between the change in anode current and a change in anode voltage at a given control grid voltage. It is given in ohms (Ω) and denoted ra (Europe) or rp (USA) in the data sheets.

Triodes have an anode resistance in the order of tens of kΩ, whereas small signal pentodes have an anode resistance of hundreds of kΩ, sometimes even higher. The RF pentode EF80, for instance, displays a ra of 650kΩ.

According to Formula 3:

$$ra = \frac{(Anode\ current\ change)}{(Anode\ voltage\ change)}$$

Formula 3

Fig 40: Visualisation of ra in a triode

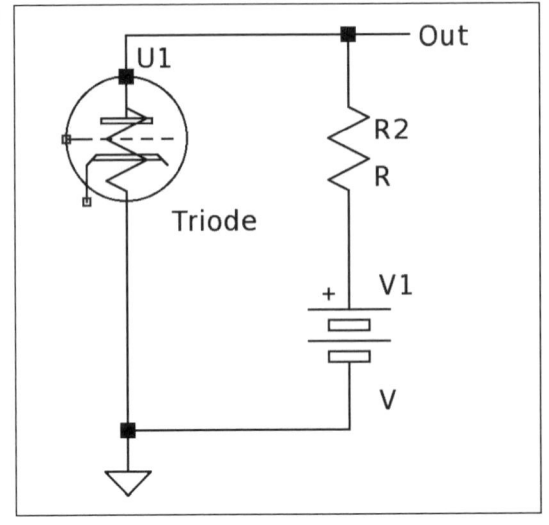

It is useful to visualise the function of ra as shown in **Fig 40**. The anode resistance (ra) can be seen as a resistor built-in into the triode and acts as one leg of a voltage divider with the load resistance R2 (or load impedance) as the other leg.

Since the high voltage source (the battery in the figure) is supposed to have a low impedance, the two resistances are effectively connected in parallel.

Finding parameters from the curves

Finding the three parameters using the Va / Ia curves is fairly straight-forward. The values you get are not exact, of course. Many variables are involved, such as the manufacturer's accuracy when the curves were drawn, the accuracy of the printing process, thickness of the lines, measurement accuracy, etc. However, you rarely need exact values - an approximation is good enough.

This is how it is done:

- Find the Va / Ia-curves for the valve you wish to use.
- Find a suitable grid bias point for your application.
- Draw a horizontal line through the bias point between two Vg-curves.
- Read the difference in Va at the X-axis and divide by the difference in Vg. This is the µ of the valve.
- Draw a vertical line through the same point between two Vg-curves. Read the difference in Ia at the Y-axis and divide by the difference in Vg. This is the Gm of the valve, expressed in mA/V.
- Calculate ra as in Formula 4 below.

$$ra = \frac{\mu}{gm}$$

Formula 4

For the sake of clarity in this example, we shall run the valve at a grid bias of about -1.3V and 210V anode voltage. This gives us an anode current of 10mA. See the marking in **Fig 41**. To find the µ, we now draw the horizontal line through the bias point between two grid voltage curves surrounding the bias point and read the anode voltage difference. **See Fig 42**.

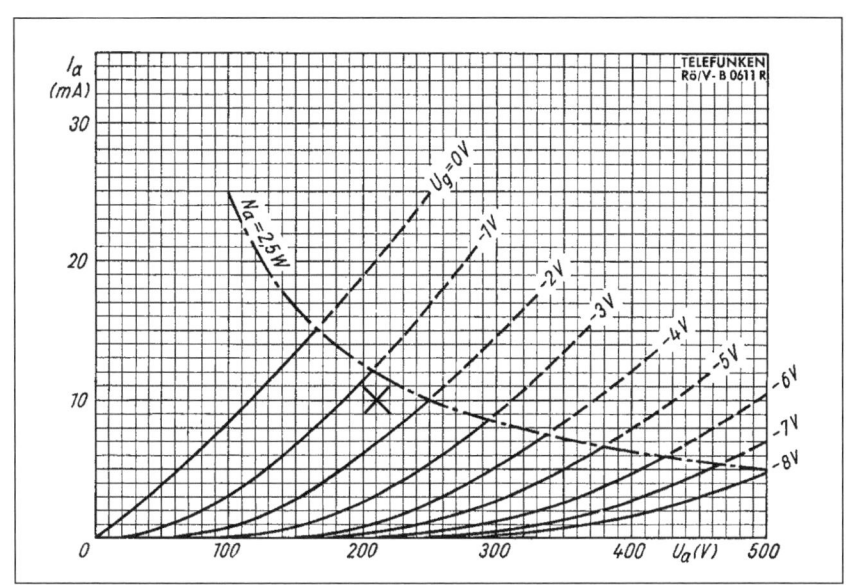

Fig 41: Finding the mu of a valve, step one

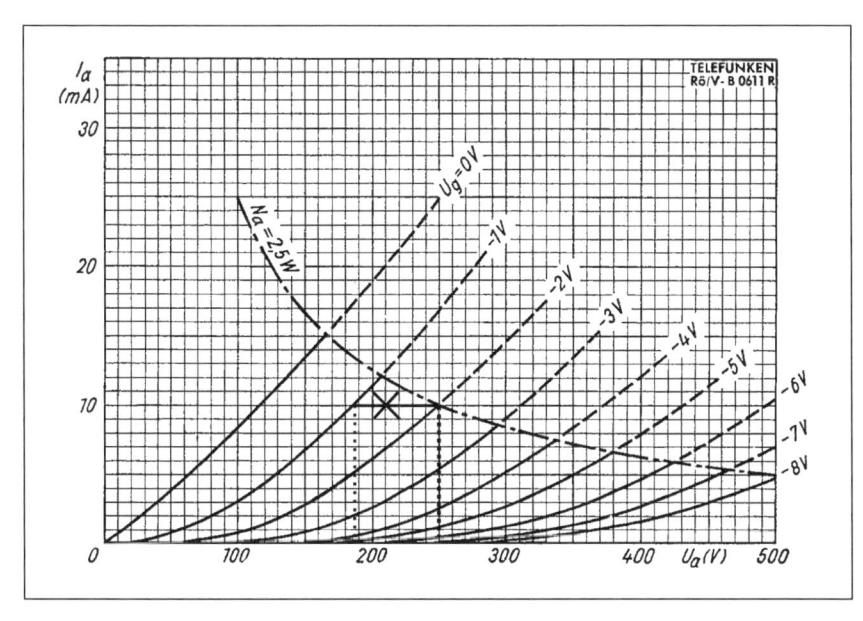

Fig 42: Finding the mu of a valve, step two

The distance between the dotted lines is slightly more than 60, say, 62. The voltage between the grid curves is 1V, so the maths are easy.

Applying Formula 1 to our numbers gives us:

$$\mu = \frac{62}{1} = 62$$

Finding Gm is just as easy. Draw the vertical line and read the Ia difference as in **Fig 43**. The Ia difference is about 5.5mA, so Gm at this point is 5.5mA/V. According to Formula 4 above, the figures we got from the curves return an ra of:

VALVES REVISITED

Fig 43: Finding the Gm of a valve

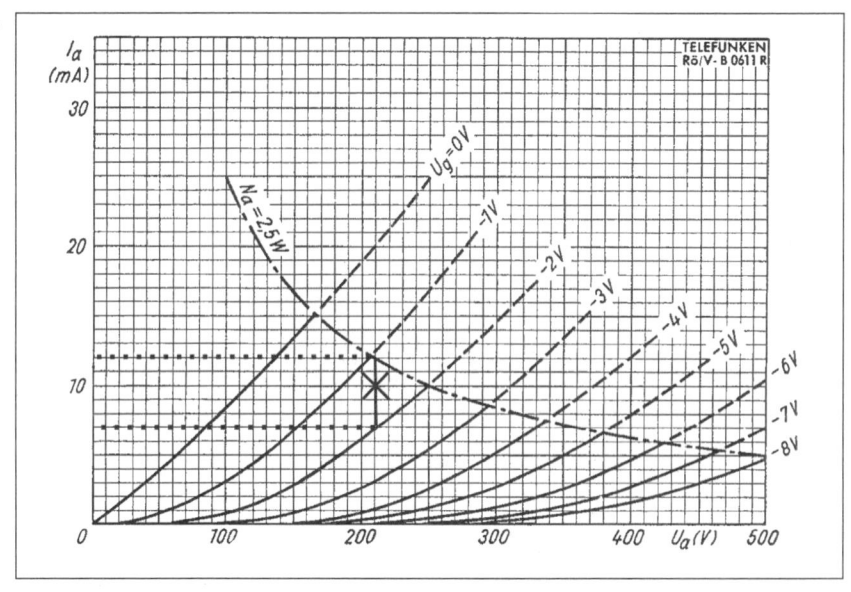

$$ra = \frac{62}{0.0055} = 11\,k\Omega$$

Reading the table in the data sheet of **Fig 44** gives us slightly different numbers. This illustrates that unless you run the valve under the exact conditions stated in the table, you will get slightly different parameters.

If you do the same exercise with the numbers given in the table, you will get a better correlation with the table. The examples are taken from the ECC81 data sheet.

The method may seem crude, but the knowledge is useful nonetheless, in case you need to plot a set of curves for an unknown valve or an unusual configuration, such as the Lunar Grid amplifier described elsewhere in this book.

Relation of the three

There is a sort of 'Ohm's Law' relationship between the three main valve parameters as shown below in Formulae 4, 5 and 6.

$$ra = \frac{\mu}{gm} \qquad \text{Formula 4}$$

Fig 44: Data sheet extract

TYPICAL CHARACTERISTICS AND OPERATING CONDITIONS (each unit)						
Anode voltage	V_a	100	170	200	250	V
Grid voltage	V_g	-1.0	-1.0	-1.0	-2.0	V
Anode current	I_a	3.0	8.5	11.5	10	mA
Transconductance	S	3.75	5.9	6.7	5.5	mA/V
Amplification factor	μ	62	66	70	60	
Internal resistance	R_i	16.5	11	10.5	11	kΩ

28

Thus:

$$\mu = gm \times ra \qquad \text{Formula 5}$$

and

$$gm = \frac{\mu}{ra} \qquad \text{Formula 6}$$

This makes it easy to calculate one parameter if the other two are given. When applying the formulae, gm must be given in A/V rather than mA/V. Thus, a valve with gm = 5.5mA/V, and μ = 20 would have an ra equal to:

$$ra = \frac{\mu}{gm} = \frac{20}{(5.5 \times 10^{-3})} = 3.6k\Omega$$

Calculating the amplification of a stage

The amplification of a (grounded cathode) stage can easily be calculated. Formula 7 shows the relation:

$$A = \frac{(\mu \times Rl)}{(ra + Rl)} \qquad \text{Formula 7}$$

where ra is the anode resistance and Rl is the load impedance between the anode and the power supply.

The Valve Parameters

Getting a valve to work is fairly simple. Ensure the various electrodes get the voltages and currents given by the data sheet, and the stage usually works.

As is the case with semiconductors, there are several parameters that tell you about the properties of the valve. All parameters you usually need are given in the valve data sheet. The following designations will be used in this book:

- Vf is the heater voltage. It is usually 6.3VAC, 12.6VAC or 1.2VDC.
- Vg is the control grid bias without signal. It is usually negative in relation to the cathode. All voltages in a valve stage are measured against the cathode. The bias can be generated in a number of different ways. The most common is automatic bias, which is achieved by a resistor between the cathode and ground. The cathode current (which, in a triode, is equal to the anode current) flows through the resistor and gives the cathode a slight positive voltage. Since the control grid is grounded (eg through a resistor), the grid becomes more negative than the cathode.
- Va is the anode voltage.
- Ia is the anode current at a given control grid voltage.
- Zin is the input impedance of the stage.
- Zout is the output impedance of the stage.

The data sheets also contain curves (of which we have seen a variety already for different valves) and the socket connections which are always shown from below. Sometimes application examples are given, and the amplification under various circumstances.

The internal capacitances are usually given for anode to grid, anode to cathode, and cathode to grid. Sometimes capacitances relating to the heater are also given.

How curves are made

You might find yourself in a situation where you have a valve of unknown origin, or for which no curves are available. Perhaps you would like to connect in a non-standard way (such as a Lunar grid amplifier stage, or a multi-grid valve in triode connection), or maybe you wish to investigate the valve's behaviour under extreme conditions (at very low anode voltages, for instance). In these circumstances, it is very convenient to be able to make the curves yourself. This is a simple procedure, but very tedious, depending on what resolution you want.

You will need two variable power supplies, one for the grid bias and another for the anode voltage. For valves with more grids, you will need to be able to connect the other grids to their respective voltages. You will also need two voltage meters (one for the grid voltage and one for the anode voltage), and a milliamp meter for the anode current.

Ia-Va-curves

Connect the heater or filament according to the data sheet, or your instinct if no data sheet is available. As a pointer, the filament/heater is connected to the pins where you can measure a small value of resistance between them. Note that some valves have a centre tap on the filament to enable operation from 6.3 or 12.6 volts. Connect the rest of the valve according to **Fig 45**.

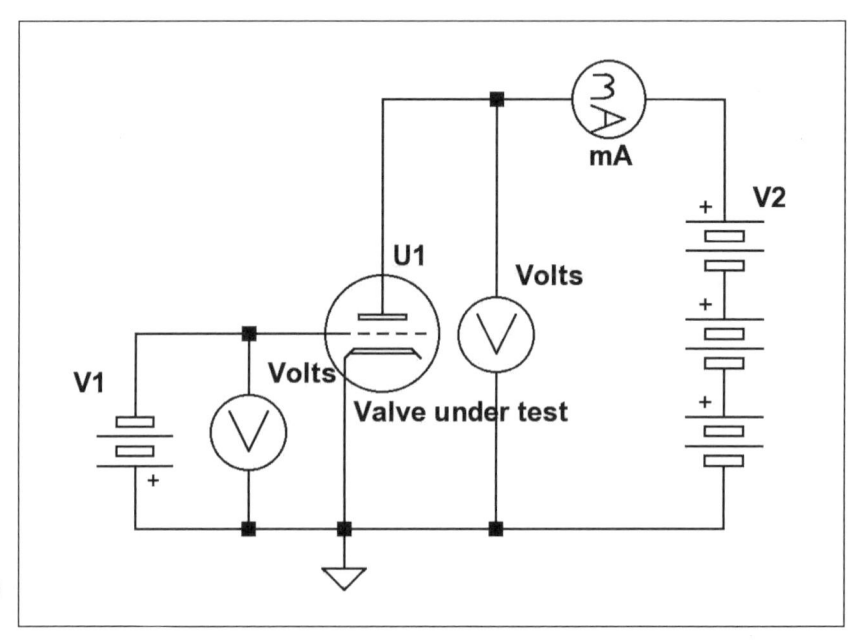

Fig 45: Measuring valve parameters

1. Set V1 to the lowest voltage you wish to plot.
2. Set V2 to 0.
3. Increase the voltage at V2 until the valve begins to draw current.
4. Note the two voltages and the current.
5. Increase V2 one step (10, 20 or 50 volts).
6. Repeat from 4 until you reach the desired highest anode voltage.
7. Increase V1 one step (0.5, 1.0 or 2.0 volts).
8. Repeat from 2 until you reach the desired highest grid voltage.
9. Plot the curves onto millimetre paper. Plot Va along the X-axis and Ia along the Y-axis.

The resulting lines will represent the Vg settings. Now you have a set of curves from which you can measure Gm, µ and calculate ra.

Ia-Vg-curves

If, instead, you wish to find the Ia-Vg curves, connect the valve according to the figure above.

1. Set V1 to 0.
2. Set V2 to the lowest voltage you wish to plot.
3. Increase the voltage at V1 until the valve begins to draw current.
4. If needed, adjust V2 to remain at the set voltage.
5. Note the two voltages and the current.
6. Increase V1 one step (0.5, 1.0 or 2.0 volts).
7. Repeat from 4 until you reach the desired highest grid voltage.
8. Increase V2 one step (20, 50 or 100 volts).
9. Repeat from 2 until you reach the desired highest anode voltage.
10. Plot the curves onto millimetre paper. Plot Vg along the X-axis and Ia along the Y-axis. The resulting lines will represent the Va settings.

Now you have a set of Ia-Vg curves, from which you can deduce the three valve parameters, and also assess a suitable grid bias. If, however, you are not interested in the curves, but still wish to find the three parameters of your valve, the procedure is much simpler. You need the same equipment and instruments as above.

Finding gm

1. Set the anode voltage to a suitable value
2. Set the grid voltage to a suitable value
3. Note the grid voltage
4. Note the anode current
5. Set a different grid voltage
6. Make sure the anode voltage stays the same. Adjust if necessary
7. Note the anode current and subtract the two currents
8. Subtract the two grid voltages
9. Divide the change in current by the change in voltage

This gives you the gm in mA/V.

Finding μ and ra

1. Set the anode voltage to a suitable value
2. Set the grid voltage to a suitable value
3. Note the grid voltage
4. Note the anode current
5. Set a different grid voltage
6. Make sure the anode voltage stays the same. Adjust if necessary
7. Note the anode current and subtract the two currents
8. Subtract the two grid voltages
9. Now, return the grid and anode voltage settings to their original
10. Change the anode voltage by such an amount that you get the same change in anode current as before
11. Subtract the two anode voltages
12. Divide the change in anode voltage by the change in grid voltage
13. This gives you the μ of the valve under the given circumstances.
14. Now you have μ and Gm. Divide μ by Gm to get ra.

Using the Curves

To illustrate how to interpret the curves for a valve, let us begin by looking at the curves for a dual triode, ECC81, shown in **Fig 46**. Since both triodes are built into the same envelope, they are very well matched, far better than usually is the case with semiconductors.

These curves show the relationship between the control grid voltage (along the X-axis) and the anode current (Y-axis) at the given anode voltages. We can see that Ia = 6mA at Vg = -2V and Va = 200V. These are the static curves of the valve.

You can see that there are no curves for positive grid voltages. The reason is that a positive grid voltage causes a grid current to flow, which loads the signal source. So, the control grid is usually kept at a negative potential.

Let us theoretically design an amplifier stage using ECC81 using the information we find in the data sheet.

First we connect the heater to 6.3VAC. Since this valve has a centre tap in the heater, pin 9 is connected to one terminal of the voltage source, and pins 4 and 5 to the other.

Then we connect the grid to ground via a resistance. The input should have a fairly high impedance, so we need to choose a fairly big resistor, say, 470kΩ. Sometimes the data sheet contains information about how big the grid resistor may or should be. The grid resistor is also called 'grid leak'.

Then we want to find out how big the anode resistor (the load of the stage) should be. This is done by means of **Fig 47**. (Note that the figure already has a load line drawn, see text below).

Looking at the data sheet, we find that they recommend -2V grid bias at 250V anode voltage. The valve then conducts 10mA anode current. But let us choose 200V instead. That gives a slightly higher amplification, according to the data sheet. The recommended Vg is then -1V.

This set of curves shows us the relationship between Va and Ia at various grid voltages. We have to stay below the bent curve, which shows the maximum

CHAPTER 3: CHARACTERISTICS OF VALVES

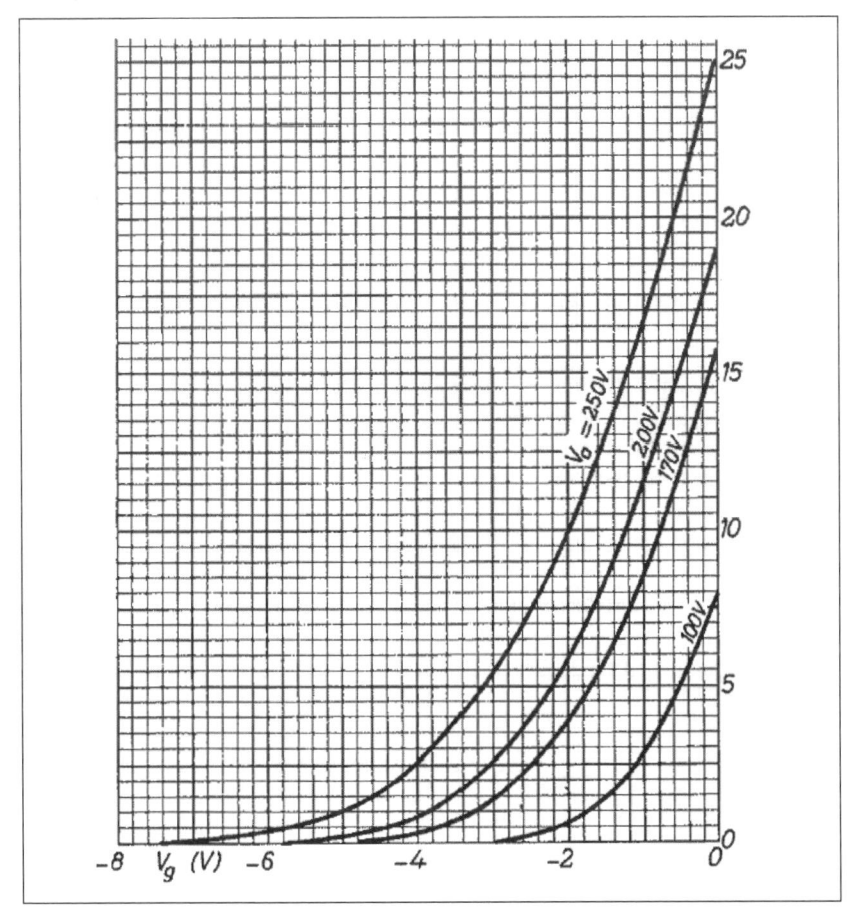

Fig 46: Ia / Vg curve for ECC81

(below) Fig 47: Finding anode resistor value

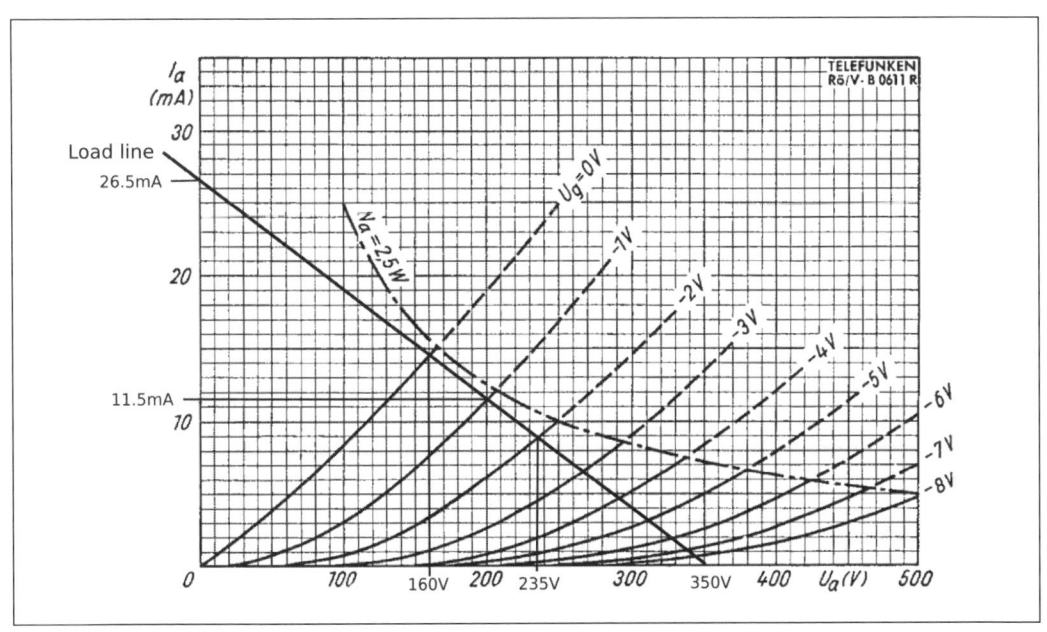

allowed power dissipation for this valve. We need to construct a load line to find out the value of the anode resistor. There are different ways of doing this.

One way is to begin by choosing a resistance and draw a line that corresponds to this resistor. Let us assume that we have 450 volts available from the power supply and that we choose a resistance of 33kΩ. According to Ohm's law (Georg Simon Ohm, 1789 - 1854), there is a relationship according to Formula 8 below:

$$I = \frac{V}{R}$$ Formula 8

or

$$I = \frac{450}{33000} = 13.6mA$$

So, we place a ruler with one end at 450V on the X-axis and the other at 13.6mA on the Y-axis. This is our load line, and its slope determines the resistor. The line can be moved up or down in parallel without the resistance changing.

Now we need to choose the bias point of the grid. The data sheet says -1V grid voltage and 200V anode voltage yields the highest amplification. Let's try that. We move the ruler, still in a parallel movement, to the intersection between -1V grid voltage and 13.6mA. We find two things that make 33kΩ an impossible choice.

- The intersection lies above maximum allowed power for the valve
- We would need over 500V from the power supply, which we don't have

So, we need to look at a different configuration. Let's instead try a grid bias of -2V. That gives us an anode voltage of 215V and an anode current of a little more than 7mA. The cathode resistor is given by Ohm's law and for 7mA and 2V we get R = 2/0.007 = 285.7Ω, which is reasonable. This configuration gives us a load line that stays below the maximum power curve, and a voltage swing at the anode from 170V to 260V for an input signal between -1V and -3V, which is a voltage amplification according to Formula 9.

$$Av = \frac{\Delta Va}{\Delta Vg}$$ Formula 9

or, inserting our numbers:

$$\frac{(260 - 170)}{(-3 - (-1))} = 45$$

Our amplifier amplifies the signal 45 times.

Here's another way: Choose a bias point for the grid in the diagram, for instance -1V grid voltage and 200V anode voltage. Then place the ruler centred to this point. Turn the ruler around the working point so that it hits a suitable voltage at the X-axis, but still remains below the power line. 350V is a good

CHAPTER 3: CHARACTERISTICS OF VALVES

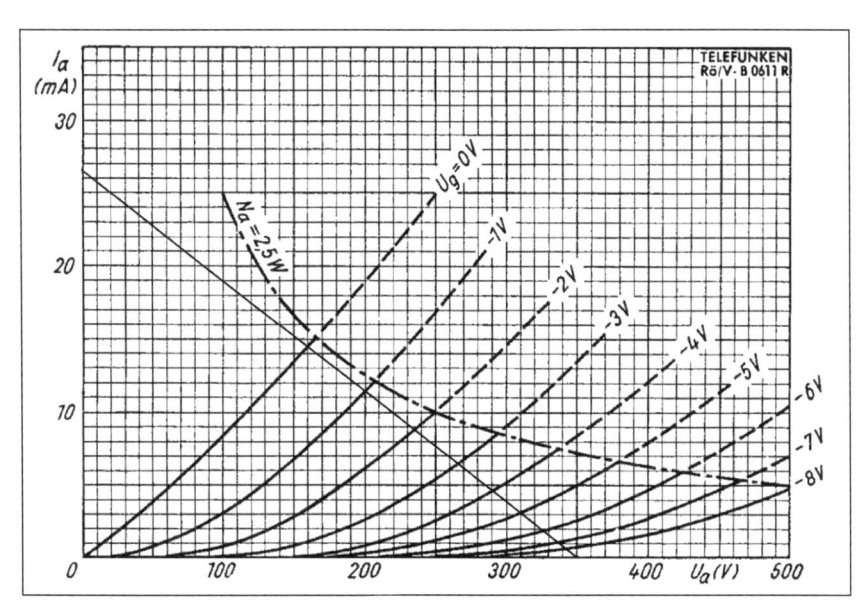

Fig 48: Load line drawn

value. At the other axis we can now read the current, 26.5mA. Calculating the anode resistor as above with the new values (350V and 26.5mA) gives us a value of 13207Ω. 12kΩ is a good standard value. The cathode resistor is calculated as above with the new current and voltage. The curve looks like **Fig 48** with the load line drawn:

From the curve it can be seen that the output at the anode should swing reasonably symmetrically around 200V and the peak-to-peak voltage should be 235 - 160 = 75V for an input peak voltage of 2V, and 75V/2V = 37.5 times amplification. The mean anode current is 11.5mA, which is the basis for calculating the cathode resistance below. So, we get a lower amplification with the new smaller anode resistor. On the other hand, we now have a power supply voltage (350V) which is easier to achieve.

Draw the load lines from both examples above and measure the distances between the grid voltage curves along respective load line. You shall find that they are more equidistant along the 33k line than along the 12k line. In this case, less slope, less distortion.

The cathode current at the working point will be 11.5mA. That gives us a cathode resistor of 86.95Ω (1V grid voltage and 11.5mA cathode current). 82Ω is a good choice.

There is one thing we haven't yet considered: The input impedance of the next stage. The anode resistor in this stage is in parallel with the grid leak and input impedance of the next stage, AC-wise. So the load line gets a slightly different slope. It is not really too important in practice, since the difference is negligible as long as the following stage has a high input impedance, higher than the output impedance of our stage.

The principles for placing load lines and calculating resistors are the same regardless what type of valve and application you use, be it triodes, pentodes, RF, AF, small signal or power. The curves may look different, but the principles are the same.

35

However, the curves provided by the manufacturer don't tell the whole truth.

Dynamic Curves

The valve characteristics are drawn by varying the grid voltage, keeping the anode voltage stable. In a grounded cathode application (the most common one), the anode has a load impedance connected to it. As we know, the anode voltage affects the amplification. Since the designer is the one determining what kind of load to connect to the anode, not the manufacturer, the final, dynamic, curve is unknown at the time of manufacture.

In order to assist in the design of an amplifier stage, a dynamic curve needs to be constructed. This curve tells us how the valve behaves when loaded, the degree of actual amplification, and how the valve behaves in the application.

Let us use ECC81 again as an example. To construct a dynamic curve for a valve, you need to begin with the Ia-Va curves as before. Draw a load line as before. Now, extend the X-axis to the left of the curves and mark as many equidistant points as the number of curves provided by the manufacturer. For the curve set used here, there are nine curves, one for each grid voltage shown (0 to -8V). Then draw a horizontal line from the intersection of the load line and the 0V curve to the first mark on your new X-axis. Then draw another line from the intersection between the load line and the 1V curve to the second marking on your X-axis.

Continue like that until all nine lines have been drawn. Finally, connect the new points at your X-axis by drawing a curve. The result will be something like **Fig 49**.

Now you can immediately see what the resulting anode voltage (as read at the anode connection) will be for a certain momentary grid voltage. So, for instance, for a grid voltage of -2V, the anode voltage will be 220V, for 0V grid voltage, the anode voltage will be 160V, etc. Naturally, this information could have been drawn from the load line alone, but as you can see, the dynamic curve is not a straight line. This implies that:

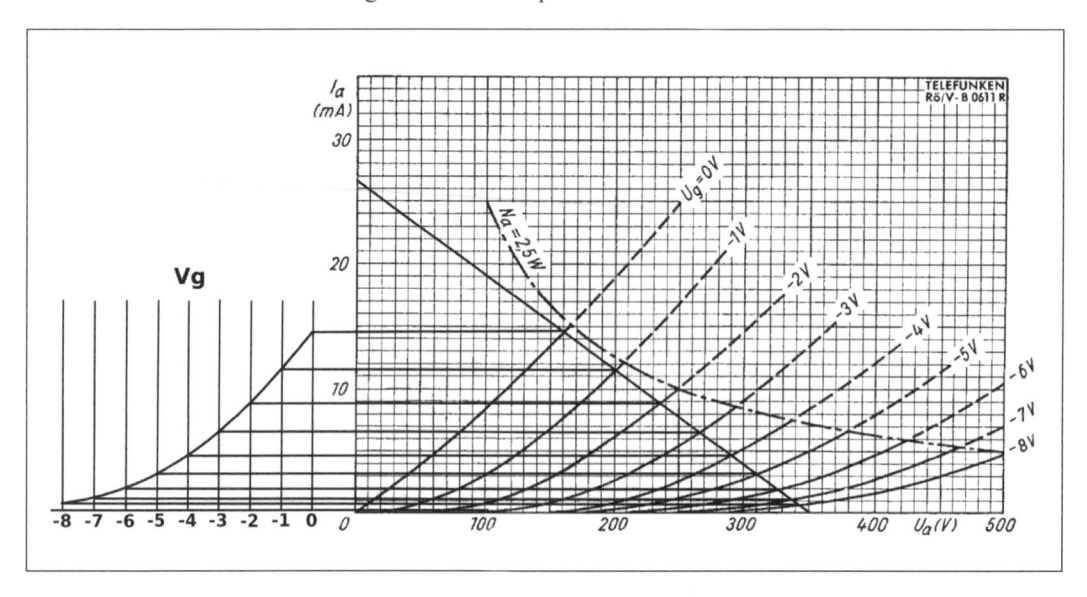

Fig 49: Dynamic curve drawn

- If you attempt to use this valve in an AF application, you will get an amount of distortion.
- Using the valve in an RF application enables you to use AVC to control the amplification.
- The valve can be used as an AM modulator.

It is quite clear that a 1V p-p input signal with a grid bias of -7.5V yields an amplification which is barely noticeable, whereas at the other end of the curve (bias -0.5V), the amplification will be about 35. This is not immediately apparent with only the load line drawn.

In radio applications, where the input voltages all are in the order of microvolts or millivolts, the curvature has no noticeable impact on the fidelity of the output signal. However, the bias voltage does have an impact on the amplitude of the output signal, ie the degree of amplification.

Distortion gives rise to harmonics or overtones. Get rid of the unwanted harmonics, and you will get rid of the distortion. On RF and IF it is easy - put a filter into the circuit and tune it to the desired frequency, and you're done.

The situation is different, of course, in AF applications and oscillators, where it is highly desirable to keep distortion at the lowest possible level and input voltages are higher.

Note, though, that it is only for triodes that the cathode current equals the anode current (see further below)!

A circuit diagram would look like **Fig 50**. R1 is the grid leak. Its purpose is to give the grid a ground potential and remove those electrons that hit the grid.

R2 is the anode resistor. Its purpose is to convert the current variations in the anode current to voltage variations.

R3 is the cathode resistor. Its purpose is to use the cathode current to raise the potential of the cathode so that the grid (which is at ground potential through R1) has a bias voltage which is negative relative to the cathode. This resistor needs to be decoupled by a capacitor to prevent negative feedback (more on that later).

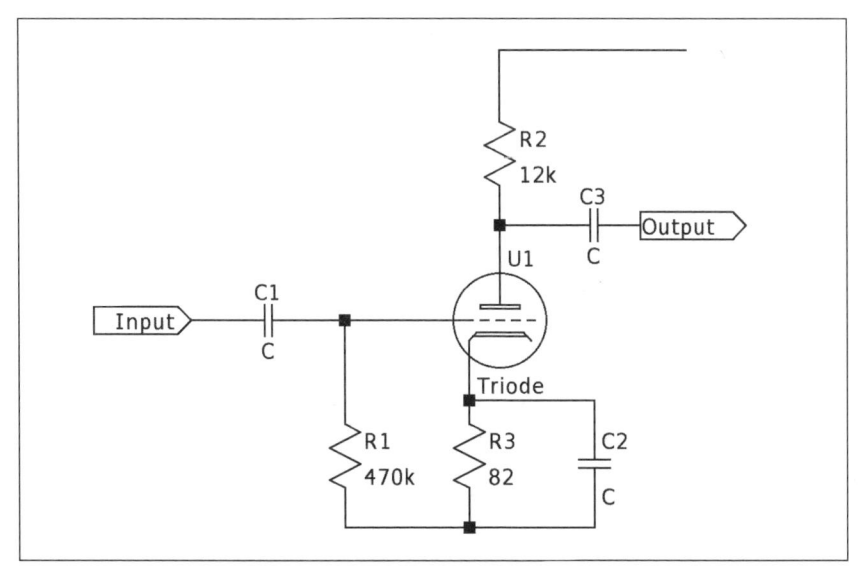

Fig 50: Schematic diagram, triode amplifier

VALVES REVISITED

Fig 51: Dynamic curve for E88CC, load 25kΩ.

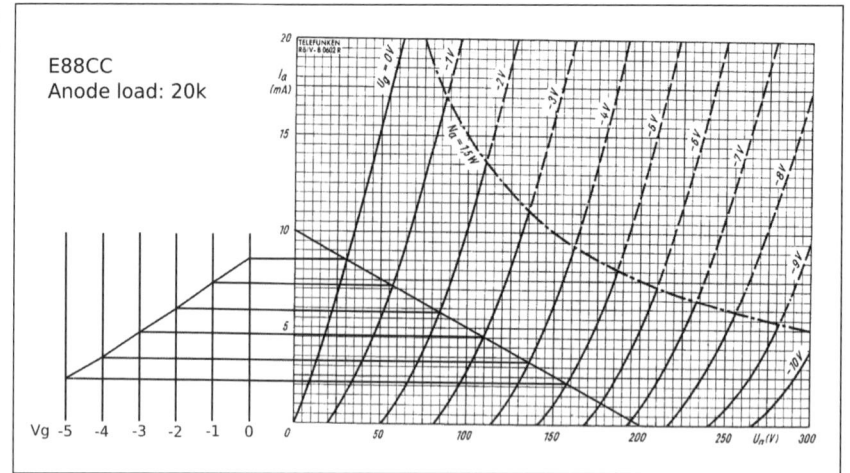

Fig 52: Dynamic curve for E88CC, load 10kΩ

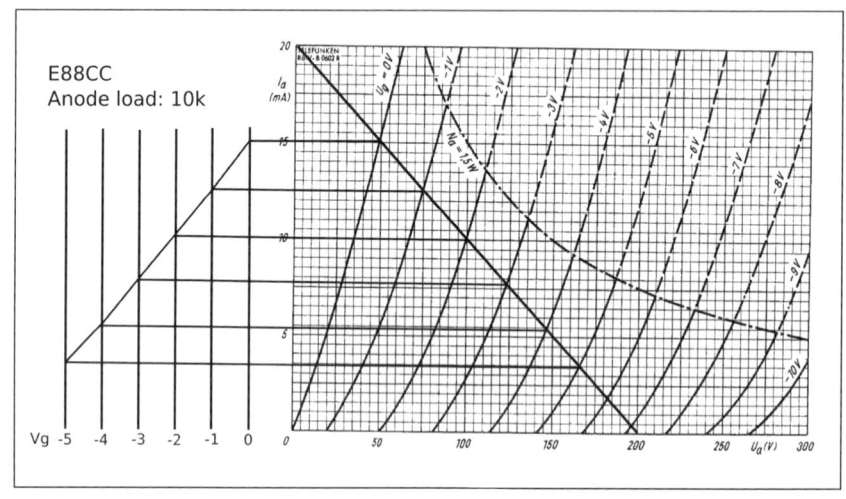

Fig 53: Dynamic curve for EF80

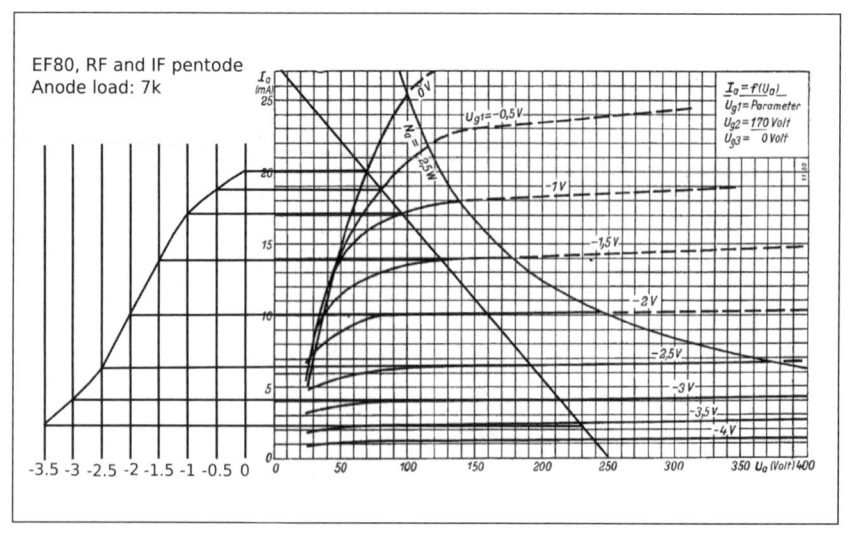

CHAPTER 3: CHARACTERISTICS OF VALVES

C1 and C3 are coupling capacitors. All three capacitors are chosen according to the desired frequency range of the amplifier.

The frequency range of a particular valve depends heavily on the Miller effect, which in turn depends on the input capacitances of the valve, including stray capacitances at or near the input. Therefore, some valves are better suited to handle high frequencies than others.

There is another double triode, E88CC, which is intended for higher frequencies than ECC81. Its dynamic curve is drawn in **Fig 51**. Here you can see that the dynamic curve is straighter than that for the ECC81 above, which implies less distortion in AF applications. Increasing the dynamic range of the amplifier is done by decreasing the anode load, as shown in **Fig 52**. It is also easy to see how straight the dynamic curve becomes.

But, what about pentodes? Can you do the same thing with them? The curves are so different! The answer is "yes". **Fig 53** shows the dynamic curve for a pentode intended for RF and IF applications.

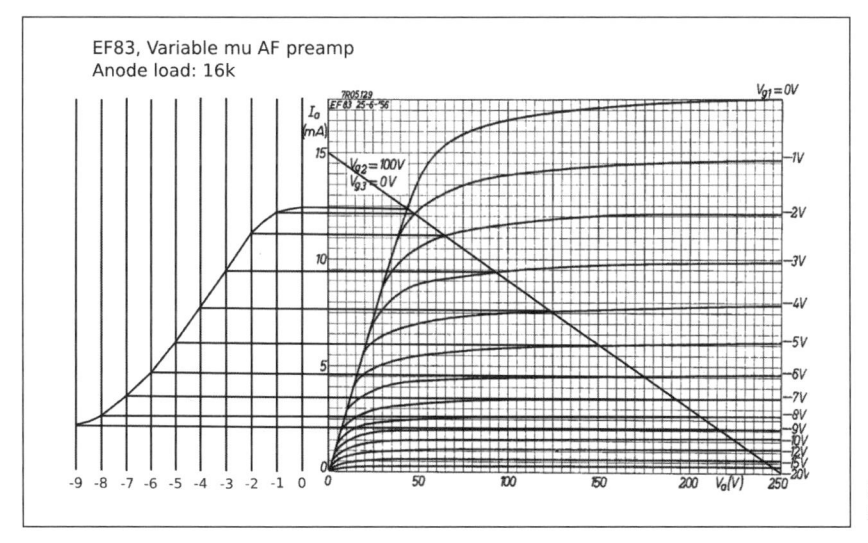

Fig 54: Dynamic curve for EF83

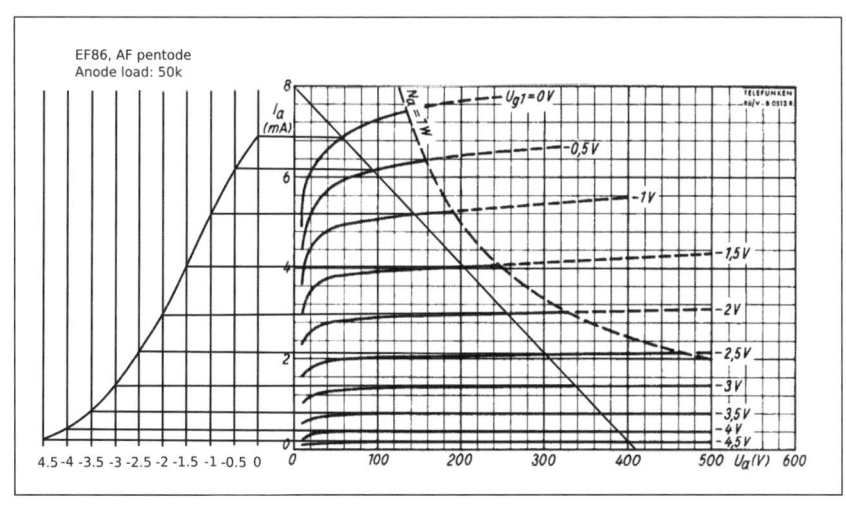

Fig 55: Dynamic curve for EF86

The curve is fairly straight at the middle, around -2V bias. In RF applications, the curvature is of less importance than in AF applications. Note also that the amplification is considerably higher than that of a triode - about sixty times.

Many pentodes are of the 'variable μ' type. Here is one, actually intended for AF use, the EF83. See **Fig 54**. The amplification is lower (about 30 times) in this valve, though the dynamic range is larger, and can be increased by reducing the anode load.

The EF86, shown in **Fig 55** is another AF pentode, without variable μ. The amplification for this stage (with a 50kΩ load) is about 50 times.

Valve Noise

Shot noise

Imagine a light summer rain. You are sitting in your rowing boat in the middle of a lake. It is calm and quiet. You can hear a light noise from the water drops as they hit the lake surface. What you hear is a noise that varies in strength with the intensity of the rain.

Now, imagine the same picture, only a couple of million times smaller. The rain cloud above your head is the cathode, the rain drops are the electrons and the lake surface is the anode. What happens is that when an electron hits the anode, a tiny voltage difference occurs. Had the electrons arrived at a fixed interval instead, you would have heard a tone, but they don't. They are randomly distributed in time, and noise is generated.

Grid noise

A grid sits in the way of the stream of electrons on their way from the cathode to the anode. A valve that is supposed to amplify a voltage is bound to contain at least one grid. Our intention is not to stop all electrons, even if the (control) grid is negative in relation to the cathode. Instead, it lets a number of electrons through, and how many that pass through depends on the momentary voltage on the grid.

Despite the facts that both the electrons and the grid have a negative voltage, and equal polarities repel each other, there are still electrons that hit the grid. Their momentum is big enough to overcome the polarity repulsion. Unless the grid is connected to ground in some way, through a coil or a grid leak (a resistance between the grid and ground), a negative charge will build up over time, which will add to the grid bias. (This phenomenon is put to use in some designs, but is not particularly common). So, the electrons have to be removed somehow. And, as in the case of the anode, each time an electron hits the grid, a small voltage difference occurs. Since the grid is the target of the electrons in this case, the tiny voltage variations will be amplified in the process. The amplitude of the grid noise depends on the impedance of the grid circuit and what frequency and bandwidth the valve works with. At low and medium frequencies it is barely audible, but can be quite serious at high frequencies.

Not only is the control grid subject to electron bombardment - all grids within the same valve contribute to the total noise. This is the reason a pentode (for instance) is noisier than a triode. The more grids, the more noise.

4

Connecting Stages Together

WHEN BUILDING EQUIPMENT that contains more than one stage, it is important to catch the variations in anode current, turn them into voltage changes and bring them to the next stage. There are several different methods to achieve this. They can be used on their own or in combination.

Transformer Coupling

At the beginning of the valve era (early 1900s), coupling between stages was nearly exclusively done with transformers. We must remember that in those days only one stage was used for RF amplification and detection. The rest of the receiver was used to amplify the AF signal given by the detector.

Transformer coupling is shown in **Fig 56**. It did have its benefits. The transformer could be used as an 'amplifier' too by transforming the voltage to a higher level. However, the drawback was that the sound quality was not very good due to the limited frequency range of the inter-stage transformer. However, in RF and IF stages, transformer coupling of some sort is still the most common method.

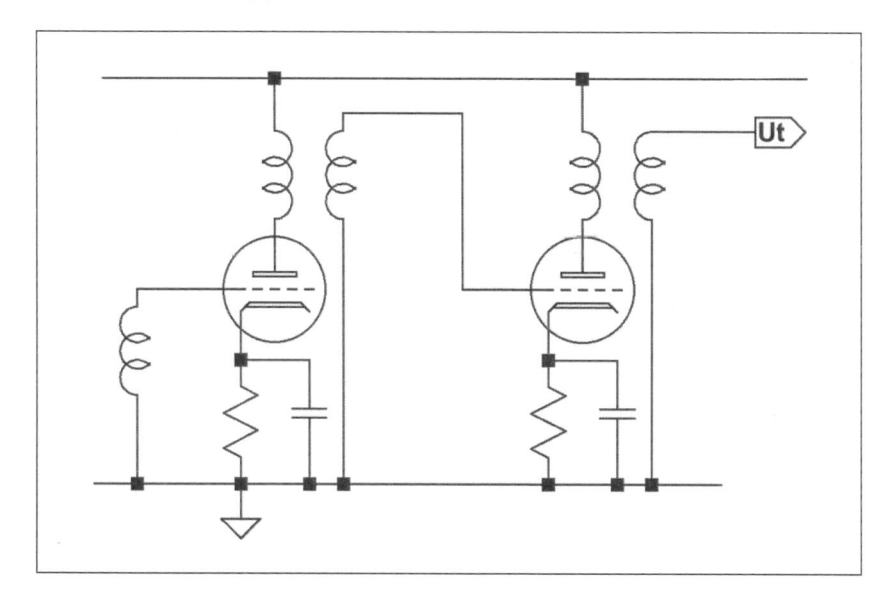

Fig 56: Transformer coupling

R/C Coupling

The most common method these days is to use resistors and capacitors to couple the signal between stages.

This applies mostly to AF amplifiers. A resistance is connected between the anode and the power supply.

A capacitor carries the signal on to the grid of the next valve. It is marked 'C' in **Fig 57**. The purpose of the capacitor is to block the high anode voltage and prevent it from reaching the grid.

Even though transformers can be built today which have a wide frequency range, they are very expensive compared to the alternative - a resistor and a capacitor.

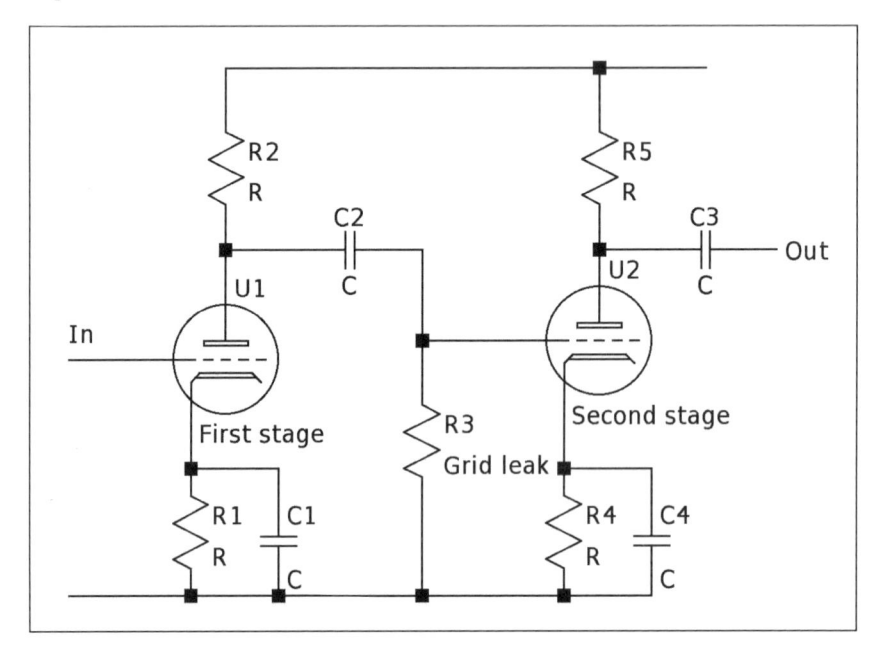

Fig 57: Resistor / Capacitor coupling

Reactive Coupling

A coil is far better at turning a varying current into a varying voltage than a resistor.

The coil needs to have a high reactance for the frequency range in question. Besides, it has to be able to cope with the anode current. However, attention needs to be given to the inevitable following capacitor. If incorrectly designed, the capacitor / coil combination could form a tuned circuit, which, in this case is undesirable.

The benefits are that the coil is more efficient in catching RF than a resistor, and that the coil doesn't drop the anode voltage from the power supply. Such a coil usually has a value of one or more millihenries, and is called an 'RF choke'. **Fig 58** shows a typical RF method of coupling, not particularly suitable for audio frequencies.

CHAPTER 4: CONNECTING STAGES TOGETHER

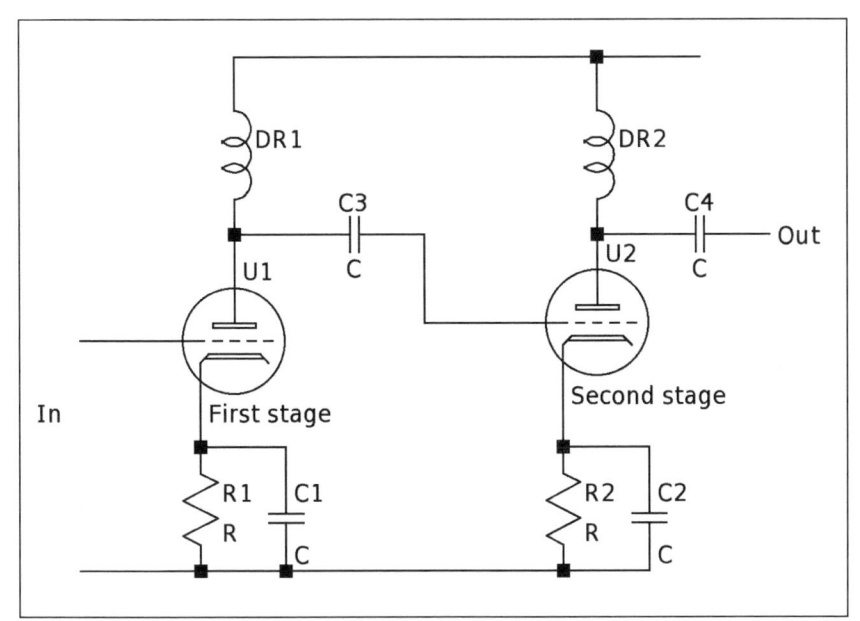

Fig 58: Reactive coupling

IF Transformer Coupling

It is common practice to connect an IF transformer between the stages of an IF amplifier as in **Fig 59**. It consists of two coils and two capacitors. Each coil / capacitor pair form a tuned circuit. These days the coils are usually adjustable to enable fine tuning. In older radios you can find trimmer capacitors to adjust the frequency. Even the coupling between the coils could be adjusted to set the bandwidth of the receiver. The coils are mechanically mounted near each other, so that the coupling between them is electromagnetic, just like in an audio transformer.

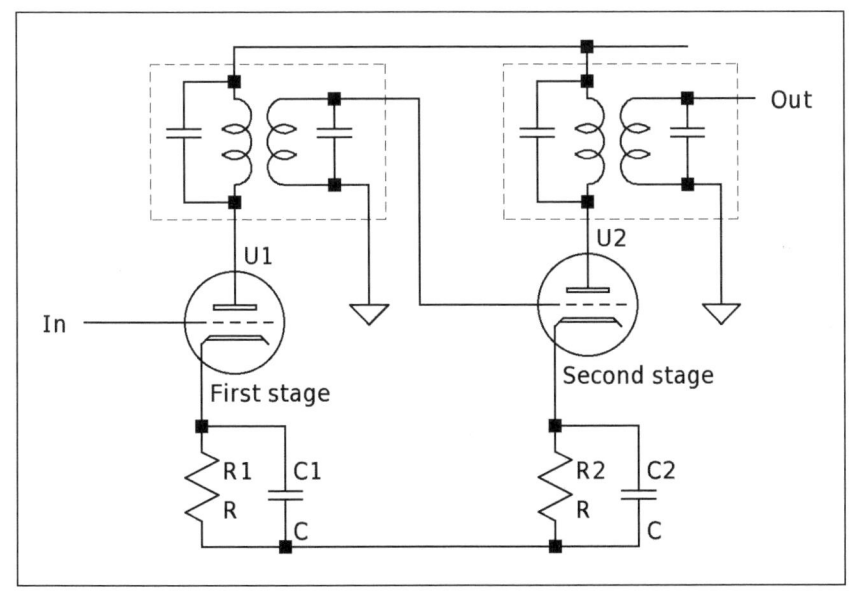

Fig 59: Coupling with IF transformers

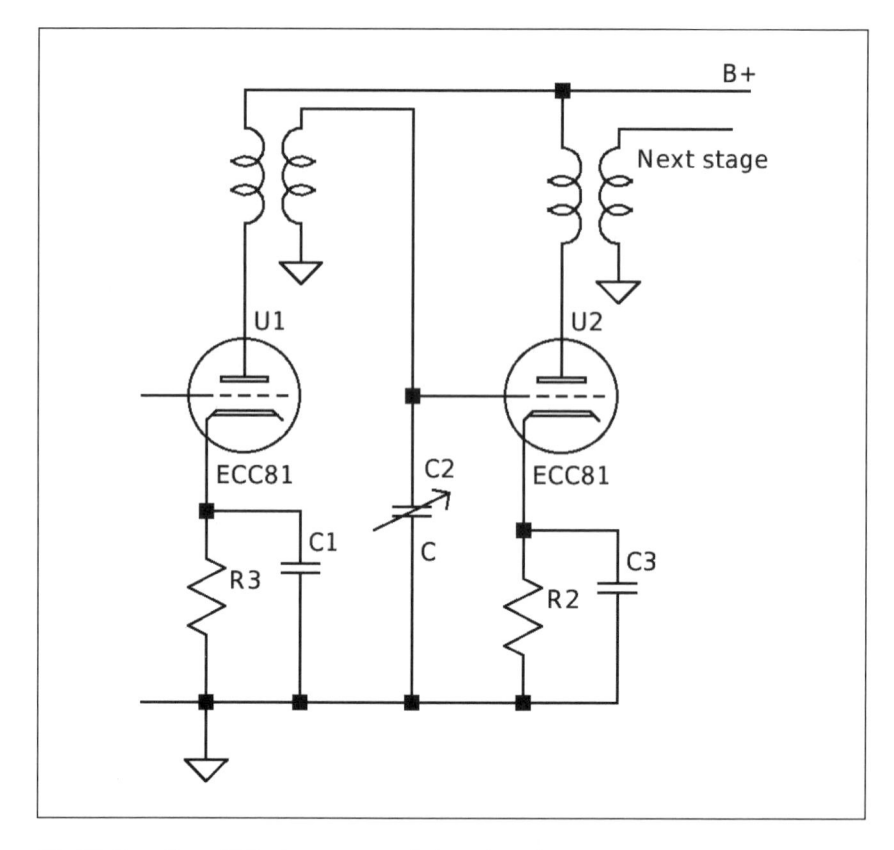

Fig 60: Coupling of RF stages, second stage tuned

RF Stages

An RF stage frequently needs to be tunable. Unless inductive tuning is used, this gives rise to some additional considerations:

- The variable capacitor needs to be grounded
- The capacitor may not tolerate the anode DC voltage

The most common (and simplest) way of solving this is using a transformer coupling with only the secondary tuned, as shown in **Fig 60**. One end of the tuning capacitor is grounded and the grid is grounded through the coil.

It is common practice to add automatic volume control (AVC) to the RF stages. This is done by connecting the AVC voltage from the detector to the bottom end of the coil in the grid circuit through a high value resistor, decoupling the voltage with a capacitor. The resistor needs to have a high resistance. AVC is described in more detail later.

Apart from the variable capacitor, this is the same coupling as the IFT coupling.

A slightly different approach is to remove the secondary winding of the RF transformer and instead connect the tuning capacitor to the anode as in **Fig 61**. However, attention needs to be paid to the fact that the capacitor will have a

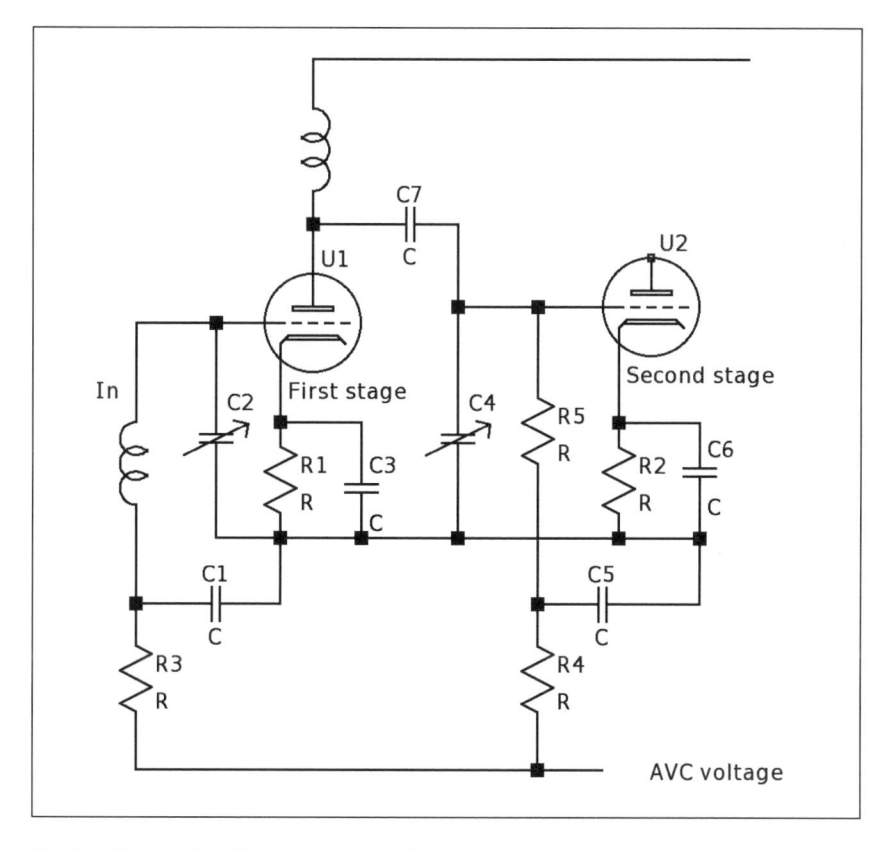

Fig 61: RF coupling, both stages tuned

high voltage on its vanes. Therefore it needs yet another capacitor (C7) in series to block the high anode voltage, and this capacitor needs to be big so as not to influence the value of the tuning capacitor. Also, the grid needs to be grounded (or connected to the AVC voltage), so a grid leak (R5) needs to be included. Other than that, this coupling works just as well. It is a form of reactance coupling.

DC Coupling

In some applications such as television, some audio applications, and in measurement instruments, such as oscilloscopes, it is important that the DC component of the input signal be preserved. Various methods can be employed to achieve this.

The problem with valves is that the input signal is negative at the grid, whereas the output signal at the anode is highly positive. Therefore, the high positive anode voltage needs to be converted to a low negative grid voltage.

One of the difficulties is to maintain balance throughout the chain of amplifiers. Small changes in DC voltages in the system are amplified in the chain and may have a significant impact on the output signal.

If the design is successful, though, the amplifier will amplify signals down to zero Hertz, and audio amplifier manufacturers are spending huge amounts of development money on solving this problem.

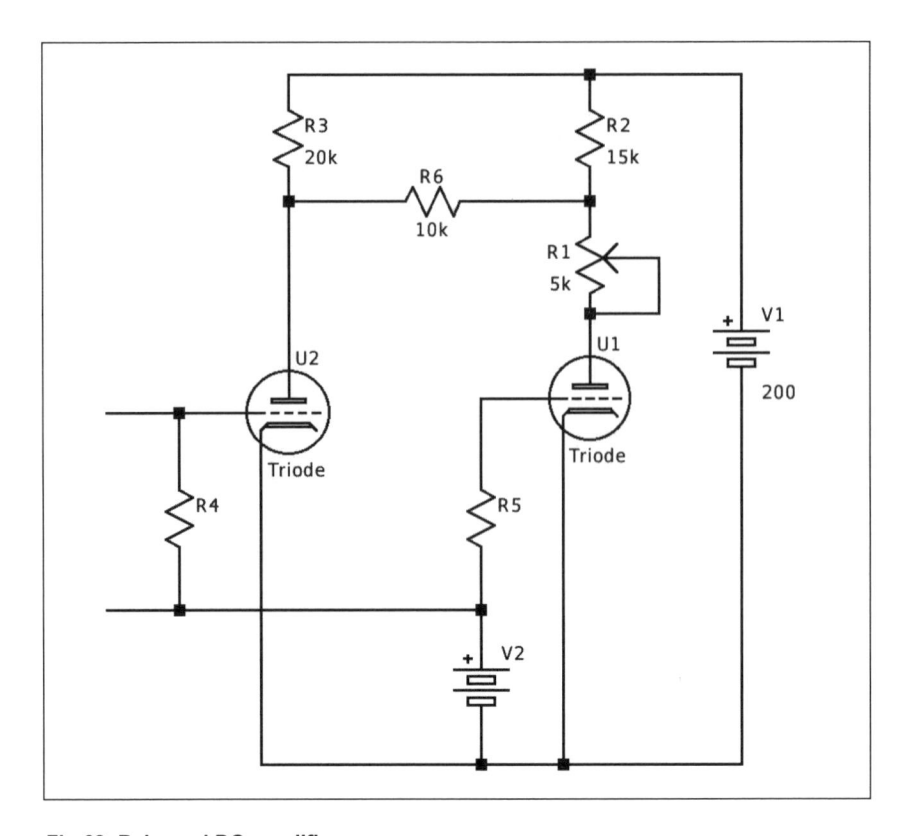

Fig 62: Balanced DC amplifier

One method of balancing the amplifier is shown in **Fig 62**.

The 10k resistor between the anodes receives the output signal. Note that the input needs to 'float', ie not referenced to ground. The potentiometer R1 is adjusted so that no current flows through the load with no signal (the two inputs connected together). When the signal is connected the balance is upset, and the signal amplified. Also note the separate grid bias battery.

In a real life implementation, the batteries would be replaced by (stabilised and well filtered) power supplies.

This is really a single stage amplifier, since both halves of the dual triode participate in amplifying the same signal. A practical example, though, is shown in the chapter on measurements.

Glow Lamp Coupling

By far the simplest solution to removing the anode DC voltage is using one or more glow lamps.

A glow lamp draws very little current, and once lit, keeps its voltage drop fairly constant.

This fact is used in voltage stabilisers, where the lamp is connected in series with a resistor between an unregulated voltage source and ground. The voltage drop across the lamp can be used to drive oscillators and other devices that require a stabilised anode voltage at reasonable currents (see further under power supplies).

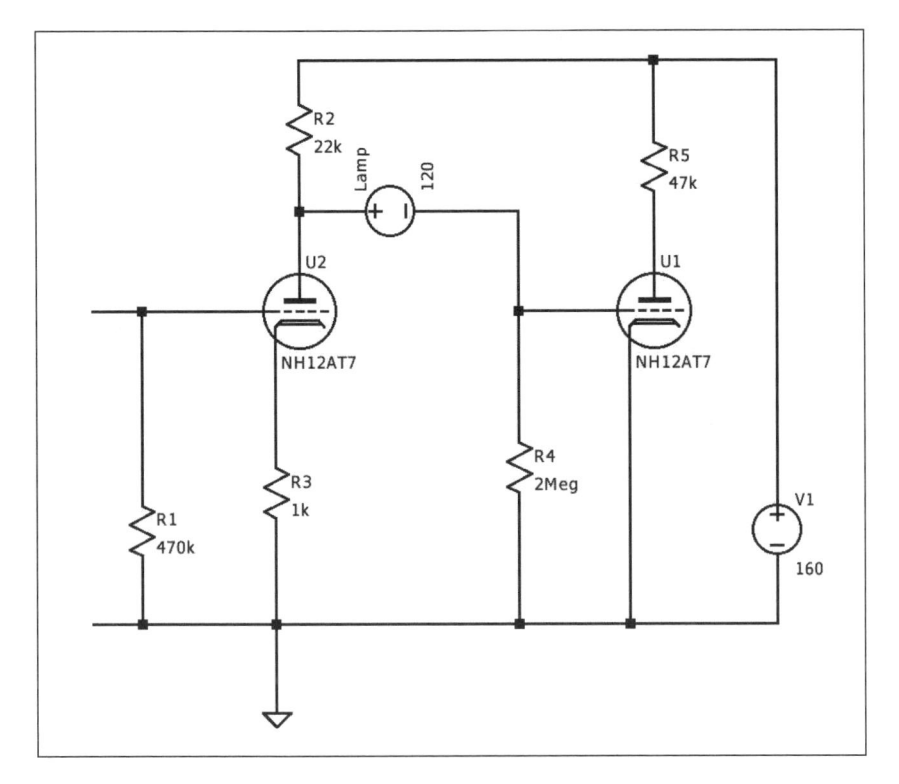

Fig 63: The use of a glow lamp to remove high voltage at the anode

In a DC amplifier, such as the one shown in **Fig 63**, the lamp can be seen as a 'voltage subtracter'. As long as the lamp is lit, the voltage at the grid of the second stage corresponds to the grid bias, provided the high voltage is correct. The result will be a DC coupled amplifier.

The schematic reveals that all capacitors have been removed from the R/C coupling schematic. Also, the coupling capacitor between the stages has been replaced with a glow lamp. Its glow voltage is chosen to be the anode voltage of the first stage plus the grid bias of the second stage. The single lamp in the drawing may represent two lamps connected in series. The anode resistor of the first stage, R2 is best a combination of a fixed resistor and a trim potentiometer.

The balance will be upset if the power supply varies, so a stabilised supply is essential. The signal is taken from the anode of the second stage. R3 is the bias forming resistor for stage 1.

The circuit acts like an R/C coupled amplifier with a leaky coupling capacitor which has the important characteristic that its voltage drop is always the same. A tiny current flows through the lamp and the grid leak of stage 2.

Dual Power Supply

If you have access to a dual high voltage power supply, one half could be used for grid bias and the other for anode voltage generation.

The anode resistor, the coupling resistor and the grid leak of the second stage form a voltage divider, as shown in **Fig 64**. With correct balance between the three, and correct voltages at the power supplies, the system achieves balance

VALVES REVISITED

Fig 64: Resistor / Resistor coupling, using a bias battery to compensate for high voltage

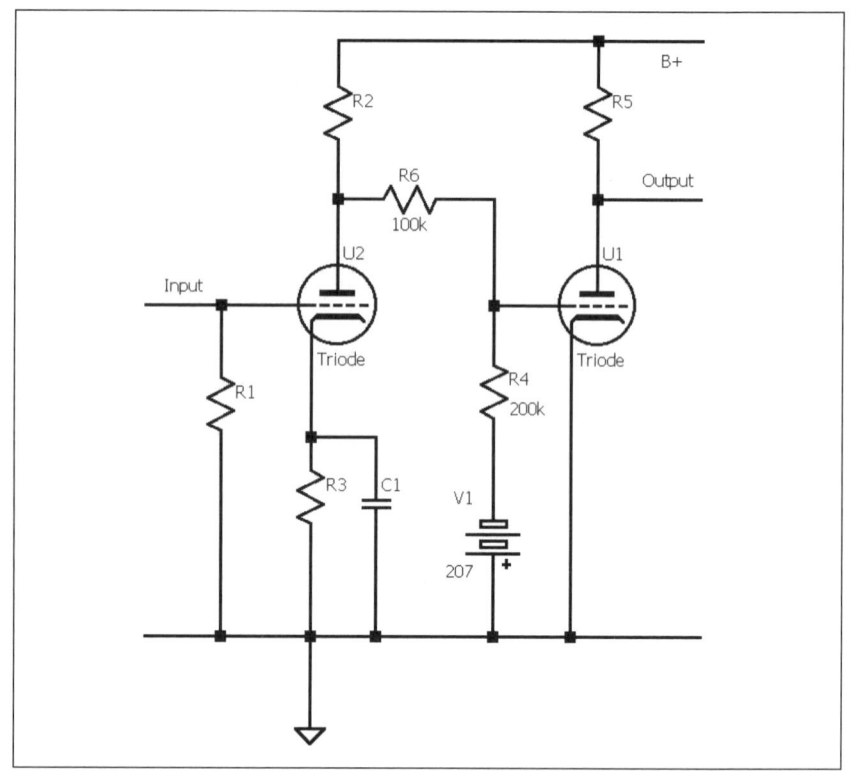

and the DC voltage at the input will be amplified and available at the anode of the last stage.

No doubt, the amplification in the first stage will diminish at a degree corresponding to the relationship between R6 and R4, but the DC component of the signal will be preserved.

Grid Bias

We have seen that the control grid of a valve needs to be slightly negative in relation to the cathode.

We have also calculated two cathode resistors. However, there are several other methods of achieving a negative grid bias.

External voltage source

This includes batteries, but could be a separate power supply too. If the cathode is grounded directly and a battery (or power supply) is connected between the grid leak and ground, with the positive pole to ground (**Fig 65**) you will get the same result as

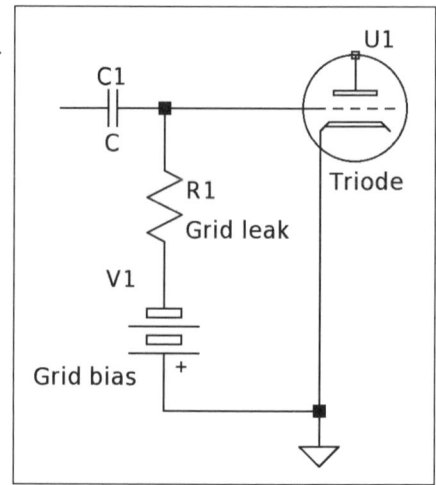

Fig 65: Battery bias

48

with a cathode resistor. The benefit is that you save the capacitor, which in audio applications needs to have a big value. If a number of valves need the same bias, the same battery can be used for all of them. The life span of the battery will be very long, because the grid circuit normally draws negligible current. This method is not very common any more.

AVC voltage
Automatic volume control (AVC) is an essential part of a receiver. It is a self controlling system, since a large aerial signal amplitude generates a larger AVC signal, and an AVC signal decreases the signal through the signal path. A normal AVC voltage can never make a signal disappear. A signal that grows too big despite AVC is a sign that the AVC control is insufficient. AVC is also known as AGC, automatic gain control.

It is a fact that the signal from different stations can vary as much as thousands of times. Even one and the same station can vary very greatly in amplitude. Consequently, if all stations were amplified by the same amount, one would constantly have to be ready at the volume control to set the volume to a comfortable level. To avoid this, AVC was invented in 1925 and patented by the American engineer Harold Alden Wheeler (1903 - 1996).

The AVC voltage is created by the detector and is proportional to the input signal level. It should be zero without a signal and go negative as the signal grows. It is easy to see why it needs to go negative, if you study one of the valve curves, eg Fig 47. Follow one of the Va-curves, and you will see that amplification decreases the more negative Vg becomes.

In principle, the AVC voltage replaces the battery in Fig 65. However, if an AVC voltage is going to be used instead of the battery, a cathode resistor needs to be incorporated as shown in **Fig 66**. The AVC voltage adds to the grid bias generated by the cathode resistor to keep the grid at its intended level without signal.

In order to keep the volume reasonably stable (which is the purpose of AVC), the AVC voltage is brought to several of the receiver's stages. The RF and IF

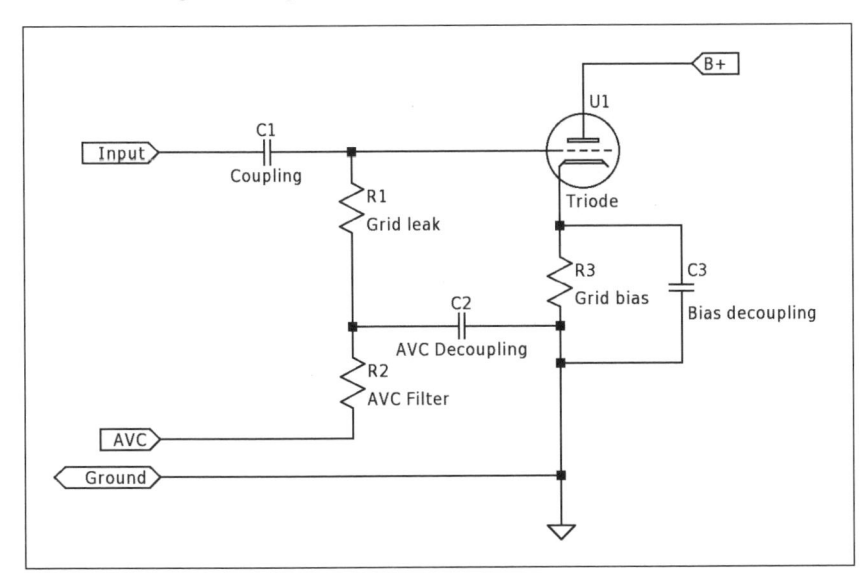

Fig 66: AVC bias

stages are natural candidates. Sometimes one or more mixers are also provided with AVC. AF stages, though, are seldom if ever AVC controlled.

Since several stages are connected to the AVC voltage, steps must be taken to avoid signals passing from one stage to another. This is done by creating 'virtual grounds' at the controlled grid (G1, mainly) of each stage, upon which the normal input structure rests. So, for instance, those parts of the input circuit of a RF or IF stage that create a DC path to the grid (represented in Fig 66 by R1), and which normally had been directly grounded, are, instead, connected to the 'virtual ground'.

This 'virtual ground' is controlled by the AVC voltage and consists of a high value resistor (R2 in Fig 66) decoupled by a large capacitor (C2). The other end of the resistor connects to the AVC detector output. The time constant of the R2/C2 combination determines the maximum rate of change the AVC chain can cope with. It needs to be long enough to stop any RF or IF from passing to other stages, but short enough to be able to follow even fast changing fading. Some communications receivers have a switch whereby the AVC function can be switched off. In those cases, the entire AVC chain is grounded, providing 'real ground' to all controlled stages. The switch allows even fast morse code transmissions to be crisp and clear. Other receivers perform this function when the mode switch is turned to CW. Yet other receivers have a separate AVC switch to select fast, slow and no AVC.

Some receivers control the volume of RF and/or IF stages through the AVC function. Rather than using the potentiometer act as a voltage divider for the signal, it is connected at the cathode and controls the grid bias by controlling the cathode voltage through its current.

Advanced receivers have an additional IF amplifier at the end of the IF chain, not to make the receiver more sensitive, but to increase the AVC voltage's sensitivity. Following that extra IF stage is a dedicated AVC detector, whose sole purpose is to generate the AVC voltage. The AF signal is generated by a different detector.

The AVC voltage begins developing as soon as there is a hint of carrier within the receiver's passband. Sometimes the term 'delayed AVC' is mentioned. This is when the signal strength of the transmitter needs to exceed a certain threshold before the AVC kicks in. Delayed AVC prevents weak stations from being additionally weakened by the AVC voltage.

Delayed AVC is achieved by offsetting the cathode voltage in positive direction of the AVC diode by a suitable amount, some four or five volts. The voltage may be taken from the cathode resistor of the next valve. Another way of achieving an AVC delay is to offset the anode of the AVC diode in negative direction. In both cases, the diode doesn't begin to conduct until the signal strength has exceeded the offset voltage, and therefore doesn't generate any control voltage.

In fact, most receivers, even domestic ones, used delayed AVC. Only the simpler and cheaper ones had 'direct AVC'.

Cathode resistor

The function of the cathode resistor has already been described. Normally, the resistor needs to be decoupled with a capacitor, whose capacitance needs to be big enough to allow for low frequencies to pass. A couple of nanofarads is sufficient at RF and IF, whereas capacitors in the order of several microfarads is needed for AF application, or the base register will suffer. The cathode voltage is not very high, so normal low voltage (10 or 16V) electrolytes can be used in this case.

In pentode and tetrode stages, the cathode current equals the sum of the anode and screen grid currents. This needs to be taken into account when the cathode resistor is calculated. The method is also called 'automatic bias' because if the cathode current increases, so does the voltage drop across the cathode resistor, which causes the anode current to decrease, etc. Automatic bias is shown in **Fig 67**.

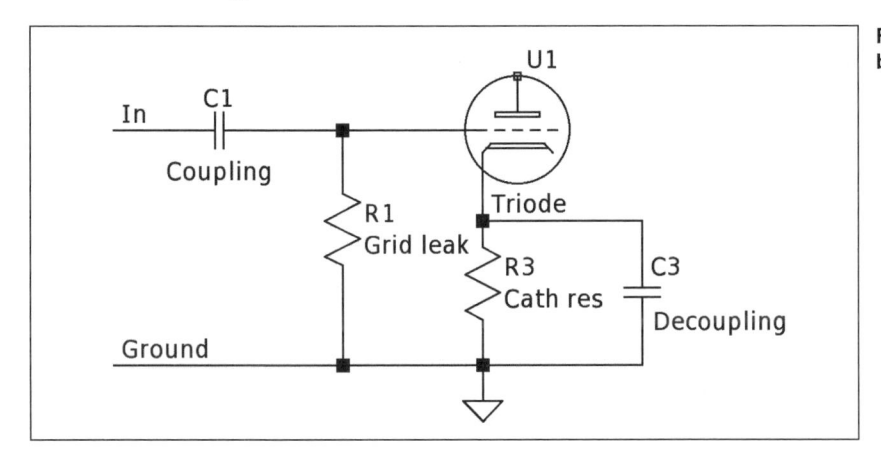

Fig 67: Automatic bias

Grid leak bias

In some detectors, the grid leak bias method is used. The principle is that, instead of the grid leak being connected to ground, it is connected in series with the grid. When the signal goes positive, the grid too goes positive and a small current will flow through the grid leak. This causes the capacitor to charge and its negative side will be facing the grid. When the signal goes negative again, the capacitor will still be charged due to the long time constant of the grid leak and the capacitor, so the grid will go negative again.

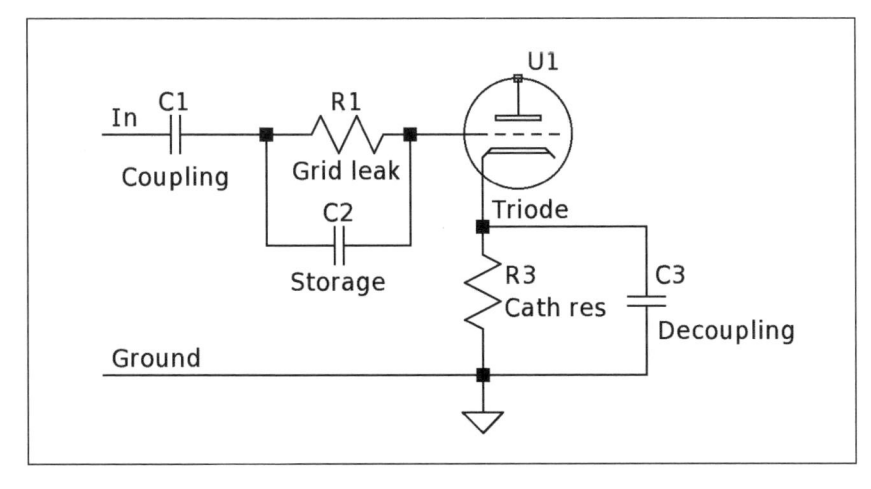

Fig 68: Grid Leak Detector bias

51

Since the method causes the grid to draw current, the grid circuit impedance decreases, so the input circuit will be loaded.

Usually the input is a tuned circuit, which eliminates the need for the coupling capacitor. The grid leak is very large, in the order of a megohm. **Fig 68** shows such an arrangement. The storage capacitor is given a value that allows for storage of the voltage during a full period of the lowest modulation frequency range.

5

Tuned Circuits

VIRTUALLY EVERY RECEIVER or transmitter, however simple or complicated, must contain a number of coils. Generally speaking, the more advanced the equipment, the more coils. This is the motivation for mentioning coils and how to make them.

Winding Coils

Winding coils is not very complicated. It requires a bit of calculation before getting started, though. There is an enormous amount of documentation on coils, and those who are into crystal receivers know a lot about coils and how to make them.

The unit Henry is a measure of the inductance of the coil (from the American scientist Joseph Henry, 1797-1878). In radio frequency (RF) applications, the unit is frequently microhenries (µH) for coils, whereas chokes are in the order of millihenries (mH).

A few general pointers:
- The larger the diameter, the more Henries (although less Q)
- The shorter the coil, the more Henries
- The more turns, the more Henries
- A coil should be at least as long as its diameter.

One formula (Formula 10) for calculating the inductance of a cylindrical single layer coil is:

$$L = \frac{r^2 \times N^2}{(228 \times r + 254 \times l)}$$

Formula 10

where:
L is the inductance in µH
l is the length of the winding in mm
r is the radius of the coil in mm
N is the number of turns.

Separating the turns decreases the inductance. Inserting a ferrous core increases the inductance (and the Q). Inserting a brass core decreases the inductance.

There are other formulae for calculating the inductance of other types of coils.

Q of a Coil

Q is a measure of the quality of the coil and is determined by:
- The frequency at which the coil is to be used
- The length of the wire
- The diameter of the wire
- The material of the wire
- The impedances of the circuits connected to the coil

Q has no unit, but it can be calculated. One way is shown in Formula 11:

$$Q = \frac{X_L}{R + r}$$
Formula 11

where:
X_L is the reactance of the coil at the given frequency
R is the ohmic resistance of the coil and
r is the equivalent resistance as a consequence of the skin effect (see below).

Another way of calculating Q is given in Formula 12:

$$Q = \frac{f}{BW}$$
Formula 12

where:
f is the resonance frequency and
BW is the bandwidth of the filter.

The reactance is given by Formula 13:

$$X_L = 2 \times \pi \times f \times L$$
Formula 13

where:
f is the operating frequency
L is the inductance of the coil
π is a constant, 3.141592

Skin Effect

The phenomenon called 'skin effect' is caused by the fact that the RF current through a coil tends to move towards the surface of the wire. Then a resistance occurs which can be represented by a resistor in series with an ideal inductance. It is a pure loss, whose value increases with the frequency. The equivalent resistance can be calculated, but it is rather complicated and involves parameters that are not generally known. Suffice it to mention its existence, because it does have an impact on coils. The depth of the conductive layer is in the order of millionths of a meter, so it is easy to see that the skin effect does have an influence.

The Wire

At low frequencies below a couple of megahertz or so, Litzendraht (which is a German word. usually shortened to Litz) can be used. It is a wire composed of a great number of very thin wires twinned together. The total surface is thereby increased (or, if you will, the losses are connected in parallel), whereby the skin effect decreases. At higher frequencies the wires act like a number of coils connected in parallel with very hard coupling, and the advantage disappears. Consequently, at higher frequencies, a thicker wire and a bigger coil diameter should be used instead.

In high power transmitters copper tubing was frequently used for winding coils. This is made possible due to the skin effect, which essentially renders the inner core of a coil wire superfluous.

The Q of the coil dominates the Q of a tuned circuit and hence its impedance and bandwidth at resonance. The Q of the coil is easy to measure, since, according to Formula 12 above, if a circuit has a resonant frequency of 5MHz and a bandwidth of 50kHz, then the Q will be:

$$Q = \frac{5000}{50} = 100$$

Increasing the frequency will decrease Q due to the skin effect. Generally speaking, the smaller the coil and bigger the capacitor, the higher the Q. However, there is an optimal point. A coil in the order of 3µH and 330pF gives a good (theoretical) Q at 5MHz.

Q can be measured by first finding the resonant frequency of the circuit, then finding the points at which the voltage across the circuit is 0.707 times the voltage at resonance. Then subtract the lowest frequency of the three from the highest, which gives the bandwidth. Finally, divide the resonant frequency by the bandwidth, and you have the Q. Use instruments with the highest possible impedance and lowest possible capacitance. A good way of removing the influence from the instrument is to connect its probe to another coil which is held at the largest practical distance from the test circuit.

Various Types of Coils and Cores

Pot cores

One way of increasing the Q of a coil is to use a ferrite core, or simply putting the coil inside a ferrite pot. Since ferrite is a material which increases the inductance of a coil, you have to remove a number of turns to maintain the same inductance. This decreases the amount of wire necessary and the skin effect and resistive losses will be reduced. A pot core, as the one in **Fig 69**, is very common these days.

This type of core is sometimes adjustable. One problem could, however, occur. Since the pot is so small, you might find that a thin

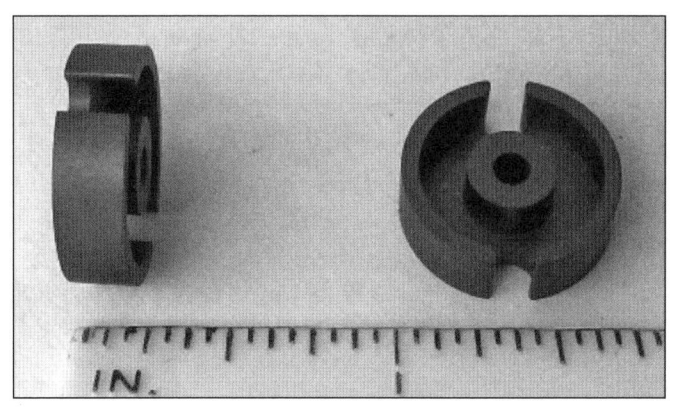

Fig 69: Example of a pot core

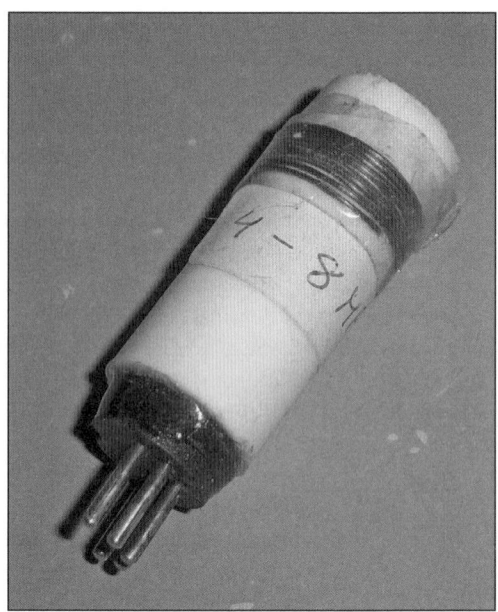

Fig 70: Inductor wound on a toroid core   **Fig 71: Plug-in cylindrical coil**

wire is needed, increasing the skin effect again, and you are back to square one. The Q of these pots may be as high as about 600. Note that the ferrite material has to be the correct one for the desired frequency range.

Toroid cores

Toroid cores (shown in **Fig 70**) are a good idea if the correct material is used and you don't need the core to be adjustable. They are a bit tricky to wind, though.

There are formulae to calculate the inductance for a toroid coil, frequently provided by the core manufacturers. Toroids can yield a Q of over 400 for certain materials and frequencies.

Cylindrical cores

Then there is the classical cylindrical coil.

One type such as the one in **Fig 71** can be plugged in. This coil is exchangeable, which was fairly common in the olden days. The principle is, however, not bad, because it rids your receiver from switches that tend to oxidise and cause other mechanical problems.

Modern cylindrical exchangeable coils can be built with a coil former and a male 3-pin or 5-pin DIN connector.

The variometer

Another older coil is the variometer. It consists of two coils, one inside the other, and they were used in TRF (Tuned Radio Frequency) receivers to vary the coupling between two coils. One coil (the large one) was connected to the grid circuit of the detector valve, and the other to the anode circuit. By varying the coupling between the coils by turning the small coil you also varied the

CHAPTER 5: TUNED CIRCUITS

Fig 72: A home-made variometer

positive feedback (the reaction) of the detector, increasing selectivity and sensitivity, but increasing the noise level of the receiver. They were very common in the 1920s and 1930s, before the superheterodyne was in general use.

The variometer shown in **Fig 72** is actually home-made.

Spider coils

Spider coils, were also used where the coupling factor of two coils needed to be varied. One coil would be fixed and the other could slide to overlap the fixed coil more or less, much like the vanes of a variable capacitor. **Fig 73** shows the components necessary to make a spider coil.

Other arrangements to vary the coupling between the coils is to mount the movable coil onto an axis. Turning the axis allowed the moving coil to approach or distance the fixed coil.

Note that there needs to be an odd number of vanes for the winding to work!

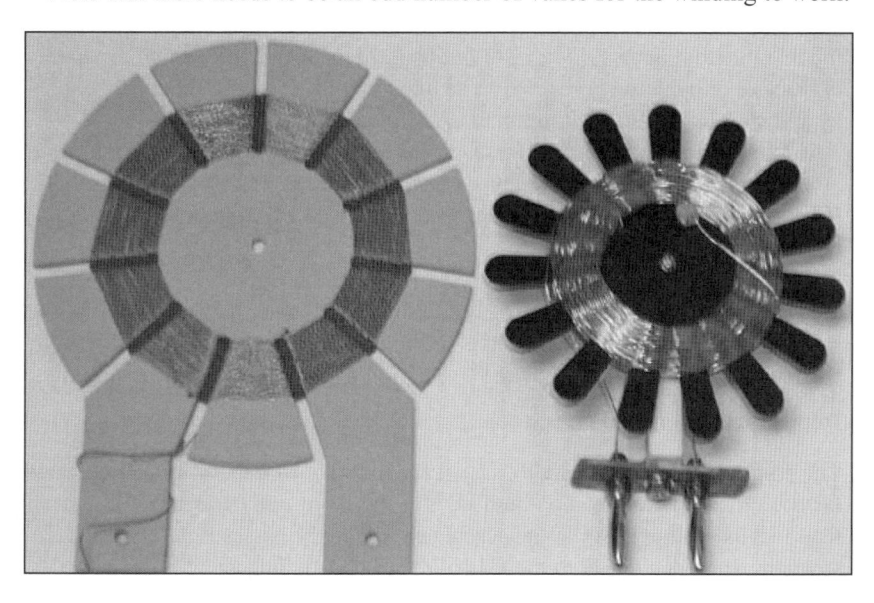

Fig 73: How to make a spider coil

Fig 74: A plug-in honeycomb coil

Honeycomb coils

A honeycomb coil (**Fig 74**) is wound in a zig-zag pattern so that no adjacent turns are parallel to the previous. This way, the capacitance between the turns is low (as opposed to, for instance, cylindrical coils, where all turns are parallel). Low capacitance of a coil means that the tuning range for a given tuning capacitor increases.

They were mounted on holders, one fixed and one movable, as described above.

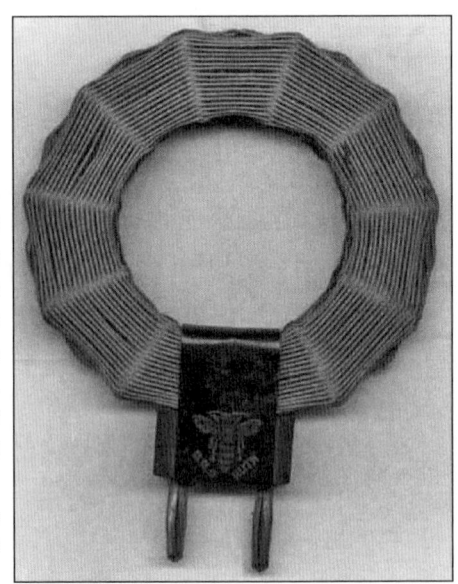

Oscillator coils

Special rules apply to oscillator coils. They need to be as mechanically stable as they can possibly be made. They should be core adjustable, since it is difficult these days to find trim capacitors. They need to have as high a Q as possible. And, of course, they need to connect to a high impedance environment.

Impedances

Tuned circuits

Tuned circuits are two-pole networks which are often used simultaneously as input and output devices. They are frequency dependent, and can be capacitively reactive or inductively reactive. The impedance varies with the input frequency and appears to their environment to be resistive at resonance.

The impedance is high at resonance for a good tuned parallel circuit, up to 50kΩ or more. Measuring the DC resistance, though, returns a low result. The RF current flowing inside a tuned parallel circuit is very high, much higher than the current that is fed into it.

A variable tuning capacitor is frequently not a very good device. The vanes are often very close together (to increase capacitance and decrease physical size), which makes them sensitive for temperature variations. Additionally, particularly in oscillators, they require a high degree of mechanical stability.

Impedance of a tuned circuit

The properties of a tuned parallel circuit are such that its impedance is high at resonance - then it goes down on both sides of the resonance frequency. This means that a signal that appears at the resonance frequency passes with nearly its full strength, whereas a signal to the side of the resonance frequency is dampened.

The Q (or Quality) factor is an important part of a tuned circuit, whether it is parallel or series. Q should be as high as possible because Q determines the bandwidth of the tuned circuit. However, a Q that is too high may result in the bandwidth becoming too narrow for the purpose of the circuit. In some applications, such as the IF amplifier of a TV, it is necessary to decrease Q.

Q was discussed in more detail earlier in this chapter.

What determines the Q of a tuned circuit are:
- How the coil is wound (how many turns of wire)
- What wire is used for the coil (wire's specific resistance, influence from skin effect)
- The Q of the capacitor (which mostly can be ignored)
- The impedance of the environment

In order for a tuned circuit to perform well, it needs to sit in an environment with high impedance. This applies to all environments, be it an oscillator or an RF or IF stage. If this impedance is too low, the quality of the tuned circuit will inevitably deteriorate, causing instability in oscillators, and bad selectivity and low amplification in IF and RF stages.

Since a parallel circuit grounded at one end is both input and output at the same time, any inputs and outputs connected to the circuit need to be high impedance. It is a matter of affecting the circuits as little as possible.

Coupled circuits

Literature mentions 'mutual inductance', or the coupling between two tuned circuits. Amongst other things, this is determined by the distance between the coils. If the coupling is too loose, the amplification of the signal will decrease. If it is too tight, a dip in the middle of the resonance curve will appear. There is an optimal point where the coupling gives the best result. This point is called 'critical coupling'. **Fig 75** shows a series of curves for three different couplings. The one with the double peak is said to be 'overcritical'. The others are 'critical' and 'under-critical'.

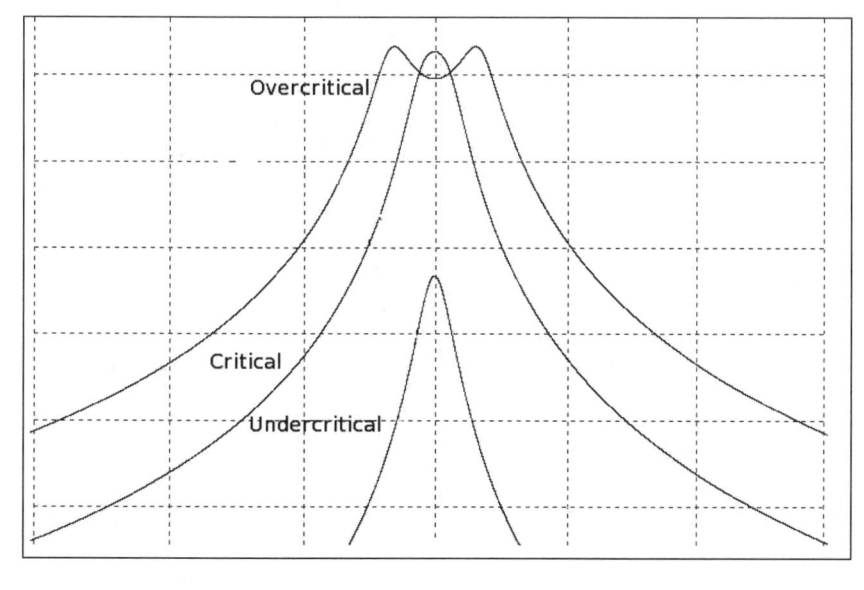

Fig 75: The results of various circuit couplings

They are all tuned to the same frequency and connected to the same inputs/outputs. The only thing that differs is the coupling. You can clearly see the twin peaks of the over-critically coupled circuits. There is no way the two peaks of the overcritical filters can be 'trimmed' away, except by decreasing the coupling.

You can also see that the amplitude of the under-critically coupled circuit is lower. This means that the circuit does not work optimally.

The best (optimal) result is achieved by the critically coupled circuit.

There are different ways of coupling tuned circuits:
- Inductive coupling
- Capacitive coupling
- Link coupling

Inductive coupling and link coupling are closely related. Inductive coupling means that the two coils are placed in each other's vicinity or even wound on the same core. The distance between the coils determines the degree of coupling.

Capacitive coupling can be done in two ways - top capacitive (**Fig 76**) and bottom capacitive (**Fig 77**). The circuits to be coupled are still in different places and isolated from each other. A capacitor is connected between the circuits and there is a risk that this coupling capacitor resonates with one or both coils at a frequency where no resonance is desired.

Bottom capacitive coupling is similar to top capacitive. As Fig 77 shows, instead of directly coupling the coils together, the coils are both grounded at one end, and the coupling capacitor is connected to the lower end of the tuning capacitors. Coupling capacitor (CC) forms a reactance at the given frequency. The higher the reactance (the lower the capacitance), the higher the degree of coupling between the circuits. In this case, an inductance or a resistor can be used instead of the capacitor, with the same result.

The coupling component may be variable to change the degree of coupling. For instance, with careful selection of the coupling capacitor in relation to the resonant frequency, the coupling between two tuned circuits can be kept at optimum (critical coupling), even though the circuits are tuned across a range.

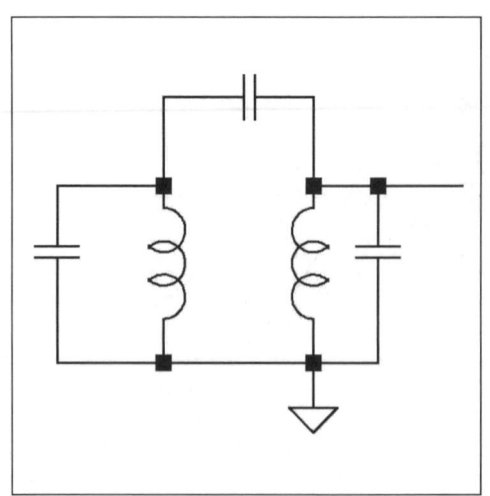

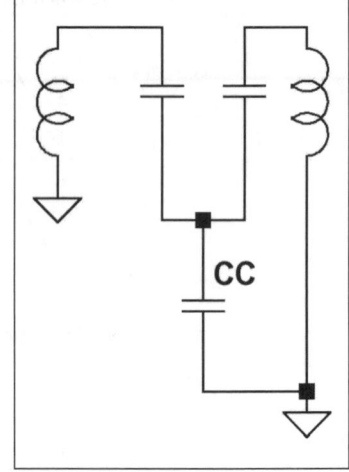

(left) Fig 76: Top capacitive coupling

(bottom) Fig 77: Bottom capacitive coupling

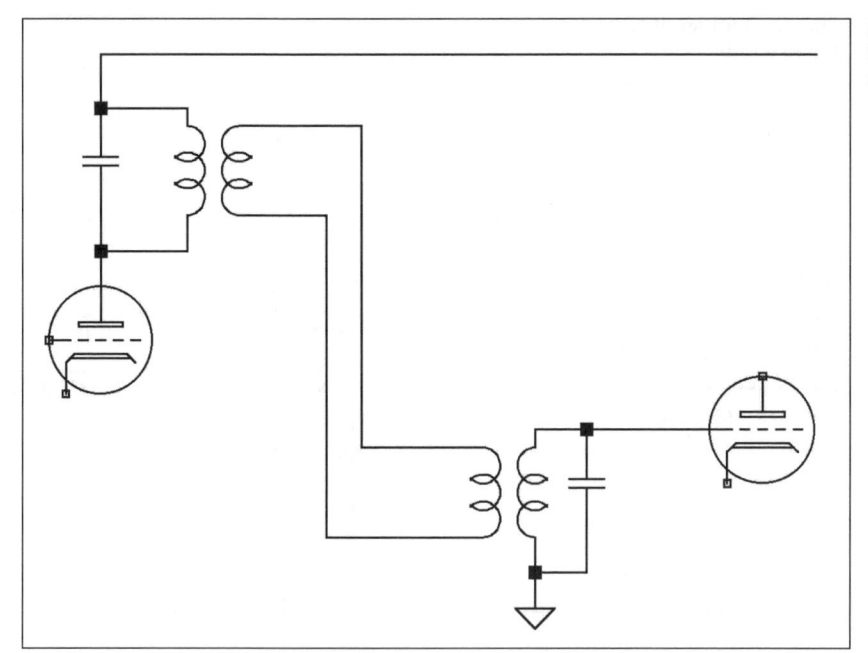

Fig 78: Connecting two stages by link coupling

Link coupling (**Fig 78**) is when the coils are located at some distance from each other. The link (coupling element) between them consists of a few turns near both coils and then the links are connected together. Few turns make the link low impedance, which is good in this case. The link turns are wound at the grounded end of the coils.

'Reading' a tuned circuit

A tuned circuit returns peak performance if it is kept in a very high impedance environment. The higher the impedance, the better the performance. It is undoubtedly true that a tuned parallel circuit only reaches a limited impedance at resonance, even if completely isolated from everything else. On the other hand, careful and thorough design of the circuit's environment does pay off. It is easy to show through practical experimentation that a tuned circuit returns a far better result at the anode of a pentode than it would at the anode of a triode. The output from the circuit increases and the tuning becomes far sharper.

The same applies to the 'reading' process of a tuned circuit. One of the best ways of protecting a tuned circuit from low input impedances is to isolate it from the next stage with a cathode follower. Those can reach impedances up to 30MΩ or more. An alternative is to connect the following stage with a link, a few turns wound around the grounded end of the coil. A tap in the coil would work in principle, but is not recommended, since the tap changes other properties of the coil.

'Writing' a tuned circuit

A parallel tuned circuit has a high impedance at resonance. It draws very little current from its environment. To avoid loading it, it needs a high impedance output. You see often the tuned circuit connected between the anode of a stage and the high voltage rail. This works in most cases, in particular if an RF pentode

Fig 79: Writing into a tuned circuit

provides the output, because a small signal pentode has generally a high output impedance. However, a pentode is noisy. The output impedance of a triode is considerably lower than that of a pentode, but is a lot more silent.

A compromise would be a grounded grid stage with the circuit in the anode. A grounded grid stage has a low input impedance and high output impedance. If you can afford it (space-wise or economically), consider using a cathode coupled pair of triodes, or a cascode pair as replacement for pentodes in an IF amplifier. The cascode gives you high output impedance and high amplification. A grounded grid (or cathode coupled) configuration gives you a lower amplification, but still a high output impedance.

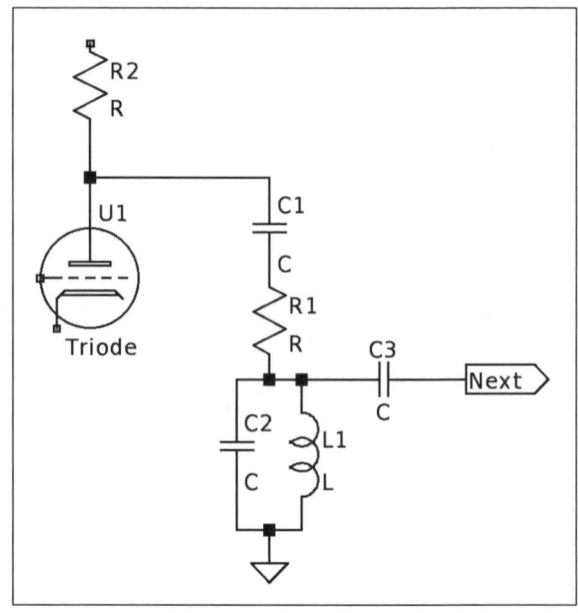

If neither of these options is available to you, you might consider connecting the circuit to ground and a capacitor and resistor in series from the anode, as in **Fig 79**. Since the circuit impedance is high, the resistor can be big. The value of C1 can be made small too.

Tuning

For a station to be heard in a receiver, it needs to be tuned in. Tuning is changing the properties of a tuned circuit such that it begins resonating on a different frequency. Since there are two components in a tuned circuit, a capacitor and a coil, changing one of them is enough to change the resonating frequency of the circuit.

For reference, Formula 14 below (also known as the 'resonance frequency formula') shows the relationship between the resonant frequency, the capacitance and the inductance:

$$f = \frac{1}{2 \times \pi \times \sqrt{(L \times C)}}$$

Formula 14

where:
f is the resonant frequency
L is the inductance
C is the capacitance

The formula is valid for both series and parallel circuits.

Capacitive Tuning

This is the most common method of tuning in receivers for long-wave (LW), medium-wave (MW) and short-wave (SW). Capacitive tuning is changing the capacitance of the circuit. It may be a Varicap (a semiconductor diode that changes its capacitance with a voltage), a variable capacitor, a reactance valve (which is a valve stage connected such that it varies the phase in proportion to a voltage, used in phase locked loops and frequency modulators) or similar. In most domestic receivers the tuning component is a variable capacitor.

A variable capacitor has very rarely a linear relationship between frequency and turning angle. Nor is it likely that the wavelength is proportional to the angle of the tuning shaft. This is why the frequency/wavelength index on a domestic receiver is non-linear.

Variable capacitors and tracking

Since you need a tuned circuit before the mixer and want to tune the oscillator and the pre-mixer circuit the same amount with one control, you will need a two gang capacitor. However, the two frequencies are different - the difference is 455kHz in a European domestic receiver (465kHz in UK receivers). This makes a big difference on the lower wavebands.

The solution may seem simple, but it is not. When using a capacitor with a linear capacitance characteristic, there will be only one point within the range which will be correct. One solution is making capacitors that cause a certain frequency difference at any setting.

Such capacitors have a slightly different mechanical size of the vanes in the two sections. Since the oscillator is tuned to a higher frequency than the mixer, it is the smaller section that controls the oscillator frequency. This is, however, not the entire solution. The capacitance relationship is correct only on one band.

There is one solution, though, that gives the correct tuning in both circuits for all bands. You can build a one-band receiver for a band where the tuning is correct and connect a converter in front of it to cover all the other bands. This is not usually done in domestic receivers, though, due to the higher production cost.

Instead a different (cheaper) technique was used, called padding. You use a tuning capacitor of the same size in both sections. Then the oscillator section is corrected using trimmer capacitors.

The relationship between the maximum and minimum frequencies in the oscillator section is slightly higher than those in the mixer section, since they both need to cover an equal amount of kilohertz, but on different bands. The minimum frequency can be decreased slightly by connecting a small capacitor in parallel with the tuning capacitor.

Then the range is increased again by connecting a large capacitor in series with the other two. By carefully selecting coils and capacitors, it is possible to make the two circuits to follow each other fairly well. Strictly speaking, the tuning will be perfect in only three points - in the middle and at both ends. The method is, however, good enough to hide the shortcomings.

In reality, the small capacitor is trimmed at one end of the band, and the coil is adjusted at the other end. This gives the same result.

VALVES REVISITED

Fig 80: Brass rod, suitable for linear tuning

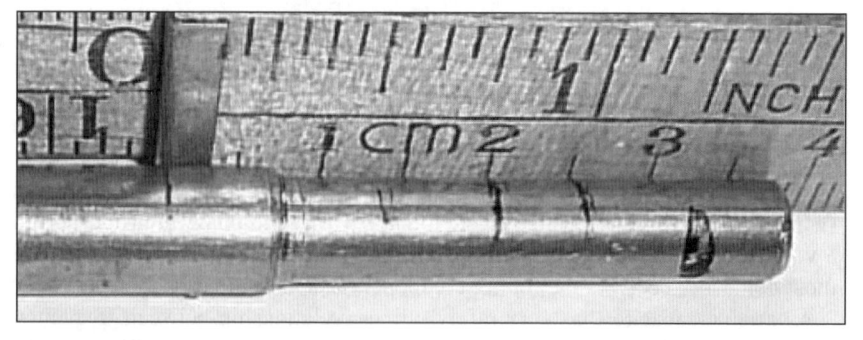

Permeability tuning

Many receivers for commercial and amateur use have frequency linear tuning, ie the distance between kilohertz markings are the same regardless of where you tune within the band. This is a desirable feature that most AM receivers are lacking. FM receivers, though, do have frequency linear tuning. This is achieved by varying the coil rather than the capacitor.

They are all based on the fact that the inductance of a coil increases (frequency decreases) if an iron powder core is inserted, but decreases (frequency increases) if instead a brass rod is inserted. Using an iron powder core decreases frequency with a factor 1:3.4 or so (slightly more than a big variable capacitor), whereas a brass rod causes a considerably smaller change in frequency, about 1:1.15 (the figures are results of experiments with an oscillator, using the same coil but inserting different materials into the coil). **Fig 80** shows a brass rod that has been used in experiments with linear permeability tuning. There is 100kHz between the markings. An inductive tuning mechanism must be very rigid to eliminate instability while tuning.

There is a number of more or less complicated designs for the mechanics behind permeability tuning. Let us look at FM tuners to get some hints.

Fig 81: Linear tuning arrangement for FM. The spiral shaped 'wheel' is clearly visible.

64

CHAPTER 5: TUNED CIRCUITS

Philips made a model of portable receiver that became popular. It had a big tuning knob with a little button that was used to set markings around the knob, so that the frequencies could easily be found again. The tuning mechanism was a spiral at the back of the knob that controlled the position of two rods, one for the oscillator coil and the other for pre-selection. Inside the spiral was inserted a little nipple that followed the spiral and which in turn moved the brass rods.

A slightly different approach is to manufacture a spiral shaped 'wheel' fixed to the tuning shaft, as in **Fig 81**. The spiral is connected to a yoke which in turn moves the rod. The yoke was spring loaded to allow for travel back when the shaft was turned the other way.

The coils are individually adjustable by means of the adjustment screws at the top of the yoke, as **Fig 82** shows. Also note the load spring fixed to the yoke.

Another solution is shown in **Fig 83**. It was based on a piece of string passing over a pulley and that was connected directly to the rods at each end. Both ends of the string were fixed to the tuning shaft around which it wounded as you tuned

Fig 82: Alternative linear tuner for FM

Fig 83: Permeability tuner for FM

VALVES REVISITED

Fig 84: Yet another tuner for FM

Fig 85: FM tuner with yoke

the receiver. The string was loaded with a spring to keep it tight at any setting. The rods could not be individually adjusted with this tuning device. Also, one coil had to be wound at one end of the former, and the other at the other end.

The pulley, string and one of the two coil formers are seen in this picture. The other coil is faintly seen behind the one in front.

The tuning shaft and the two coils are shown in detail in **Fig 84**. Note how the string is attached to the shaft and how the coils are mounted asymmetrically. The spring wound around the shaft keeps the string tense.

Yet another solution, shown in **Fig 85**, was a hinged yoke, connected to the rods. A string was wound around the tuning shaft, went up to a pulley fixed to the yoke and down again to a fixed point. As the shaft was turned, the string wound or unwound from the shaft.

As the yoke was moved down, it inserted the rods into the coil formers. At the top of the yoke you can see the inductance adjustment screws that allow for individual trimming of the coils. Here is clearly seen that the two coils have different inductances (fewer turns on the left hand coil) but are wound along the same length of the former.

A similar solution is found in Drake (the RL Drake company was founded in 1943 by Robert L Drake) R-4a receivers, where the yoke's movements were controlled instead by a friction coupling to the tuning shaft.

The VFO and associated circuits in more elaborate receivers (Collins, Drake, Heathkit and others) were permeability tuned with a frequency linear scale. The tuning shaft was connected to a gear which turned a long screw. The screw was inserted into inner threads, like a nut, so that when the shaft was turned, the 'nut' moved out and in, carrying the rod out and in into the coil former. This is the standard mechanism for tuning VFOs on short wave. Again, it must be very stable, and provisions made to avoid backlash while tuning.

6

Amplifiers

AS IS THE CASE with transistors, valve amplifiers come in three categories: Grounded cathode, grounded grid and grounded anode. Combining basic amplifiers can result in an amplifier with completely different properties.

Considerations for RF, IF and Hi-fi Stages

There are many things to consider when building for RF and IF. These considerations are valid for valve equipment and semiconductor equipment alike. Oscillators have particularly high demands (see later in this book).

- Mechanical stability is important
- Use only one ground point per stage
- Make sure the components' leads are kept short
- Use short leads between the stages
- Separate the output circuit components from the input circuit components
- If you must use a coax to connect a stage to another, the signal should be buffered by a cathode follower (see later in this chapter)
- Pay special attention to the solder joints

Feedback

One important aspect when building RF and IF stages is feedback, both positive and negative. A stage with positive feedback may begin to oscillate and its performance will be compromised. If the stage is close to oscillating, the bandwidth of the stage will be smaller than intended. The degree of feedback may vary with mechanical and/or thermal influences.

Certain valve holders have a metallic pin in the middle. It is intended to be used as a screen between different parts of the valve and a grounding point for the stage components, and should be well grounded, even if not used for components. The higher the frequency, the more important screening becomes.

The components are best mounted directly onto the holder, even though it may not always be very neat. Neatness is less important here than functionality.

Components mechanically mounted in parallel may be neat, but every component generates an electro-magnetic field, which may cause problems. The various components of a stage have different functions, and some of them are more sensitive to influences from their neighbours than others. Ideally, the components should be mounted at a 90 degree angle, but this is not always practical.

VALVES REVISITED

Solder lugs may constitute ground or mounting points, but they have to be properly mechanically fixed.

Temperature

Components that generate heat should be kept as far away from oscillators as possible. If they are not, the heat will spread into the oscillator circuits and cause instability and frequency drift. You should also make sure that the equipment is well ventilated. Overheating is not good. If you are using a closed chassis, you should drill ventilation holes here and there to allow air to flow through the box. Make sure the air is allowed to circulate. This is particularly important in transmitters and audio amplifiers, where final amplifiers and power supplies can generate a lot of heat. PC fans are cheap these days and could be used in many cases.

Grounded Cathode

The grounded cathode configuration (**Fig 86**) is by far the most common one. The cathode is, as the name implies, grounded, the grid is the input and the anode is the output.

The component values and valve type to choose have to be determined by the frequency range for the stage. For audio, the capacitors need to be high value and the triode may be one to use (eg ECC82). Note that C3 is the coupling capacitor that leads to the next stage. One end is connected to the high voltage anode. Therefore, this capacitor needs to be of a high voltage type, 400V - 600V. Resistor R1 is the grid leak and would be in the order of 100k - 470k. R2 is determined by the valve type and stage function and would be in the order of 30k - 100k. R3 is calculated from the grid bias and cathode voltage, usually less than a couple of kilohms.

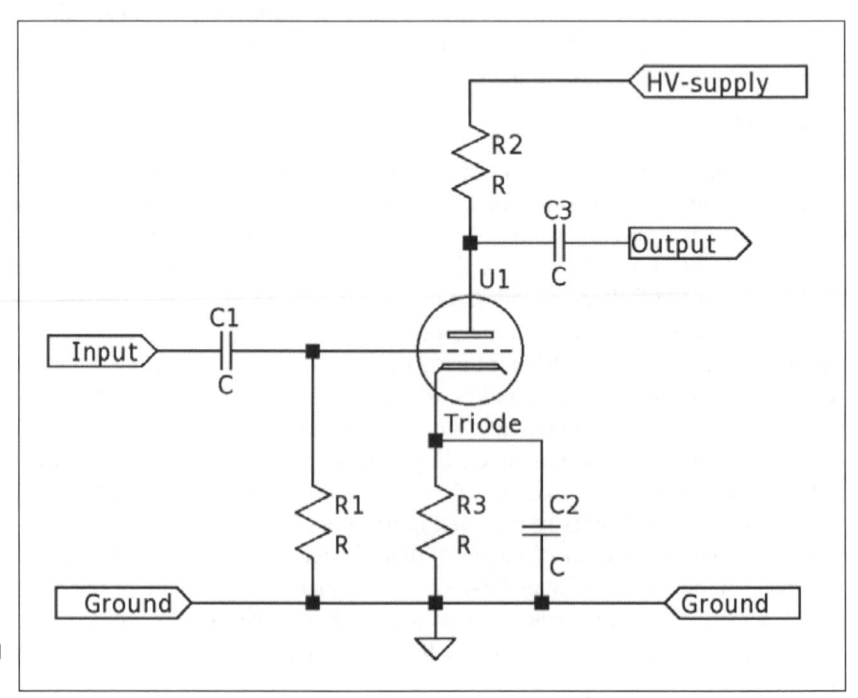

Fig 86: Grounded Cathode

As the grid voltage changes in the positive direction, the valve current increases, increasing the voltage drop across the anode resistance and decreasing the anode voltage. Therefore, the phase shift from input to output is 180 degrees.

The input impedance at low frequencies for a triode in a grounded cathode stage approximates the value of the grid leak.

The amplification of the stage is calculated according to Formula 15 or Formula 16:

$$A = \mu \times \frac{Ra}{ra + Ra}$$ Formula 15

$$A = Gm \times \frac{ra \times Ra}{ra + Ra}$$ Formula 16

where
μ is the amplification factor
R_a is the anode load resistance
r_a is the internal impedance of the valve
A is the amplification
Gm is the transconductance

The output impedance at low frequencies can be calculated as approximately ra in parallel with Ra.

See **Table 7** for a summary.

> **Fact sheet - Grounded Cathode Stage, Triode**
> **Input impedance:** Moderate
> **Output impedance:** Moderate
> **Voltage amplification:** Moderate
> **Capacitance:** Cag high due to the Miller Effect
> **Phase shift input-output:** 180 degrees
> **Applications:** AF, IF, RF

Table 7: Grounded cathode amplifier fact sheet

Grounded Grid

A grounded grid stage, shown in **Fig 87**, has a low input impedance and a high output impedance. It is frequently used as an impedance converter, and in the final stage of a transmitter. Further, since the grid is grounded, it also acts as a screen between the anode and the cathode, eliminating the anode-cathode capacitance. The cathode acts as the input and the anode is the output.

When the cathode voltage changes in the positive direction, the grid becomes more negative relative to the cathode, decreasing the valve current and the voltage drop across the anode resistance. Consequently, the anode voltage increases. The phase shift from input to output is therefore zero degrees.

Fig 87: Grounded grid stage

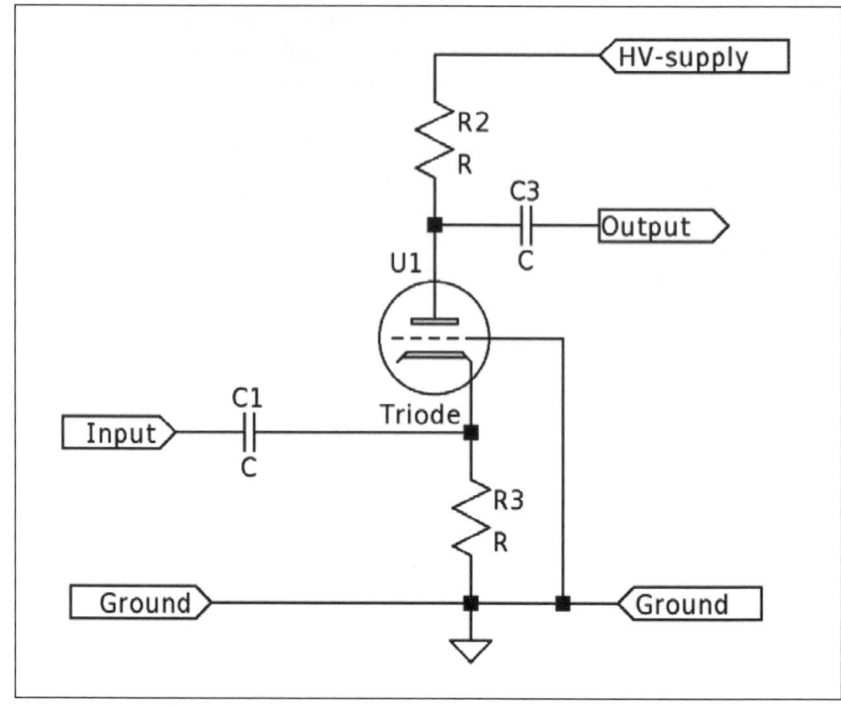

For a summary of the Grounded grid amplifier, see **Table 8**.

Note that the decoupling capacitor is missing from the cathode. Using it would short circuit the signal to ground. The same component calculations apply.

The input impedance for a triode in a grounded grid stage is calculated as shown in Formula 17 or Formula 18:

$$Zin = \frac{ra + Ra}{(ra + Gm) + 1} \qquad \text{Formula 17}$$

$$Zin = \frac{ra + Ra}{\mu + 1} \qquad \text{Formula 18}$$

where
Zin is the input impedance
Gm is the transconductance
Ra is the anode load resistance
ra is the internal resistance of the valve

If ra x Gm is much bigger than 1, and Ra much bigger than ra, the impedance calculations can be simplified as shown in Formula 19:

$$Zin \approx \frac{1}{Gm} \qquad \text{Formula 19}$$

Fact Sheet - Grounded Grid Stage, Triode

Input impedance: Low
Output impedance: High
Voltage amplification: Moderate
Capacitance: Cag low due to the grid being grounded
Phase shift input-output: 0 degrees
Applications: RF, more recently also AF, impedance conversion

Table 8: Grounded grid amplifier fact sheet

The amplification of a grounded grid stage is shown in Formula 20:

$$A = (\mu + 1) \times \frac{Ra}{ra + Ra} \qquad \text{Formula 20}$$

where
A is the amplification
μ is the amplification factor
Ra is the anode load resistance
ra is the internal anode resistance

The output impedance of a grounded grid stage is given by Formula 21:

$$Zout = ra + (1 + \mu) \times Rk \qquad \text{Formula 21}$$

where ra is the anode impedance of the valve,
μ is the amplification factor,
Rk is the cathode resistance.

A Compromise

The input impedance of a grounded grid stage is relatively low. This means that if the input is connected to a tuned circuit which needs a high Q, the Q will be decreased by the input of the stage.

To overcome this, a stage according to **Fig 88** was invented, where the ground point is determined by a couple of capacitors. That way, the input impedance is controllable (within limits, of course).

The ground point is determined by the relationship between the capacitors C2 and C3 according to Formula 22:

$$X = \frac{C1}{C1 + C2} \qquad \text{Formula 22}$$

If X = 1, the ground point is at the cathode, whereas X = 0 places it at the grid. Across the tuned circuit is a resistance as shown in Formula 23,

$$R = \frac{ra + Ra}{X \times (\mu + X)} \qquad \text{Formula 23}$$

Fig 88: Intermediate grounding

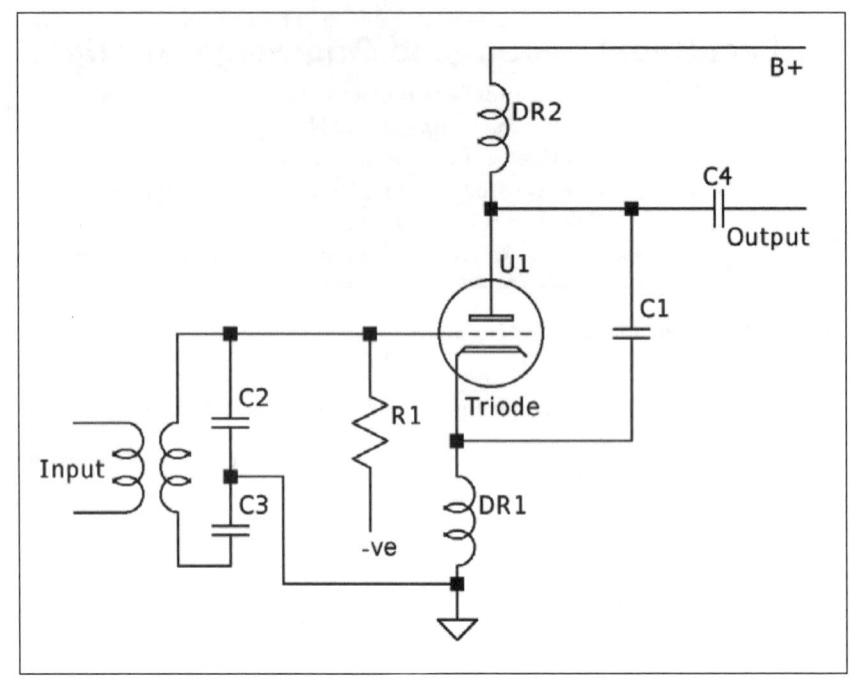

So, when $X = 0$, R equals infinity, ie the input circuit is loaded only by the input impedance of a grounded cathode stage, whereas $X = 1$ yields an input impedance of a grounded grid stage. Similarly, the amplification of the stage is dependent on X.

The greatest benefit of this stage is that designing it for maximum amplification also gives lowest noise. At the same time, a low noise triode can be used without the need for elaborate neutralisation at high frequencies. The design was frequently used as UHF amplifiers in FM receivers.

Grounded Anode (Cathode Follower)

The grounded anode configuration is also called 'cathode follower'. A cathode follower has very interesting qualities, but also a couple of drawbacks. It is very useful in many applications.

Its prime characteristic is that it transforms a high or very high input impedance to a low output impedance. It is therefore very useful to drive long transmission lines, for 'reading' tuned circuits, as an input stage in measurement instruments etc.

A cathode follower has the following characteristics:
- High or very high input impedance
- Low output impedance
- Large bandwidth
- Low input capacitance
- Voltage amplification less than one
- Power and current amplification
- Phase shift from input to output = 0°

CHAPTER 6: AMPLIFIERS

> **Fact Sheet - Cathode Follower**
> **Input impedance:** High or very high
> **Output impedance:** Low
> **Voltage amplification:** Close to one but always less than one
> **Capacitance:** Cag low due to the negative feedback and low voltage amplification
> **Phase shift input-output:** 0 degrees
> **Applications:** Impedance conversion, high impedance inputs

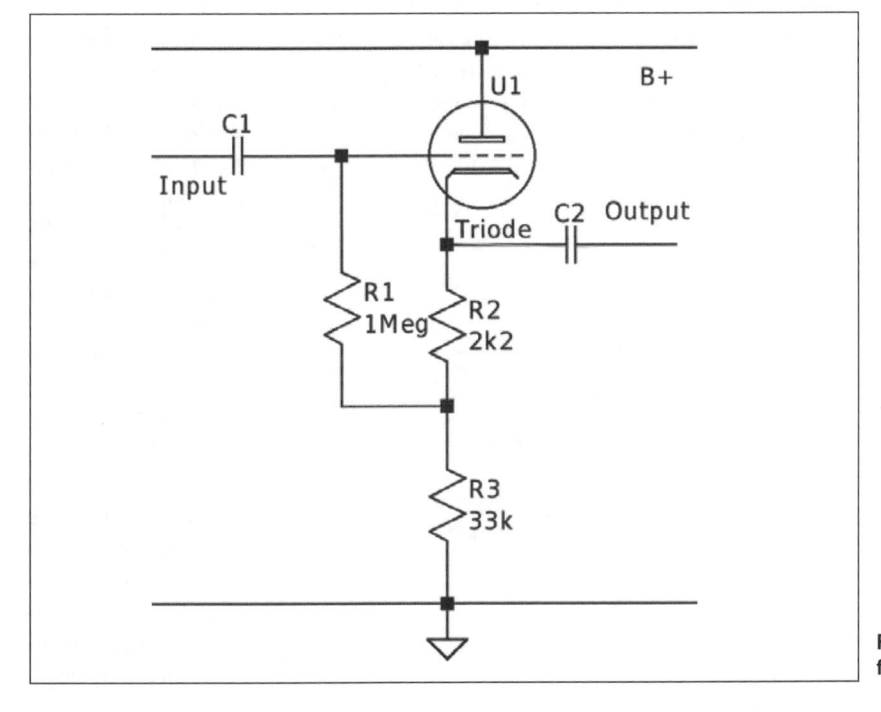

Table 9: Cathode follower fact sheet

Even though the output impedance is low, and even though the cathode follower has current and power amplification, there are still limits to how much current and power it can give. It is therefore not suitable for driving heavy loads.

The principle of a cathode follower is an amplifier with very heavy negative feedback (near 100%). The feedback yields the wide bandwidth, the low amplification and the low input capacitances. See **Table 9** for a summary and **Fig 89** for illustration.

The principle for building a cathode follower are:
- Select a valve with a high amplification factor and low anode resistance ('rp' or 'ra' in the data sheet) to get a low output impedance.
- Connect the anode directly to the power supply, or decouple the anode efficiently with a capacitor.

Fig 89: Cathode follower

73

- Select a cathode resistor which with the given current at a selected bias point gives the desired grid bias (R2).
- Connect the grid to the lower end of the resistor via a grid leak in the order of 1MΩ (R1).
- Connect R3 from the intersection of R1 and R2 to ground. R3 can be 20 - 30kΩ or more.

Done! The signal is connected to the grid and is taken from the cathode. The input impedance can be calculated as in Formula 24:

$$Zin = \frac{R1}{1 - \frac{R2}{R2 + R3}} \qquad \text{Formula 24}$$

This shows that the input impedance is determined only by the resistance combination at the input.

The voltage amplification is always less than, but close to, one. This is because of the high degree of negative feedback (100%). How close to one it comes depends on the amplification factor of the valve used and the configuration at the cathode (see below).

We know that negative feedback in an amplifier increases its bandwidth. The cathode follower has (very close to) the entire signal fed back, so the bandwidth will be very wide indeed. The low input capacitances are due to the low amplification. The Miller effect does not come into effect here.

According to simulations, the bandwidth is well over 100MHz. The 3dB points are at 100Hz and 250MHz. The coupling capacitor C1 determines the lower limit. 100nF reduces the lower limit to less than 1Hz. The output signal follows the input signal very well (hence the name 'follower') due to the feedback.

Another good quality is that the output impedance is independent of the input impedance, as opposed to, for instance, a transformer. A cathode follower can therefore be beneficial as the input stage of a receiver. The impedance of an antenna depends strongly on the wavelength in relation to the electrical length of the antenna and can vary within wide limits. Using a cathode follower at the input makes the receiver independent of the antenna size (up to a limit).

Some people receive long wave or ultra long wave signals with an aerial only a meter or so long successfully with a cathode follower (or its semiconductor equivalent) as the input stage.

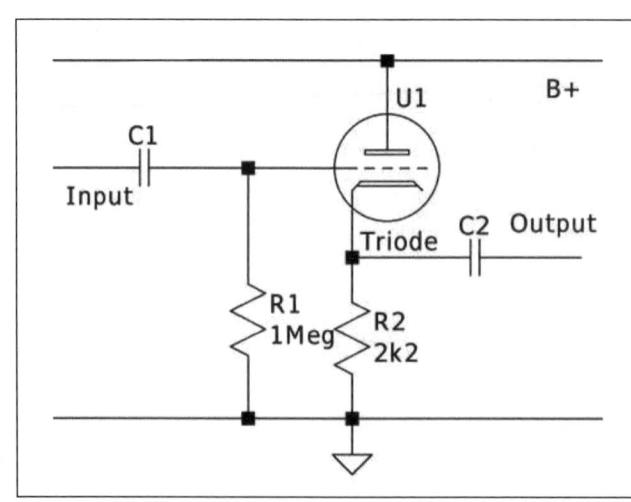

Fig 90: Simple cathode follower

The above example is the slightly more complicated cathode follower. The simple version has only one cathode resistor, as shown in **Fig 90**.

This is the simplest cathode follower imaginable. Two resistors, two capacitors, and a valve is all it takes. This version has a few drawbacks in comparison to the above example, though. These include a lower output level and a lower input impedance. The bandwidth is about the same, though.

The output impedance is calculated according to Formula 25:

$$Zout = \frac{ra}{1 + \mu}$$ Formula 25

where ra is the anode resistance as given in the data sheet
μ is the amplification factor of the valve

Some sources gives the output impedance as in Formula 26:

$$Zout = \frac{1}{Gm}$$ Formula 26

There is a way to additionally increase the bandwidth and output amplitude. Connecting a choke in series with the cathode resistor does the trick. The resistor needs to be there to generate the appropriate grid bias.

A choke is a more efficient stopper for RF than a resistor, which means more bandwidth. **Fig 91** shows how this can be done. Since the output is taken from the cathode, the amplitude will also be higher.

It is less beneficial to connect a choke at the cathode of the more complicated follower above. The reason is that R3 already plays an important part in the feedback, so very little would be gained.

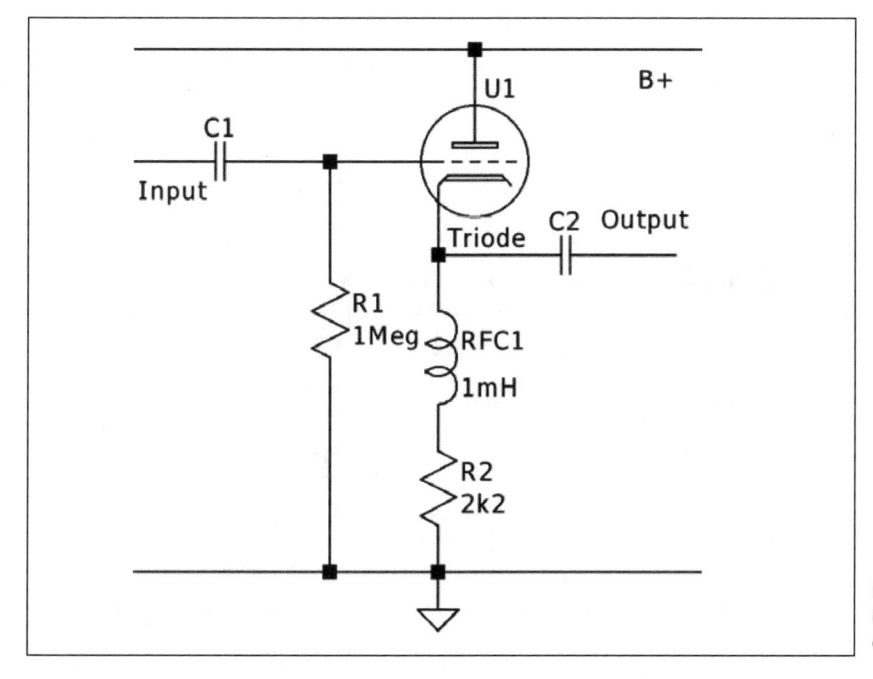

Fig 91: Cathode follower with choke output

VALVES REVISITED

The cathode resistor is calculated from the desired grid bias and the cathode current.

If you build this cathode follower, make sure that the choke can withstand the cathode current. Some modern chokes tend to break in valve applications. A suitable inductance value for RF is a couple of millihenries.

There are many applications for a cathode follower. Whenever you have a source with a high impedance that you want to match to a load with a lower impedance, a cathode follower would be the ideal solution. For instance, a microphone with a high impedance needs to be connected to a long coaxial wire. A cathode follower reduces or eliminates the load on the microphone and reduces or eliminates hum and noise that could otherwise be picked up along the way.

An amplifier with a low impedance input, such as a grounded grid stage, can be driven by a cathode follower (see below).

There are additional cathode followers, some involving more than one valve and are a lot more complicated. It is beyond the scope of this chapter to deal with them. An entire book could be written on cathode followers alone.

Anode Follower

The name 'anode follower' is not strictly correct. Whereas the cathode follower got its name from the fact that the output follows the input, the anode follower inverts the signal, and so does not follow the input signal.

The main difference from the standard grounded cathode stage is that there is a feedback chain between the anode and the grid. **Fig 92** shows how.

The input impedance is determined by a fairly low value resistor in series with the grid. The amplification is only moderate, albeit adjustable. The bandwidth is limited by the Miller effect. A summary is shown in **Table 10**.

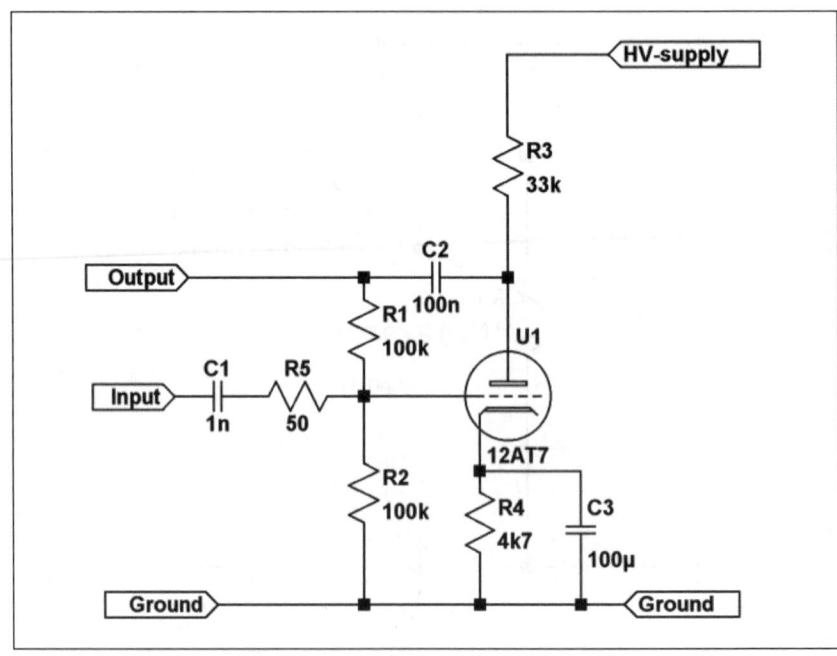

Fig 92: Anode follower

CHAPTER 6: AMPLIFIERS

> **Fact Sheet - Anode Follower**
> **Input impedance:** High (variable)
> **Output impedance:** Low
> **Amplification:** Moderate
> **Capacitance:** Cag high, due to the Miller Effect
> **Phase shift input-output:** 180 degrees
> **Applications:** Mostly AF
> **Misc:** Not very common in RF applications

Table 10: Anode follower fact sheet

Lunar Grid Amplifier

The lunar grid connection (also know as the 'Inverted amplifier') is very interesting indeed. It displays the following characteristics:

- The anode has a negative bias
- The control grid has a positive bias
- The anode is the input
- The control grid is the output
- The input impedance is very high
- The output impedance is very low
- It displays good linearity
- Wide bandwidth

The amplification factor is low, approximately 1/μ, where μ is the normal amplification factor for the valve as given in the data sheet.

Due to the grid being closer to the cathode, the current controlled is high, ie the power amplification is high.

Fig 93 shows a lunar grid amplifier.

The stage works in the following manner:

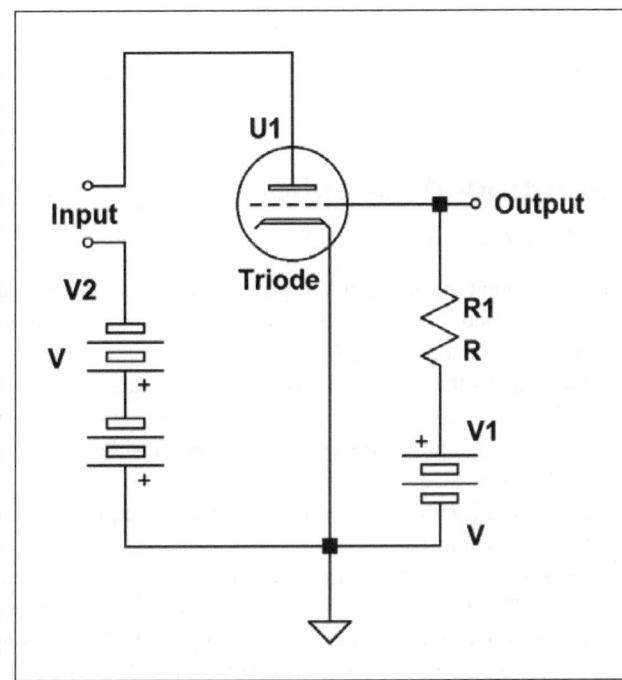

Fig 93: Lunar Grid amplifier

77

The space charge current of a valve depends on the potential gradient at the cathode, which in turn depends on the relative potential between the grid and the anode. In the lunar grid configuration, the potential difference between the grid and the anode is still great, although they have opposite signs compared to a normal configuration. But, since the grid is positive and the anode negative, the grid becomes the electrode catching the electrons.

The negative voltage on the anode controls the flow of electrons to the grid. Because the anode is so much farther away from the cathode, the influence on the current is much less than if the grid had been the input electrode, thence the low amplification factor of the stage.

R1 is the load impedance across which the output signal is taken.

The grid is closer to the cathode than the anode. Therefore, the positive grid bias does not need to be as high as a normal anode voltage.

Table 11 shows a summary of the lunar grid.

A complete lunar grid amplifier description can be found in a later chapter of this book.

Fact Sheet - Lunar Grid Configuration

Input impedance: Very high
Output impedance: Very low
Amplification: Very low, basically $1/\mu$
Capacitance: Cag very low, due to the low amplification
Phase shift input-output: 180 degrees
Applications: AF
Misc: Good linearity. Can be used for OTL power amplifiers. Requires high amplitude drive voltage.

Table 11: Lunar grid fact sheet

Compound Amplifiers

Cathode coupled

The combination of cathode follower and grounded grid amplifier is very useful. It is called 'cathode coupled amplifier'. The input impedance as well as the output impedance are high.

This means that the high impedance terminals can be connected to tuned circuits, whereas the two low impedance terminals - the cathodes, can be used for low impedance purposes, eg other filters. Besides that, the combination yields amplification, albeit not very high.

The phase relationship between input and output is zero degrees.

See also **Table 12**.

Fig 94 shows such an amplifier. It demonstrates a way of arranging for grid bias for the first valve half which yields a very high input impedance.

The first half of the valve forms the cathode follower. Its low impedance output is connected to the low input impedance cathode of the grounded grid stage.

CHAPTER 6: AMPLIFIERS

> **Fact Sheet - Cathode Coupled Amplifier**
> **Input impedance:** High or very high
> **Output impedance:** High or very high
> **Amplification:** Moderate
> **Capacitance:** Cag low due to the high feedback at the first stage
> **Phase shift input-output:** 0 degrees
> **Applications:** RF, oscillators
> **Misc:** The impedances between the stages is low

Table 12: Cathode coupled amplifier fact sheet

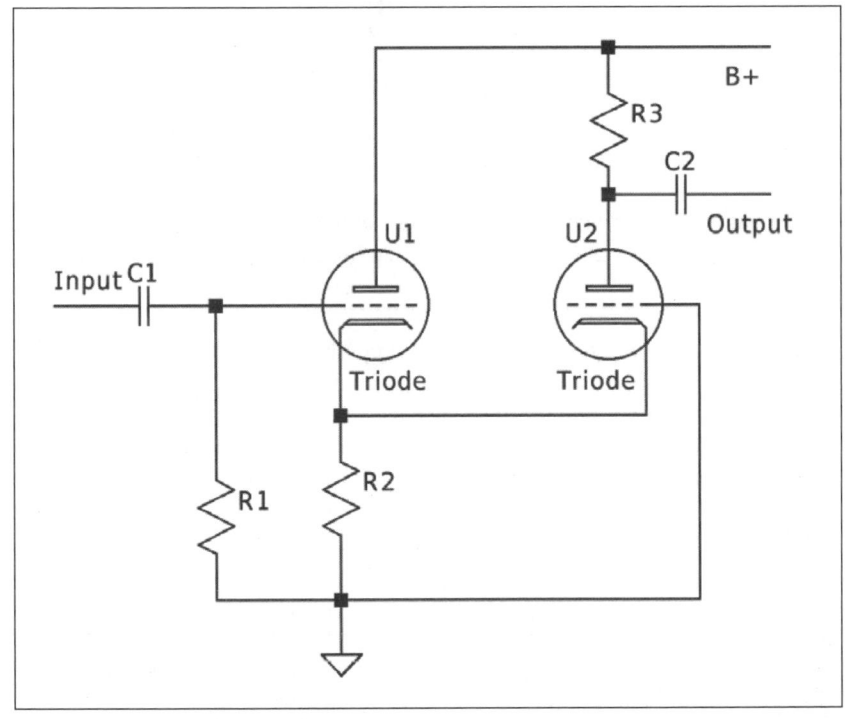

Fig 94: Cathode coupled amplifier

C1 and C2 are coupling capacitors chosen from the desired frequency range. R1 is a normal grid leak. R1 is chosen from the desired voltage drop (grid bias) at the two valves, and the sum of the cathode currents.

Those resistors need to be chosen so that the voltage equals the voltage drop across R2 minus the desired grid bias. The output impedance will be the same as that for a grounded grid stage.

Low impedance device matching

A pi-filter is a low impedance device, and there are more applications where the filter needs low impedances. In such an application it is valuable again to use the cathode follower - grounded grid combination. Simulations show that the filter in

Fig 95: Pi filter matching

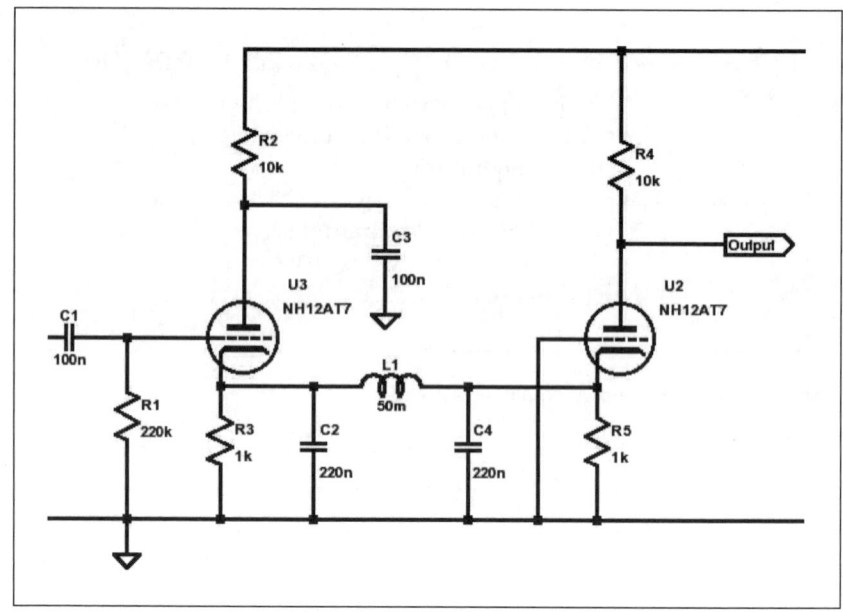

Fig 95 has a corner frequency of 2kHz and that the slope is 20dB per octave. The amplification is about 20dB. A bigger decoupling capacitor (C3) is beneficial.

Cascode

The cascode stage as shown in **Fig 96** is a combination of a grounded cathode stage and a grounded grid stage. The combination yields a moderate input impedance and a high output impedance as well as a high degree of amplification. It has the good qualities of a pentode, high amplification and high output impedance, and none of its bad sides, noise and the need for high voltage at the screen grid. Besides, it requires fewer components. The cascode is an inverting amplifier. Cascode amplifiers are designed for RF as well as AF usage.

The resistors R3 and R4 determine the grid bias for the two grids, and C3/C4 are the decoupling capacitors. The upper grid is grounded via C2. R5 and C5 form a filter to prevent RF from the anode to spread through the power supply.

The grid bias of the top grid is taken from the bottom of its cathode resistor. However, the top bias components may be omitted without detrimental effects. If omitted, the anode of the bottom valve functions as a bias generator.

Component values are not critical.

A 100Ω - 1kΩ resistor in series with the input nearest the grid would serve to reduce the risk of parasitic oscillations.

The output is taken from the junction L2/C6. The very high impedance at the output and the loose coupling (K = circa 0.1) between the primary and secondary coils cause the tuned circuit bandwidth to be very narrow, which eliminates the need for another circuit at the input.

The cascode constitutes an excellent input amplifier for a receiver, in particular at higher frequencies (SW and UHF).

Substituting the tuned circuit in the anode with a resistor and increasing the values of the decoupling capacitors turns this amplifier into an excellent AF

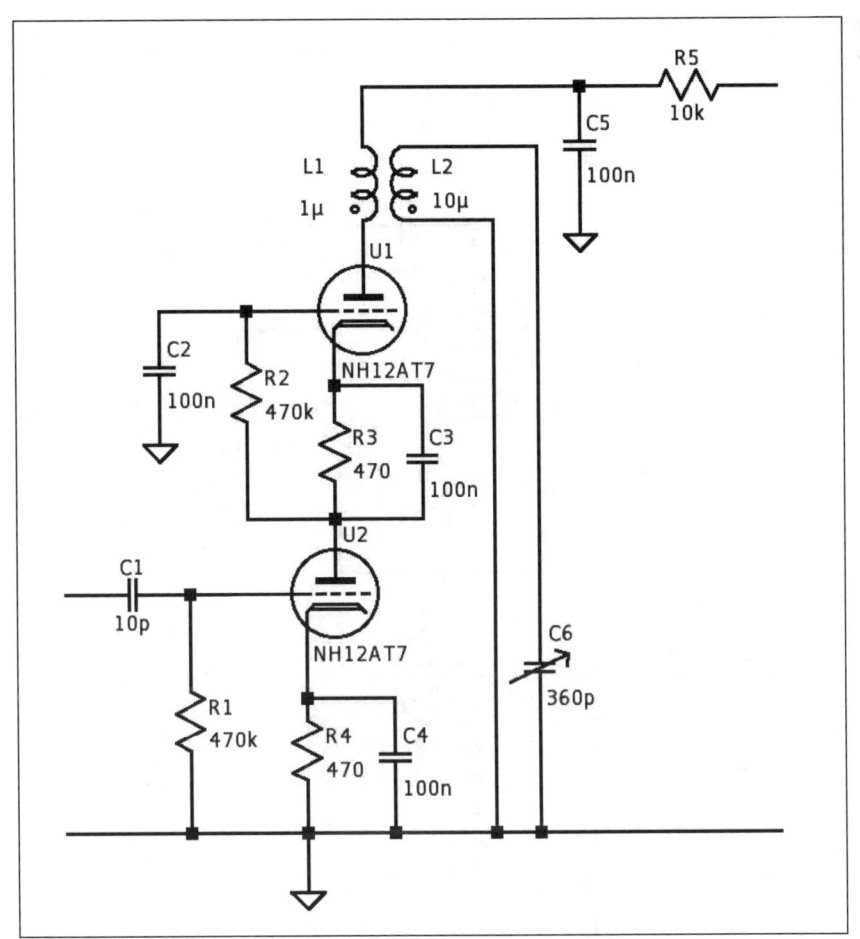

Fig 96: Cascode amplifier for RF

input amplifier. To achieve a high input impedance, connect a cathode follower in front of the cascode.

When connecting two valves in a cascode configuration, attention must be paid to the voltage restrictions between the filament and cathode voltages of the top valve. This voltage is generally given in the data sheet. There is no need to drive a cascode stage with higher voltages than would otherwise be used for a normal triode or pentode stage.

A cascode stage for AF would look something like **Fig 97**. This is a pre-amplifier for an AF amplifier. Apart from the tuned circuit in the previous schematic, the main difference lies in the capacitors. These have much higher values to enable the stage to cope with the low frequencies involved.

R6 is the grid resistor in the next stage (assumed to be 470k). The only capacitors that need to be high voltage types are C2 and C5. The others can be of a type normally used with semiconductors. C2 prevents the grid voltage from varying with the signal. The cathode is the controlling electrode in the upper stage, so the grid must short-circuit all signals reaching the grid to ground.

According to simulations, this stage is capable of an amplification of about 37dB between 10Hz and 1MHz. See the fact sheet in **Table 13**.

VALVES REVISITED

Fig 97: Cascode amplifier for AF

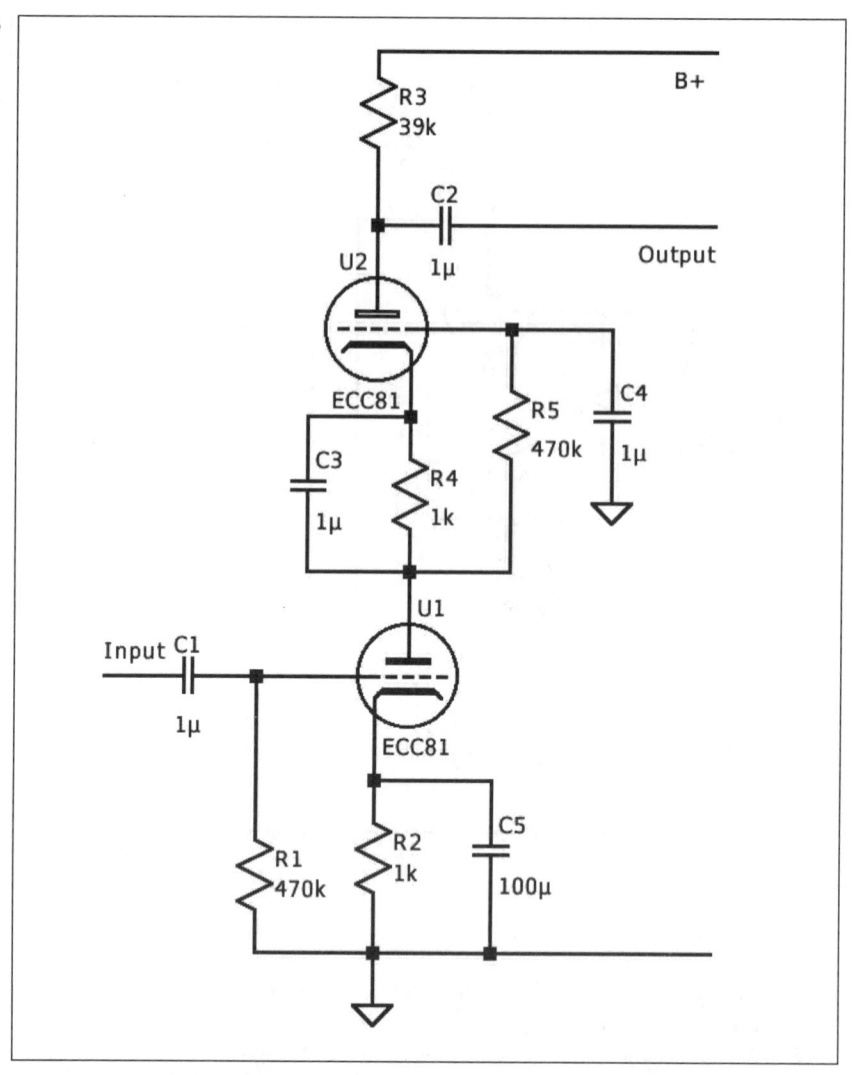

Fact Sheet - Cascode Amplifier

Input impedance: Moderate
Output impedance: High or very high
Amplification: High
Capacitance: Cag low due to the grid or the top triode being grounded, and due to the low amplification of the bottom triode
Phase shift input-output: 180 degrees
Applications: RF, AF, IF
Misc: The entire stage can be regarded as a low noise pentode stage

Table 13: Cascode amplifier fact sheet

CHAPTER 6: AMPLIFIERS

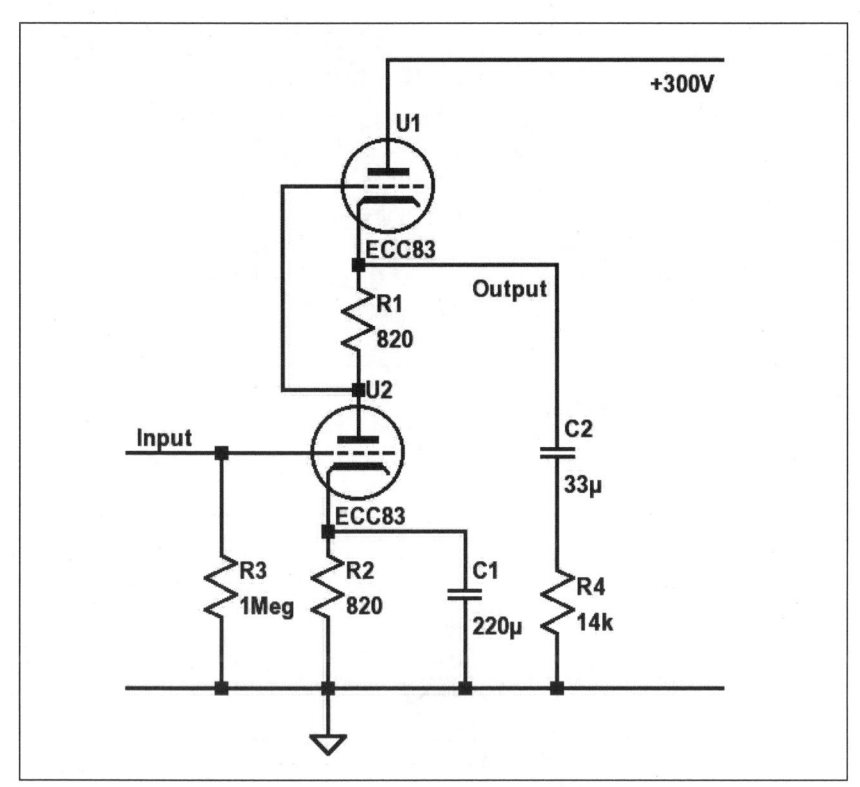

Fig 98: SRPP Amplifier

There is a dual triode especially designed for cascode amplifiers, the ECC84. It is intended to run as RF cascode amplifiers in TV receivers up to 220MHz.

SRPP, µ-follower etc

There is a Swedish saying: "Beloved child has many names". The saying is highly applicable to this design. It has been called the 'SRPP', the 'SEPP', the 'µ follower', the 'cascoded cathode follower', the 'totem-pole amplifier'.

SRPP stands for Series Regulated Push-Pull. **Fig 98** shows an example. At first glance it doesn't look like a push-pull stage as we know them (see the chapter on hi-fi amplifiers). However, the fact is that it is actively controlling current in both directions. Most of all, it looks like a cascode stage. If the signal had been taken from an anode resistor at the top valve, it would have been. What it is, though, is a cathode follower placed on top of a grounded cathode stage.

C2 is the coupling capacitor, and R4 represents the load. The fact is that an SRPP works best into a specific load, which happens to be 14kΩ with the values given. The cathode resistors should be equal in value, as shown in Formula 27.

$$Rk1 = Rk2 = \frac{2Rl + Ra}{\mu - 1}$$ Formula 27

Where
Rk1 is the bottom cathode resistor,
Rk2 is the top cathode resistor,
Rl is the load resistance (impedance)

83

Fact Sheet - SRPP

Input impedance: Moderate
Output impedance: Low to very low, fixed
Amplification: Moderate
Capacitance: Cag normal for a grounded cathode amplifier
Phase shift input-output: 180 degree
Applications: Audio, measurements within bandwidth restrictions
Misc: Wide bandwidth

Table 14: SRPP amplifier fact sheet

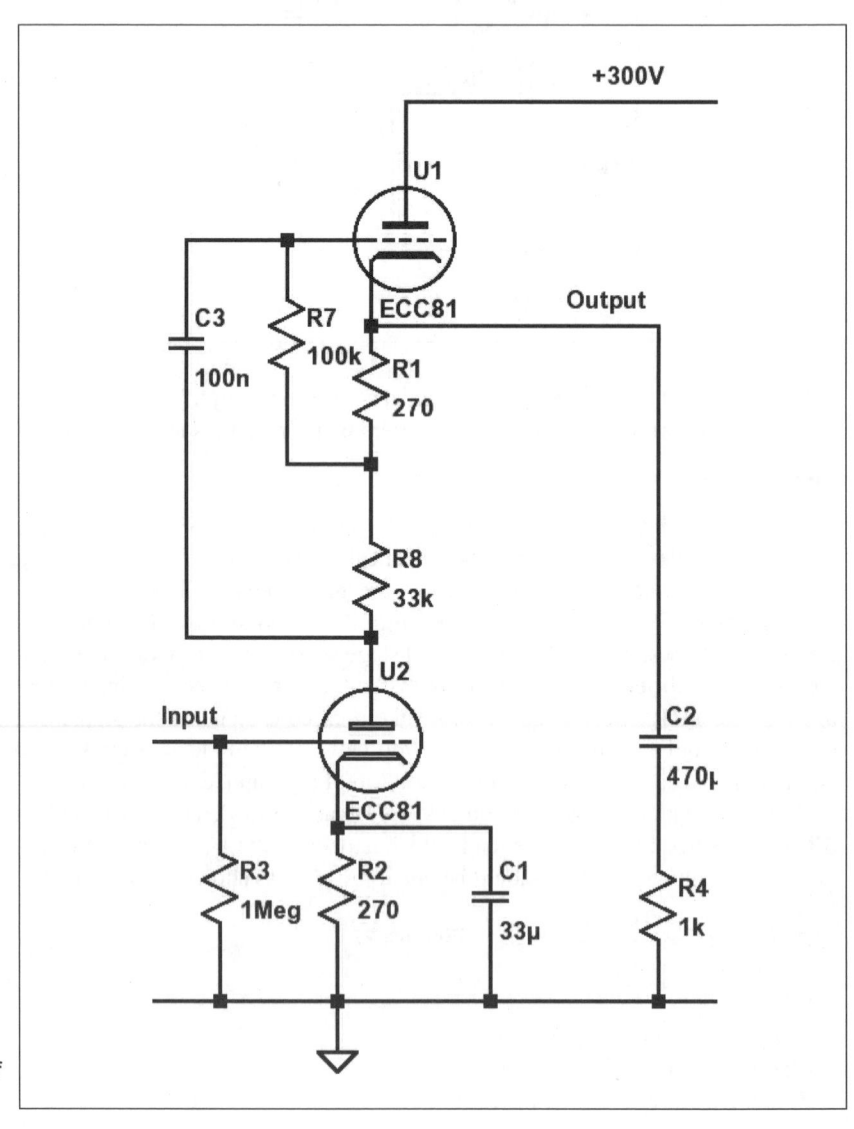

Fig 99: A variant of the SRPP

The output impedance is shown in Formula 28:

$$Rk1 = Rk2 = \frac{2Rl + Ra}{\mu - 1} \qquad \text{Formula 28}$$

since the bottom cathode resistor is by-passed (through C1), which in this case computes to about 14.3kΩ.

Simulations reveal that the given component values yield a bandwidth of nearly 2MHz, which is fine for an instrumentation amplifier but too much for audio use. Also see the fact sheet in **Table 14**.

The output impedance can be lowered to a couple of hundred ohms by re-designing the cathode follower on top. The type shown is a simple one, but the high-impedance type improves the output performance of the stage. The function *per se* is still the same, though.

The SRPP (first designed by Maurice Artzt and patented in 1943) is a very popular design in audiophile circuits. **Fig 99** shows one solution. However, remember that, according to some sources, the bandwidth of one stage should be narrower than the following stage for optimum performance, and the SRPP is a device with a very wide bandwidth.

The bandwidth can be restricted at the upper end by connecting a capacitor of a couple of hundred picofarads between the cathode of the top valve and the anode of the bottom valve without compromising the remaining performance of the stage (according to simulations).

Here, the grid leak of the cathode follower (R7) is connected to the junction of R1 and R8, providing the correct grid bias, whereas the signal is still taken from the anode of the bottom triode. Due to the greater voltage difference between the anode of U2 and the grid bias point, a capacitor, C3, must be inserted. C3 does not have to be very large, though, due to the high impedance at the grid of U1. Nor does the value of the grid leak make much difference. R8 cannot be much larger than 33kΩ. C1 can be omitted, at the cost of a couple of dB amplification. R4 represents the load, about 1kΩ, and it is fed through the capacitor C2.

In both cases, the heater voltage needs to be elevated or floating so as not to compromise the heater-to-cathode voltage restriction. Check with the data sheet before you build the stage! For the ECC81, the maximum permissible heater-to-cathode voltage is 90V.

Anode follower - cathode follower

In some cases it is necessary to be able to deliver a signal to a very low impedance device. A typical example is the power amplifier which drives a loudspeaker. There are other examples too, for instance long coaxial leads from one point to another.

One interesting combination to try is the anode follower - cathode follower combination. As the heading implies, it consists of two halves of a dual valve. The first half plays the role of an anode follower, the other is a cathode follower. See **Fig 100** for details.

The anode follower in the loop receives a signal from the output of the cathode follower and amplifies it, assisting the cathode follower and hence lowering the output impedance.

VALVES REVISITED

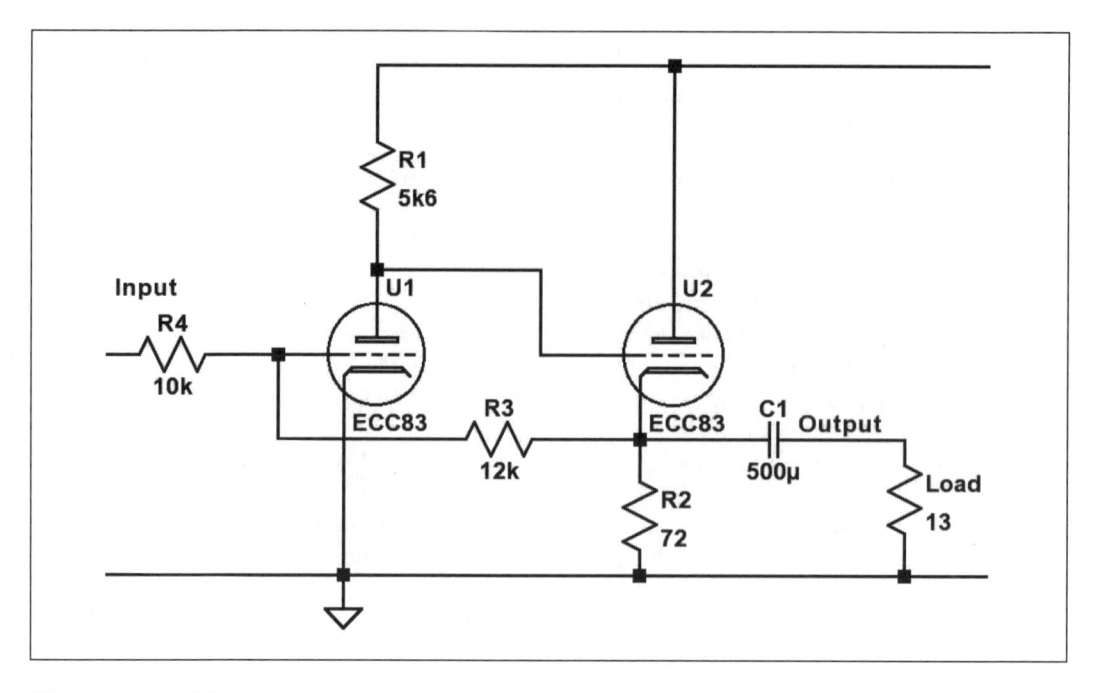

Fig 100: Combination of anode follower and cathode follower

The cathode follower is DC coupled to the anode of the anode follower. The cathode resistor of the cathode follower is lower than it would normally have been.

The design above has been simulated to drive an astonishing 13Ω resistive load. However, don't be fooled into thinking that you could drive a loudspeaker directly from the output! Even though the load is heavy, it still doesn't mean that the design can deliver a lot of current.

The simplest way to design this is to connect everything up except for the load. Then the relationship between R3 and R4 is adjusted to give a 0dB amplification, ie the output has the same peak voltage as the input (before R4). Then connect the load and check that the output signal now has half of the amplitude of the input signal. Some adjustments may have to be done to R1 and R2 for optimum results. These can be done under load. C1 may be omitted if a small DC level can be tolerated at the output, in which case the load could replace R2. This would give a better response for low frequencies. In fact, the whole arrangement would then become DC coupled all the way through.

The simulated bandwidth of the arrangement in **Fig 100** is about 2MHz.

An even lower output impedance is achieved if R2 is replaced by a choke. It is worth trying.

Operational Amplifiers

Do not, even for a second, believe that things like operational amplifiers were unknown to the electronics engineers in the days of yore! **Fig 101** shows a valve based one. Even though operational amplifiers didn't get their name until 1947, one of the first patents was granted in 1941.

As always, the design evolved, and the example shown dates from the early 1960s. It is capable of amplifying signals between DC and about 60Hz. It has two inputs, one inverting and the other non-inverting. The open loop amplification

Fig 101: Op-amp (Operational Amplifier) with valves

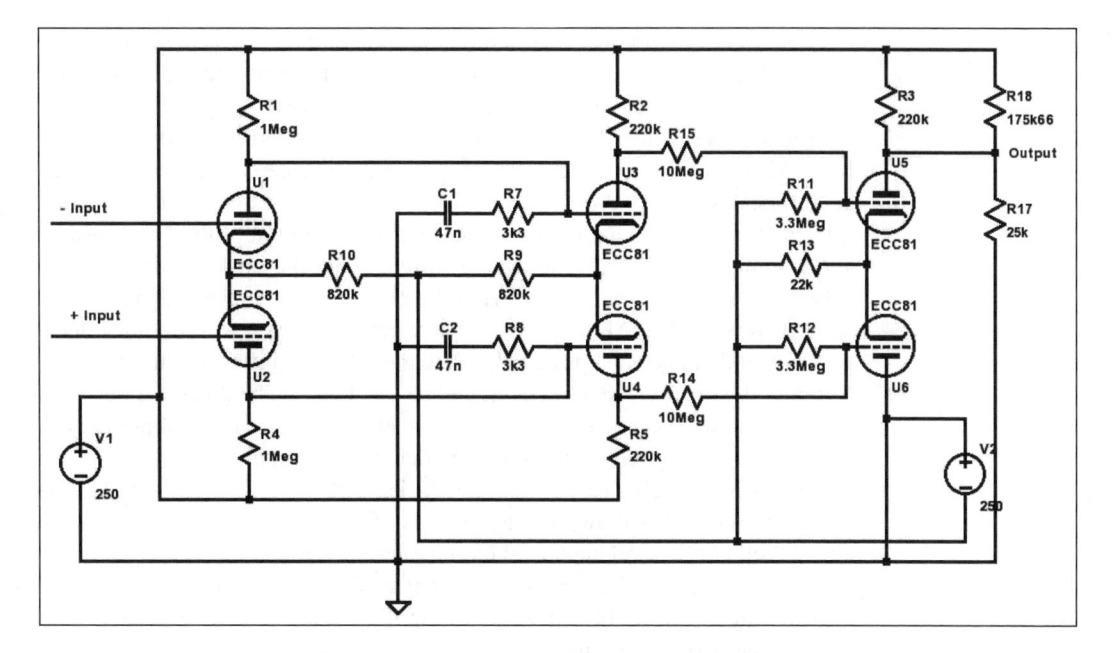

factor is around 77 dB with the valves selected here (according to simulations) and two-phase input.

The design is actually simple. The device is driven by a dual power supply and consists of three cascaded amplifiers. The cathodes of U1/U2 and U3/U4 should not contain any signal, since the valves are connected in a push-pull configuration. The output is symmetrical around zero, and the balance is set by selection of the resistor R18. R17 is the load, which should be 25kΩ or more. Connecting the output to the inverting output via a voltage divider, as for a semiconductor op-amp, sets the amplification factor. Note that the output stage is asymmetrical - there is only one anode resistor.

The design is a good example of DC-coupled amplifiers. It does not appear to be overly sensitive to voltage changes in the power supplies. Decreasing the values of C1 and C2 will increase the bandwidth to an apparent maximum of 2kHz with both capacitors removed.

The fact that it is differential all the way through means that it handles CM (Common Mode) signals with lower errors and improves drift characteristics. There is only one (dual) level shift stage, U3/U4, which minimises gain loss and improves its overall performance.

The unit could easily be built into a suitable box and used as a general purpose amplifier (within the given constraints). Or, combine it with a VTVM and a RF probe (see elsewhere in this book), and you have an instrument to measure DC and AC voltages in the microvolt range. Alternatively, hook it up to an oscilloscope, and you have an instrument to monitor muscle activities.

Classification of Amplifiers

So far we have dealt with amplifiers that work on the straight part of the Ia/Vg curve. However, for various reasons, the bias point can be placed elsewhere on the curve, and even outside the curve.

VALVES REVISITED

The reason for working on the straight portion of the curve is to minimise distortion. There is a linear (or near linear) relationship between the control grid voltage and the anode current. This is known as a class A amplifier. As long as the signal stays within the linear part of the curve, distortion is kept at a minimum. There is a drawback, though - the valve draws a certain amount of current at all times, with or without signal. This current is called quiescent current.

If, instead, the working point were placed at the bottom of the curve, near or at the point where the valve draws no current at all, the 'cut-off point', then the valve would draw current only during the positive halves of the signal. This is class B. It is used in AF amplifiers, where two valves are working together in such a way that one valve is cut-off while the other amplifies its half of the signal; then the other valve takes over, and so on. This can be done in a push-pull arrangement which is frequently used in AF amplifiers (for music reproduction) and PA systems as well as in more elaborate domestic radios.

Class B has benefits as well as drawbacks. The major benefit is that no (or very little) current is drawn during periods of no signal. The drawback is that the method requires yet another stage - a phase inverter.

A phase inverter takes the input signal and turns it into two signals 180 degrees out of phase. In a push-pull amplifier, a phase inverter is necessary because the two power valves each requires a signal, 180 degrees out of phase with the other.

The simplest phase inverter is a transformer. Many transformers today yield a bandwidth sufficient for most purposes, and they can amplify a signal voltage as well. However, they are bulky, heavy and expensive. Another very simple phase inverter is a valve with equal resistances in the cathode and anode, such as the one shown in **Fig 102**.

Resistors R8 and R9 are the grid leaks of the next stage. The outputs of this phase inverter are (relatively) high impedance.

The curves in **Fig 103** show both outputs of the phase inverter.

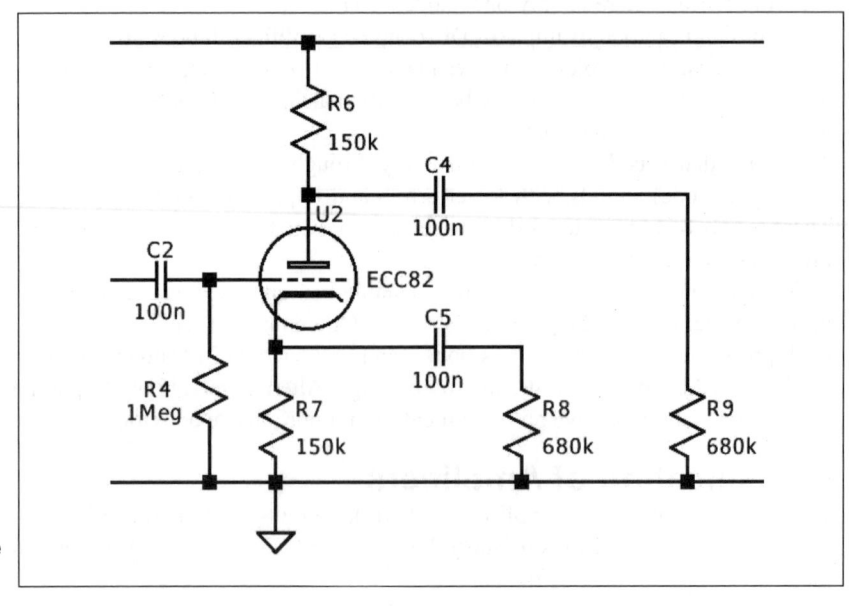

Fig 102: Phase inverter

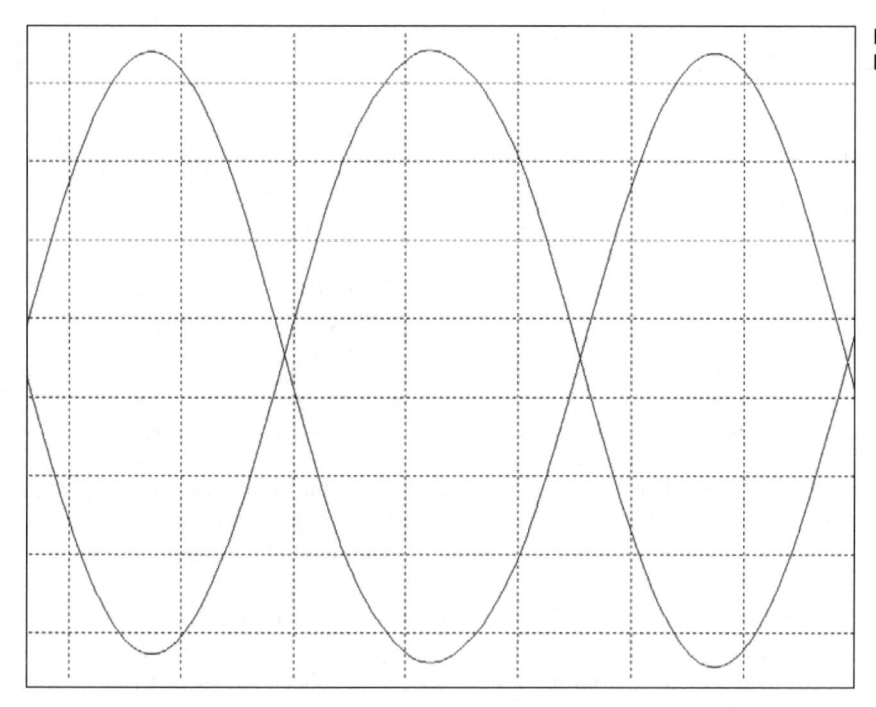

Fig 103: Phase Inverter output

Running the power amplifier in class B push-pull sometimes, but not always, means that some sort of dedicated grid voltage supply needs to be provided. Class B amplifiers sometimes suffer from a type of distortion called 'transition distortion'. It happens at the moment one valve takes over from the other and is caused by dissimilarities in the valves or bias set-ups.

Fig 104 shows a push-pull stage with EL84, capable of delivering some 10W.

Fig 104: EL84 in push-pull

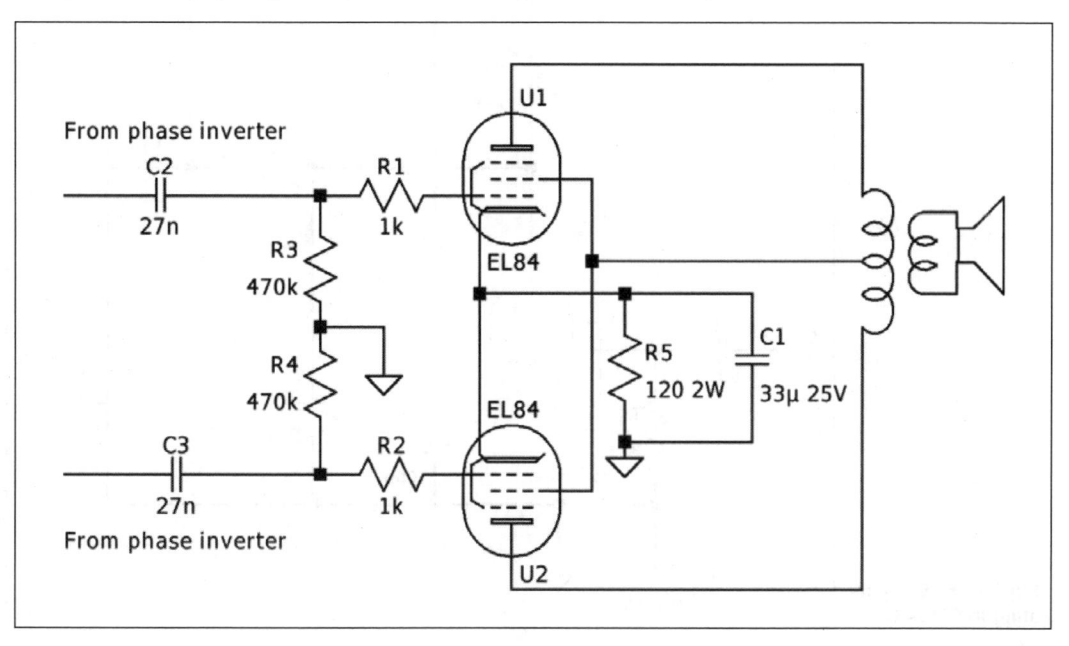

89

VALVES REVISITED

Fig 105: Extract from EL84 data sheet

Class AB, two tubes in push-pull						
Anode voltage	V_a	250		300		V
Grid No.2 voltage	V_{g2}	250		300		V
Common cathode resistor	R_k	130		130		Ω
Load resistance	$R_{aa\sim}$	8		8		kΩ
Grid No.1 driving voltage	V_i	0	8	0	10	V_{RMS}
Anode current	I_a	2x31	2x37.5	2x36	2x46	mA
Grid No.2 current	I_{g2}	2x3.5	2x7.5	2x4	2x11	mA
Output power	W_o	0	11	0	17	W
Distortion	d_{tot}	-	3	-	4	%

The 250V HT line is connected to the centre of the output transformer. R5/C1 delivers a grid voltage suitable for class AB, which lies between class A and B. A grid bias supply capable of feeding the cathodes 11.6V places the stage in class B. With the given value for R5 yields about -9.6V control grid bias. Class A operation requires -8.4V bias. The anode resistance is circa 8kΩ. Other classes are AB1 and AB2.

All data are taken from the Philips EL84 data sheet. An extract is shown in **Fig 105**.

There is an additional class, class C, which is used in RF applications only. In this class, the control grid bias is so high that only the peaks of the signal pass through the valve. It is used in conjunction with radio transmitters, where a tuned circuit is connected to the anode. Each peak of the signal gives the tuned

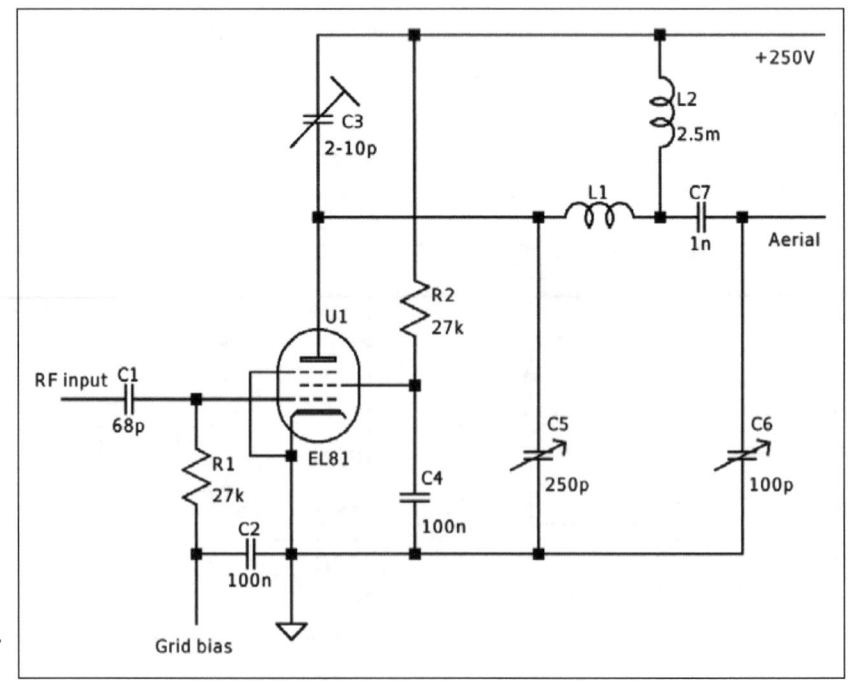

Fig 106: EL81 operating in Class C

circuit a push. The gaps between are 'filled in' by the Q of the tuned circuit. The tuned circuit works like a flywheel - once kicked, it maintains the oscillations for long enough to keep them going without noticeable loss until the next kick, which normally occurs at the latest one period later.

A class C amplifier is shown in **Fig 106**. It requires an external grid bias supply. The supply is connected to the grid via the grid leak (much like an AVC voltage). The cathode of a normal class C stage is directly grounded.

The EL81 is a pentode intended for line time base generation and audio output in TV sets. This implies that the valve has a limited power handling capability. However, the mean value of the current counts, so when a valve is driven in class C, the peak output power can be much higher than the nominal values for the valve.

The grid bias is taken from a potentiometer in the bias power supply and can reach -100V.

Note the pi-network at the anode of the valve. Since the stage works in class C, overtones (harmonics) of the signal are bound to appear at the anode. These need to be suppressed to avoid interference to other services and other disturbances. This is a very important feature of a class C amplifier, so a pi-filter is frequently used. It is a low pass filter, tuned to the working frequency of the amplifier, so that it suppresses overtones as much as possible. The amplifier in Fig 106 above is claimed to deliver 100 watts.

These days, other classes exist. Class D is emerging as a means of amplifying audio signals (in hi-fi amplifiers). This is, however, a complicated and therefore expensive task. Essentially, class D is a sampled, pulse duration modulated system, where an output filter restores the original sound. The main application is control of electrical motors, etc.

Amplifiers have been classified up to class I.

RF Design Basics
By John Fielding, ZS5JF

RF Design Basics is the latest book by acclaimed author John Fielding, ZS5JF. This book is a practical guide to Radio Frequency (RF) design rather than the more usual text book written for post-graduate electronics engineers. Aimed at those who wish to design and build their own RF equipment, this book provides a gentle introduction to the art and science of RF design. The fourteen chapters of RF Design Basics cover subjects such as tuned circuits, receiver design, oscillators, frequency multipliers, design of RF filters, impedance matching, the pi tank network, making RF measurements, and both solid-state and valve RF power amplifiers. One chapter explains the meaning of S parameters, while another is devoted to understanding the dual gate Mosfet. Much attention is given to the necessity of cooling valve PAs and there is even a practical design for water cooling a large linear amplifier, a subject overlooked by most other publications.

RF Design Basics neatly fills the gap between a beginner's 'introduction to radio' and RF design text books. Written for the average radio amateur, this book is an accessible and useful reference work for everyone interested in RF design.

Size 210x297mm, 192 pages
ISBN 9781-9050- 8625-2

ONLY £17.99

RSGB shop

Radio Society of Great Britain
3 Abbey Court, Fraser Road, Priory Business Park, Bedford, MK44 3WH
Tel: 01234 832 700 Fax: 01234 831 496

www.rsgbshop.org

E&OE All prices shown plus p&p

7

Modulation

HUMANS HAVE WANTED to communicate with each other throughout the ages. But it turned out that grunting and groaning was not enough to give detailed information about the details of everyday life. So we developed the ability to speak. However, speech could not travel very far.

Many inventions have been made over the aeons, smoke signals, jungle drums, the optical and electrical telegraph, the telephone, radio, television, mobile telephones, etc. The list could go on and on.

These days, transfer of information over long distances is done electronically. The optical telegraph is no longer in use, nor is fire used as a transfer medium. Over the last few centuries development has accelerated enormously.

Although the following chapter is not specific to valve technology, it is helpful to understand modulation before seeing how a receiver or transmitter works.

Modulation

A transmitter radiates electrical energy. This energy consists of a carrier, ie the magnetic and electrical fields around the aerial vary with a certain periodicity. This is the carrier's frequency.

All kinds of information, be it TV, mobile phone calls, satellites, data, radio, text or pictures are all transferred using a carrier. Whether the carrier is transmitted via an antenna or a line is irrelevant.

However, a carrier alone is not enough to transfer information. The only information the carrier contains is that of its mere existence.

In order to transfer a meaningful message, one or more of the carrier's parameters must be varied somehow. What first springs to mind is the amplitude and the frequency.

For the transmitter and receiver both to understand the message, they have to 'agree' which parameter is varied and how. If you try to receive information which is transferred by varying the amplitude and you are using a receiver for signals that vary in frequency, you won't hear anything. Each modulation type requires its own demodulator.

Varying the Amplitude

The amplitude of a carrier is its signal strength, the momentary output power of the transmitter. The strength is varied by an amplitude modulator.

The pattern of the variations is determined by the information which is to be transferred. Speech, for instance, is variations in air pressure caused by the vibrations of the vocal cords. The microphone converts these variations into a

Fig 107: Microphone signal

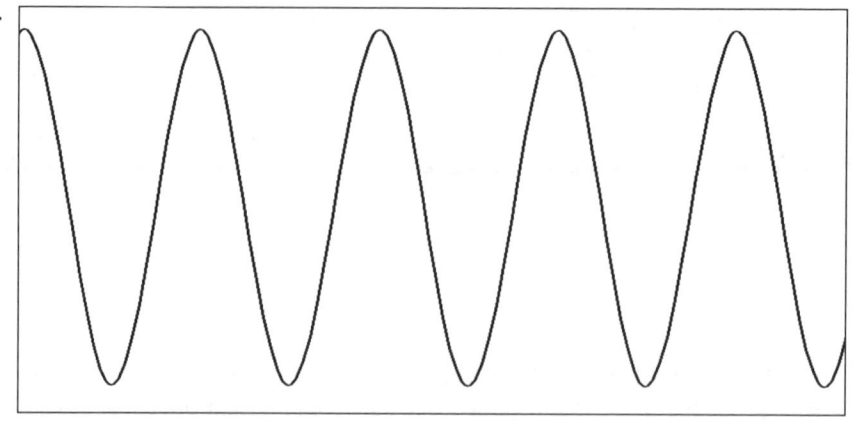

Fig 108: Amplitude modulated carrier

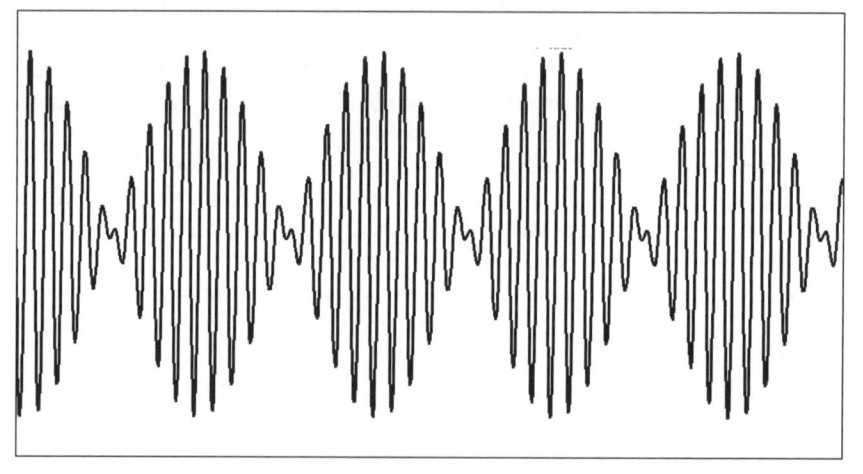

varying electrical current or voltage. These are then amplified and connected to the modulator, which varies the signal strength of the carrier.

Fig 107 shows the signal from the microphone, and **Fig 108** the resulting RF after modulation. Amplitude modulation is the most common modulation on BC (broadcast) transmitters on long, medium and short wave.

The process is quite the reverse at the receiver end. The modulated carrier is received, demodulated, and converted to sound by the loudspeaker. Ideally, you should get an exact copy of the original information. This is, however, not within reach, due to the limited bandwidth at the transmitters' disposal. In fact, even with today's technology, it is impossible to reproduce sound, be it recorded or broadcast material, with 100% accuracy.

There are also sub-categories of amplitude modulation. They will be described in more detail later.

Every change of the carrier causes sidebands to develop. They can be displayed on a spectrum analyser. You could not dispense of the side-bands - this is where the information is.

It is both theoretically and practically possible to make a receiver which receives only the carrier, but since the carrier contains no information, such a receiver would be useless.

On the other hand, the carrier can be dispensed of, leaving only the sidebands. This is a good thing because there is a lot of energy transmitted in the carrier. Only one side-band is necessary to transfer the entire information. The method is called SSB (single sideband) and is extensively used by amateurs and other stations the world over.

There is also a method which transmits both side-bands, called DSB (double sideband), but is hardly used these days. A mixture of SSB and DSB is called ISB (independent sideband), where one information is transmitted by one sideband and another on the other.

The transmission of Morse is sometimes called CW. This stands for Continuous Wave referring to the carrier. It is not totally descriptive because the carrier is not continuous - it is interrupted according to a certain pattern, an on-off method. The pattern may form letters and numbers, and is in most cases according to the Morse alphabet, invented and patented 1833 by the American Samuel Finlay Breese Morse (1791 - 1872).

Varying the Frequency

The frequency is another property of the carrier which can be varied. Every time you tune in an FM transmitter to listen to music or speech, or, indeed, watch TV, you listen to a frequency modulated signal. Recent TV transmissions, though, are most likely to be digital, which is an entirely different kettle of fish.

The frequency of the carrier is varied according to the variations of air pressure picked up by the microphone. The FM system was invented by Edwin Howard Armstrong (1890 - 1954), for which he was granted the patent in 1933.

The method is an improvement of sound quality, but takes up far more space in the spectrum (100 - 250kHz per transmitter, as compared to 9 or 10kHz for AM). It has low sensitivity to amplitude variations.

A frequency modulated signal may look like in **Fig 109**.

FM is, however used on higher frequencies too, in a modified form that does not take up all that much space. This is called NBFM (Narrow Band Frequency Modulation), and its bandwidth is limited to a few kilohertz.

Voice communication is not the only type of information transmitted with FM. There is also Fax (telefacsimile, press pictures or weather maps), RTTY (radio teletype) and other digital transmissions use FSK (frequency shift keying with a

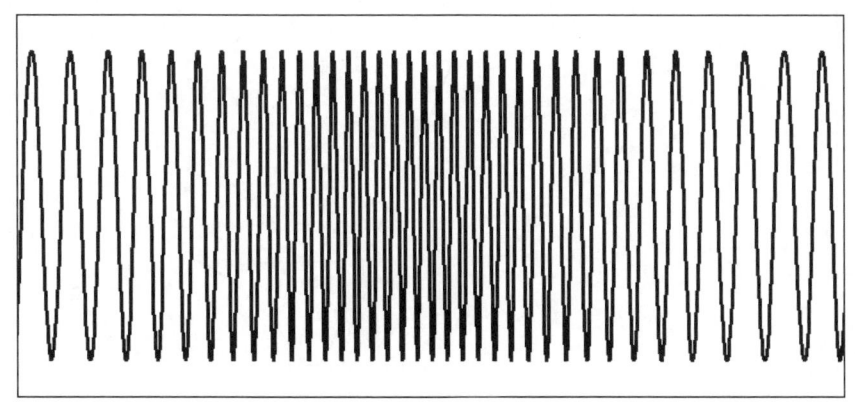

Fig 109: Frequency modulated signal

Fig 110: Resonance curve

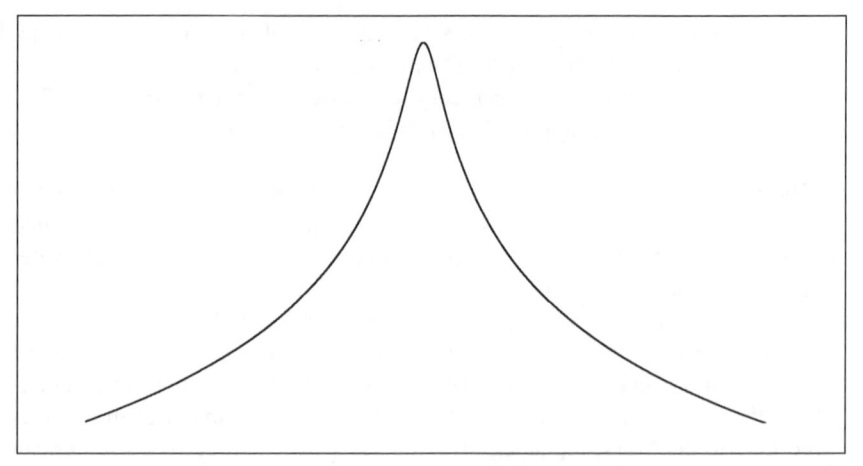

large variety of modulations) where the carrier jumps between two or more frequencies to transfer the information.

There is a large number of FM detectors, also called Frequency Discriminators. Many of them create a frequency dependent filter flank where the carrier is allowed to move back and forth, thereby creating an amplitude modulated signal, which is rectified in a couple of diodes.

The simplest way of receiving an FM signal is to tune the receiver so that the carrier is received on one of the flanks of the IF curve (which, naturally, must not be too steep). However, it causes significant distortion of the modulation.

Fig 110 shows a resonance curve for a single LC filter: As you can see, it is not linear. The method does work, however, if you can accept the distortion.

A proper FM discriminator works in much the same way. It uses two filters which have been combined - one is tuned slightly too high and the other slightly too low. The resulting curve is shown in **Fig 111**.

As you can see, the linearity of this curve is better than the one shown above. However, it would be fixing a fault with another fault. The curve is still not straight, which would be needed for good reproduction.

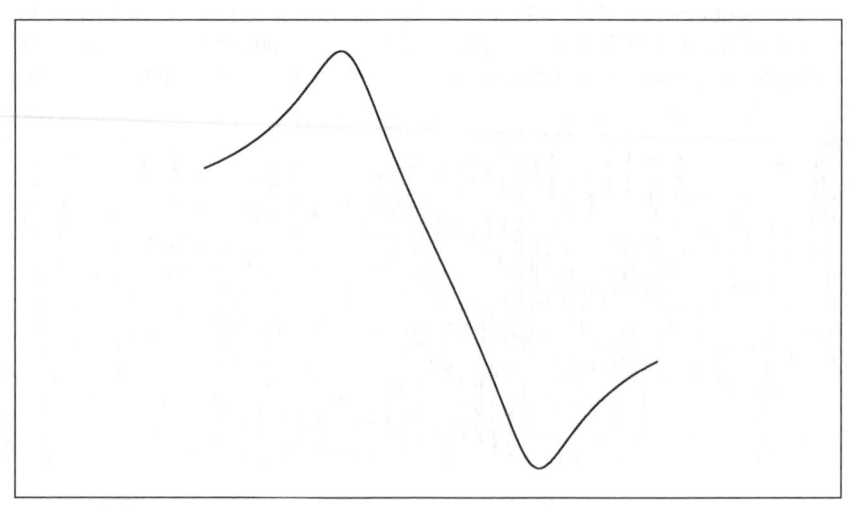

Fig 111: Curve for double circuits

Fig 112: Foster-Seeley detector

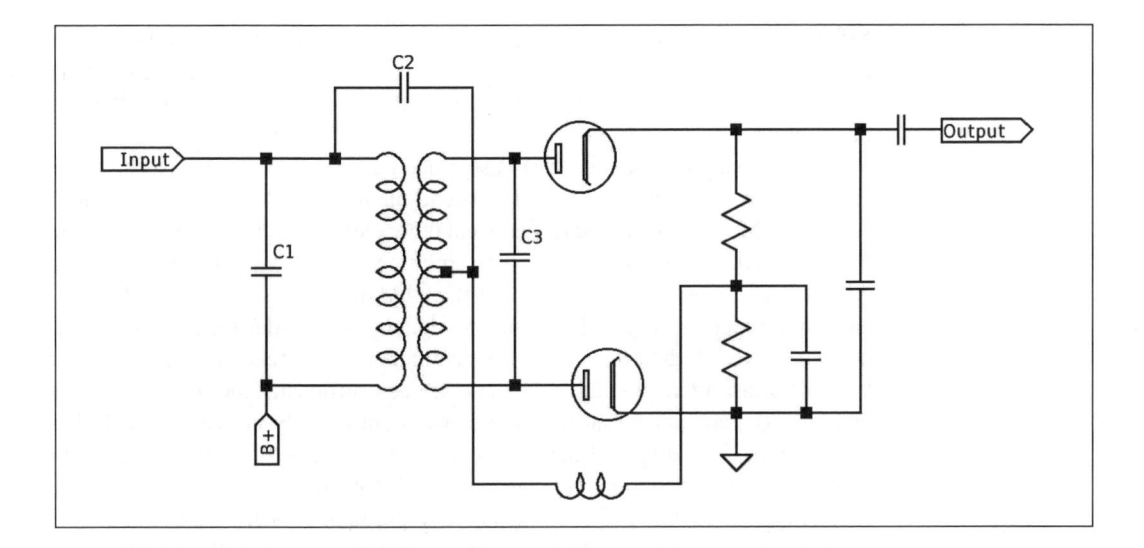

There is a discriminator called Foster-Seeley (Stuart William Seeley, 1901 - 1978 and D E Foster). The principle is built on phase comparison between the same signal in two points. At the nominal frequency, the signals are out of phase and cancel each other out. When the frequency is changed, the balance is upset and the amplitude of the output signal is proportional to the deviation of the frequency from the nominal frequency.

This discriminator also uses two filters, but they are tuned to the same frequency. The signal is fed from one filter to the mid point of the second and is brought to two diodes at the ends of the second filter. The capacitor C2 in the schematic is **Fig 112** is the coupling element.

The diodes must have separate cathodes for the system to work. The coil at the bottom is an RF choke. C1 and C3 are the tuning capacitors, and C2 is the coupling capacitor. Sometimes it is necessary to preserve the DC component of the signal, in which case the output capacitor must be omitted.

The resulting curve is shown in **Fig 113**. As you can see, this method is better than the two described above, from a distortion point of view.

There are advantages with FM. The possibility of locking the receiver to the transmitter frequency is given for free. A low pass filter which takes the signal

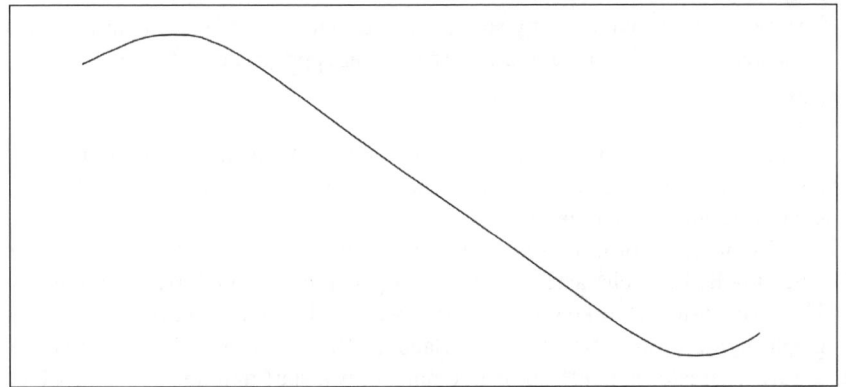

Fig 113: Resulting discriminator curve for Foster-Seeley Detector

before the output capacitor produces a DC signal which can be (and is) used to lock the local oscillator. The filter needs to have a cut-off frequency lower than the lowest modulation frequency to prevent the receiver tuning from following the modulation.

It is frequently claimed that FM is so much better than AM. This is true, to an extent, but it comes with a price. The sound quality of FM is better than that of AM, and FM is less sensitive to static and other electromagnetic interference. On the other hand, since FM requires such an immense bandwidth in comparison to AM, it needs to use higher frequencies than AM does. Consequently, an FM signal can not reach nearly as far as an AM signal can. In order for an entire country to be covered, thousands of FM transmitters may be needed, whereas one or two AM transmitters would have otherwise been sufficient. Another drawback is that an FM transmitter transmits the carrier continuously at peak level. FM is therefore an expensive and power hungry system. Besides, an FM system still doesn't reach hi-fi requirements. The baseband transmitted covers 30Hz - 15kHz. For comparison, a CD player is capable of reproducing 10Hz - 22.5kHz.

Recent work and tests suggest that digital modulation on MW and SW could yield a high quality signal. Few digital SW receivers have as yet reached the market, and there is probably still work to be done. When the system has been fully developed and agreed upon internationally, high quality radio will be able to be heard all over the globe. The digital transmissions take no more space than the allowed 10kHz. On the other hand, their upper audio bandwidth limit is 10kHz.

Telefacsimile

Telefacsimile or, shorter, fax, is heard here and there on the short wave. It is a way of transmitting pictures over the air, such as press photos, weather maps, etc, using frequency modulation. The picture or map is scanned in lines from left to right and from top to bottom. Each line takes in general one or half a second to scan. Then the next line is scanned, and so on, rather like a TV picture, only much slower.

Fax is frequency modulated with a narrow bandwidth, where one frequency corresponds to maximal white, and another maximal black. Everything in between is the grey scale. There is also a 'sync frequency' where one pulse is transmitted every second or half second to tell the receiver that a new line begins.

Fax can be received on any stable receiver with a BFO and the information presented on paper or a computer screen using appropriate software.

FSK

Frequency shift keying uses two or more discrete frequencies used to transmit text or digital data in one or more channels. It may therefore be regarded as a kind of frequency modulation.

In its simplest form, two frequencies are involved, one is called 'space', the other 'mark'. Each character is made up of a combination of spaces and marks. The early systems worked with an alphabet called Baudot after the French telegraph engineer Jean-Maurice-Émile Baudot, (1845 - 1903). It is a five unit code, ie each character is made up of five units plus a start unit and 1½ unit, which constitutes a stop unit. It was in frequent use on shortwave, in particular by news agencies, until most agencies changed to satellite. There are other codes in use too, such as seven-bit or eight-bit ASCII and others. Other modes use phase shift keying (PSK) rather than frequency shift.

Pulse Modulation

A pulse is, by definition, a DC level of brief duration in relation to its repetition rate. A pulse train is a continuous train of pulses. A single pulse is illustrated in **Fig 114** and **Fig 115** shows a pulse train.

These days, many communication systems use some form of pulse modulation, such as mobile phones, digital TV, digital radio, etc. One of the reasons is the amount of power that can be saved. A pulse modulated transmitter sends out the carrier during the duration of the pulse, and is quiet the rest of the time. If the pulse is very short, say 1μs, and the pulse repetition frequency is, say, 1kHz, then the medium power transmitted will be 1 mW, given a pulse peak power of 1W. This is a considerable power saving, compared to AM or FM. The shorter the pulse and the lower the pulse repetition frequency, the more power you save.

A pulse modulated signal requires a very wide bandwidth indeed. Therefore, the carrier frequency of a pulse modulated transmitter needs to be very high. This is a drawback, since the higher the frequency, the shorter its reach along the surface of the earth. This in turn means that a large number of relay stations is required for national coverage.

One of the benefits is that the quality of the transmission can be made very high. Another is that several channels of information can be transmitted through the same transmitter and on the same frequency.

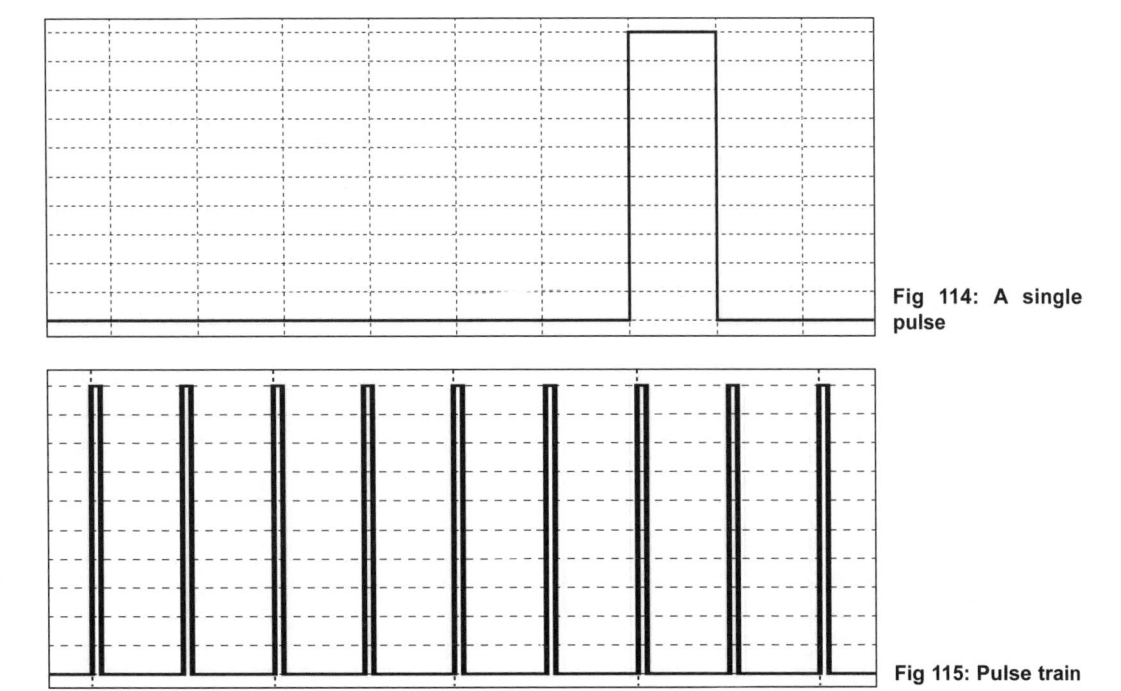

Fig 114: A single pulse

Fig 115: Pulse train

A pulse contains several parameters that can be varied - the PRF (pulse repetition frequency), the amplitude, the PW (pulse width), etc. Also, a pulse train can be coded (PCM, pulse code modulation), ie a combination of long and short pulses may represent characters (similar to RTTY) or discrete analogue levels, such as the CD system. A pulse modulated transmitter offers, in other words, a large variety of possibilities.

It is impossible to go through pulse modulation in detail in a book of this kind. Suffice it to mention that pulse modulation has been used for a considerable amount of time, and its use is still spreading.

8

Receivers

THE DEVELOPMENT OF RECEIVERS started during the first few years of electronics. Several types of receivers were developed according to different principles. Each new type was an improvement over the previous one.

Crystal Receivers

In the very beginning, there were reception devices built on mechanical or chemical principles (the 'coherer' being one of them) that were used to detect radio signals. Generators or spark gap transmitters were used to generate radio frequencies. However, those seemed impractical and forced further development.

Crystal receivers were far more practical devices. They often had one or more tuned circuits to filter out the desired station.

This type of receiver had several drawbacks, though. It was insensitive and difficult to adjust, since the crystal was not equally sensitive over its surface. It was a simple device, however, and became a popular object for home construction.

Those crystals are collector's items these days, and are very difficult to find. There is a modern device that can be used instead - a semiconductor diode.

A crystal receiver consists of a tuned circuit connected to the aerial and that passes its energy to the diode (an OA47 has successfully been used), followed by a filter and a pair of headphones.

To load the tuned circuit as little as possible, the headphones should have a high impedance. Crystal earplugs should work fine.

A switch was frequently incorporated to connect a larger or smaller portion of the coil in order to cover a larger frequency range.

Fig 116 shows a simple crystal receiver. After the diode is a capacitor. Its purpose is to short-circuit the RF. The remaining AF is connected to the high impedance earphones. Pi-filters too have been used in this position.

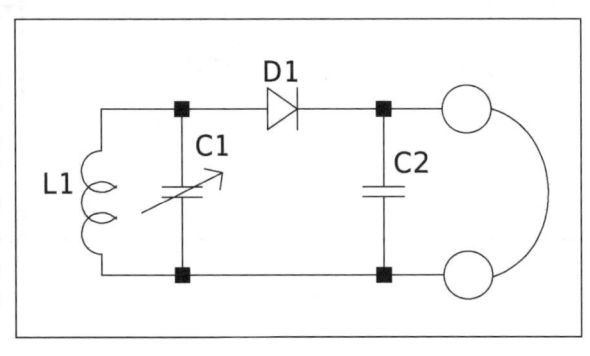

Fig 116: Crystal receiver

TRF Receivers

TRF stands for tuned radio frequency. It was patented in 1916 by the Swedish-American engineer Ernst Frederick Werner Alexanderson (1878 - 1975).

As a consequence of the invention of the valve, receivers became more sensitive and stations further away from the receiver could be heard. The signal could be amplified in one or more stages before brought to the 'phones. There were drawbacks too.

The bandwidth of a filter depends on its resonant frequency, so several filters would have to be tuned, which made the receiver more complicated to operate.

Fig 117 shows a simple TRF receiver, and **Fig 118** a block diagram of one. It has one valve, a dual triode. The first stage is a straightforward detector and the second stage an AF amplifier. It drives a loudspeaker through the output transformer T1. The receiver has two bands, LW and MW, switchable by the two-pole switch S1. It is tuned by C1. Coil details are shown opposite.

The detector is a leaky-grid detector, where the grid bias is developed across R1 and C4. C5 forms a filter to give a pleasant tone in the speaker. R7 gives a slight negative feedback for the output stage to improve the quality of the sound. R6 is the volume control. For best performance, all leads connected to U1a should be as short as possible.

Fig 117: One valve TRF receiver

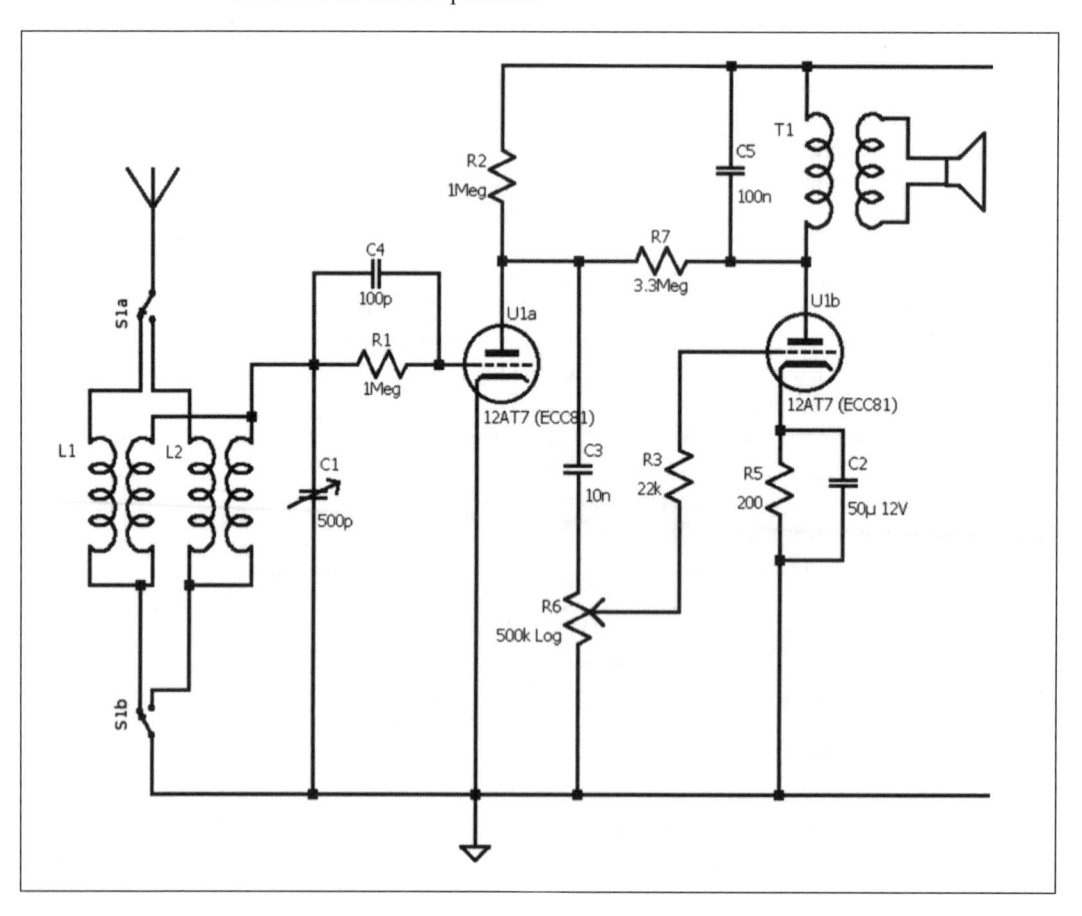

CHAPTER 8: RECEIVERS

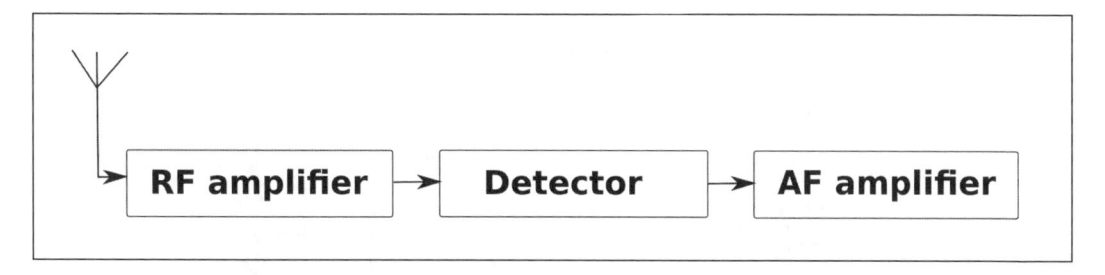

Fig 118: Block diagram of TRF receiver

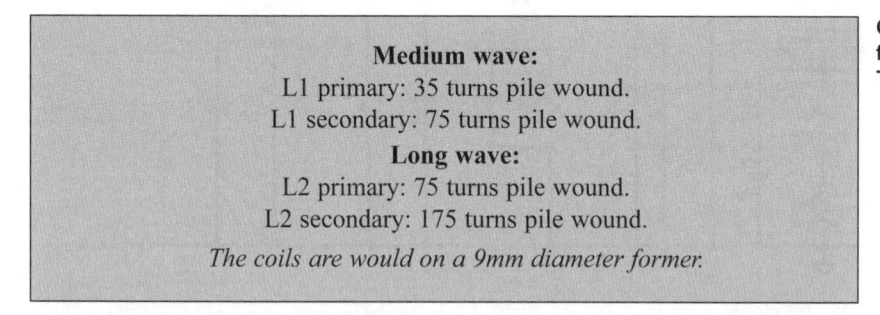

Coil winding data for the one-valve TRF receiver

Medium wave:
L1 primary: 35 turns pile wound.
L1 secondary: 75 turns pile wound.
Long wave:
L2 primary: 75 turns pile wound.
L2 secondary: 175 turns pile wound.

The coils are would on a 9mm diameter former.

Fig 119: Front panel of a TRF receiver from the 1920s

The exterior of a more advanced, commercial TRF receiver is shown in **Fig 119**. There were several filters to tune; each had its own knob on the front panel. In some models the valve filament voltages could be varied as a volume control.

TRF receiver with pentode RF stage

This is a receiver with a pentode in an anode bend detector configuration. As you can see in **Fig 120**, it has two tuned circuits and should yield a decent sensitivity and selectivity. In the olden days, only triodes were available, of course, which meant lower sensitivity.

L2 and L4 determine the frequency band along with C1 and C4. U1, an EF89, is a variable-μ pentode. This means that the grid bias determines the amplification

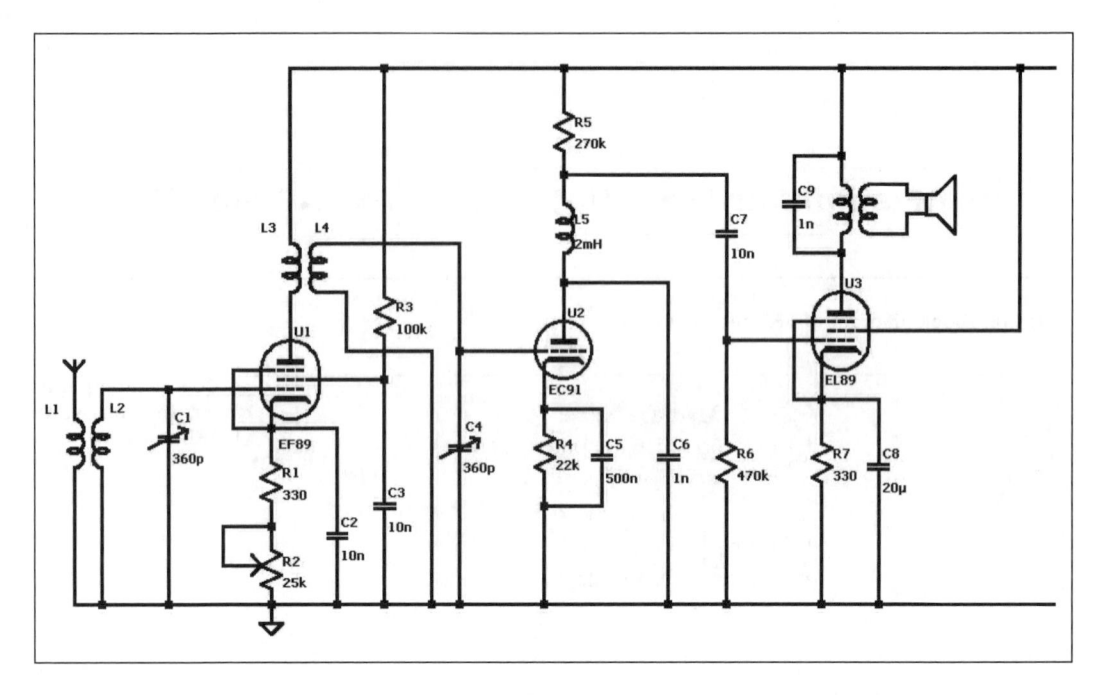

Fig 120: A three-valve TRF receiver

of the stage. Since the cathode resistors determine the grid bias, R2 functions as the sensitivity control by varying the grid bias up and down the curve. The bottom of the curve gives less amplification than the linear portion.

Note the components at the anode of U2. The choke L5 stops most of the RF signal, which is grounded via C6. However, the AF signal passes the choke to the anode resistor and is connected to the control grid of the power pentode EL89. Therefore, the decoupling capacitor, C6, cannot be too large.

Other valves could, of course, be used. There is, for instance, a triode-pentode called an ECL82, where the triode section could be used as the detector, and the pentode section to drive the loudspeaker.

The coils L2 and L4 are best wound on formers with an iron core, so they can be trimmed to match each other. If no iron core is at hand, a piece of plastic tubing can be used and trimming can be done by separating the turns slightly. They need to have the same number of turns, though. If C1 and C4 are ganged, only one control is needed for tuning.

Feedback

Eventually it was found that a positive feedback could compensate for losses in the tuning components. However, too much feedback could turn the stage into an oscillator, and that could interfere with other receivers nearby.

The principle is used in a device called 'Q-multiplier', which was a link in the IF chain of superheterodyne receivers.

The Homodyne

The superheterodyne principle was invented by Edwin Armstrong and patented in 1918. In 1932 a British team of scientists took a different approach - the homodyne.

An oscillator that oscillates on the signal frequency and is mixed together with the input signal acts as a detector for AM signals. It is easy to picture a TRF receiver with feedback inserted into the detector to such a degree that is begins to oscillate. The detector is connected in series with the signal path, the input signal and the oscillating detector signal mixed together and produced a demodulated signal, if the detector circuits were correctly tuned. The reason the detector produced an AF output is that its frequency is affected by the incoming signal such that it locked onto it. The phenomenon is named 'pulling' and may occur if a portion of a signal on the oscillator frequency finds its way into the oscillator.

Certain types of mixers are prone to cause pulling. In such cases in a superheterodyne, the oscillator should be followed by an insulating stage, such as a cathode follower.

Due to its function, the homodyne is a type of DC (direct conversion) receiver.

Later, work was carried out to improve the performance of the homodyne. The carrier of the transmitter was filtered out and re-used as a carrier insertion oscillator. This eventually led to the development of the synchrodyne, which has a built-in frequency converter.

The Synchrodyne

This is basically a homodyne, but the oscillator is not connected into the signal path. Instead, a separate frequency converter mixes the received signal with the oscillator. A block diagram is shown in **Fig 121**.

As stated above, the output of a mixer contains among other signals the difference between the aerial signal and the oscillator signal. If the two signals are equal and the transmitter is amplitude modulated, the transmitter signal will be demodulated. An AF filter at the output stops the RF and passes the AF, which is amplified and brought to the loudspeaker.

The principle was first published in 1947 and is called 'synchrodyne reception' and it is still in use in some 'world receivers'. As the name implies, the oscillator needs to be synchronised with the received signal to function properly. The benefits are less distortion of the AF and higher sensitivity.

As the name and working principle imply, a synchrodyne receiver involves a phase lock of the oscillator to the signal.

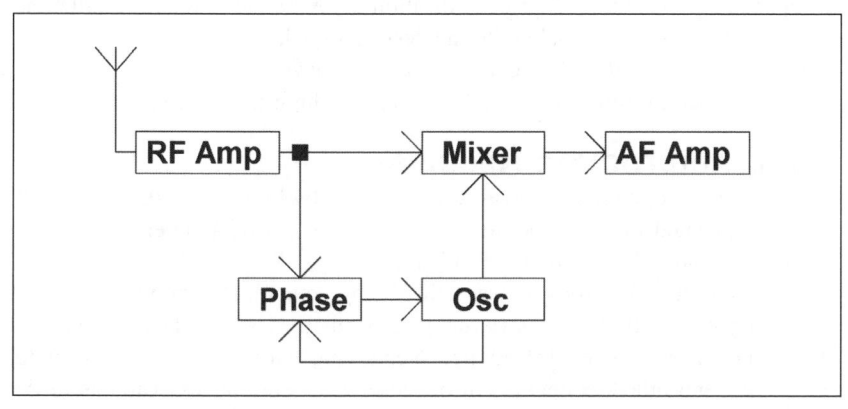

Fig 121: The block diagram of a Synchrodyne receiver

Fig 122: The RF stage of a Neutrodyne

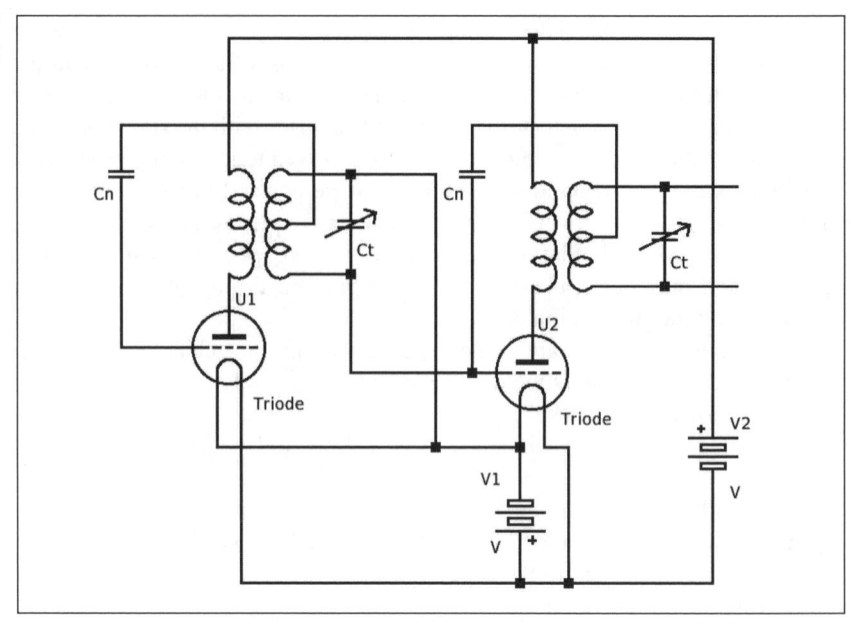

The principle also lends itself to a detector after the IF chain, which is the case in today's receivers with a synchrodyne detector.

Synchronous detection can be made very sensitive and has a better immunity to distortion due to fading than a 'normal' amplitude detector has.

The Neutrodyne

Triodes, in particular those of the early days of valve technology, suffered from large inter-electrode capacitances. This caused instability in RF circuits. The American Louis Alan Hazeltine (1886 - 1964) invented the 'neutrodyne' in the early 1920s to overcome this problem.

A neutrodyne is a receiver in which a portion of the signal is capacitively fed back to a previous stage to neutralise the valve capacitances.

A neutrodyne was cheap to build and easier to manipulate than the early superhets. An RF stage of a neutrodyne could look like **Fig 122**.

The capacitors denoted Cn are the neutralising capacitors, and those denoted Ct are tuning capacitors. Bad neutralisation gives rise to parasitic oscillations and squeaks which don't belong to the desired signal.

These days the inter-electrode capacitances have been reduced to such a degree that neutralisation is needed only on very high frequencies.

Superheterodyne Receivers

The superheterodyne receiver was first described by Edwin Armstrong in 1918. He also invented the 'Fremodyne', which is a variant of a super-regenerative receiver, intended for use on the FM bands.

While the superheterodyne was still young, manufacturers went to great length trying to find exclusive names for their designs. Names such as Autodyne, Fremodyne, Infradyne, Negadyne, Novodyne, Selectrodyne, Tetrodyne, and others came up. All of those were more or less variants of the

CHAPTER 8: RECEIVERS

Fig 123: Block diagram of superheterodyne receiver

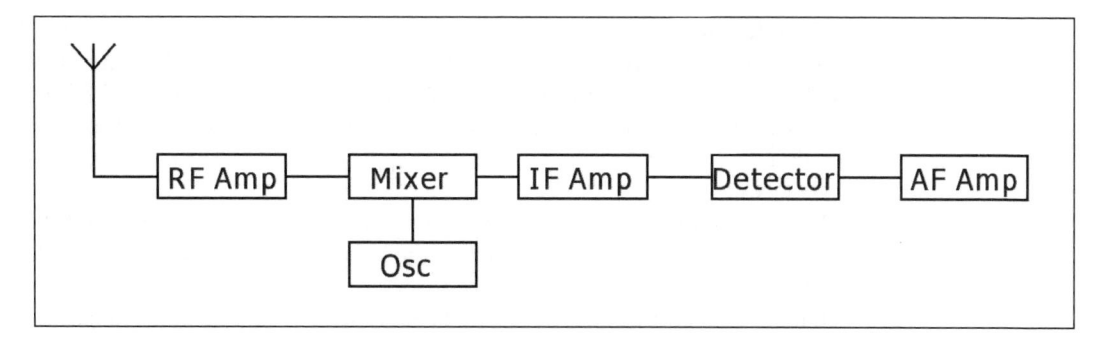

superheterodyne receiver. They never brought any new principles to life, but all of them participated in the improvement of radio reception. Even today, radio receivers are superhets. The ultimate step has been taken.

One of the problems with TRF (and similar) receivers is that the bandwidth of a tuned circuit depends on the frequency to which it is tuned. So, the bandwidth is smaller at the low end of a band, whereas it increases towards the top of the same band.

Additionally, the more tuned circuits, the more controls are needed on the front panel and the more complicated it is to operate the receiver. Consequently, if all tuned circuits could be tuned to the same frequency, these drawbacks could be eliminated.

The way to do this is to convert the signal frequency to a fixed frequency. To do this, a frequency converter and an oscillator are needed.

If the oscillator operates at a fixed offset from the desired signal frequency, the resulting mixer product will be the same, regardless of what frequency the receiver tunes in. This principle, shown in **Fig 123**, also eliminated the need for feedback, because if selectivity or sensitivity was poor, all you needed to do was to increase the number of IF (intermediate frequency) transformers and IF amplifier stages.

Usually, the oscillator works on a frequency higher than the signal frequency. This decreases the risk that the oscillator interferes with other signals and causes unwanted conversion products. In some receivers, though, the oscillator frequency is lower than that of the incoming signal.

Such a receiver is called 'infradyne'. A drawback is, among others, that overtones from the oscillator can find their way into the IF amplifier and seriously damage reception.

The following signals appear at the output of the mixer:

- The signal from the aerial
- The oscillator signal
- The difference between the two
- The sum of the two

The aerial signal isn't wanted, since it varies with the tuning of the receiver. The oscillator signal is of no use once it has done its job. The sum signal also varies with the tuning of the receiver. The difference is at a fixed frequency, so this is what we want.

107

Consider this:
You want to receive the signal from a transmitter on 1.5MHz. Your IF is 500kHz. If you tune the oscillator to 2MHz, the difference will be 500kHz, which is what you want. There is also a signal on 3.5MHz, but this falls outside the IF.

Case two:
Your desired signal is on 2.5MHz. Tune the oscillator to 3.0MHz, and the difference will still be 500kHz, whereas the sum will be 5.5MHz, again outside the IF.

If there is no filter before the mixer, the same station will appear in two places, one where the oscillator is tuned to 500kHz lower than the signal, and one where it is tuned to 500kHz above the signal. The correct signal is the one where the oscillator is tuned to 500kHz above the desired signal. The other signal is named 'image frequency', and the purpose of the filter(s) before the mixer is to get rid of the image.

Additionally, tuning a signal on a higher band, say, 20MHz, may allow the image to slip through the IF filters if the IF is chosen too low. This is caused by the width of the mixer filters - a 500kHz image is close enough to the wanted signal to fall within the filter bandwidth at higher frequencies. The same problem appears again - the higher the frequency, the wider the bandwidth. On the other hand, a high IF means a wider IF bandwidth.

One way of solving this problem is to use more than one IF - one high IF for image elimination at high frequencies, and one low IF for selectivity. An IF should also be chosen in a band with few transmitters. There is otherwise a risk that transmitters find their way directly into the IF amplifier, and then there is no way of getting rid of them. Also, the IFs need to be chosen so that they fall outside the bands that the receiver is supposed to cover.

Some advanced receivers, such as Collins 51-S1 (which covers the range 0 - 30MHz continuously), re-arrange the IFs according to the received band.

The Drake R4 receiver, which covers the amateur bands, is a double super with the second IF as low as 50kHz.

The superheterodyne (or, shorter, 'super' or 'superhet') is, in other words, a compromise.

Tuned IF superhet

One type of receiver that may be seen as a mix between a TRF and superhet is a receiver with tunable IF. It consists of a frequency converter, an IF, detector and low frequency amplifiers, as is usual in a superhet. The difference is that the oscillator in the frequency converter is fixed, whereas the IF is tuned like a TRF receiver.

The benefit is that it can be made very stable if the oscillator is crystal controlled. The building requirements of a crystal controlled oscillator are far lower than those of a VFO for good stability. On the other hand, such a receiver can not be used for signals that require a BFO, unless a VFO is introduced. It is a simple variant of a superhet, though.

The filter is frequently a tunable double filter, arranged to give a fairly flat-topped curve so as not to affect the 'higher' modulation frequencies (frequencies above 5kHz or less on the AM bands).

The IF can be arranged to be fairly low, 1MHz or lower, which means that another frequency converter can be built, bringing the last IF down to a couple of hundred kHz or lower. A VFO on such frequencies would be very stable, as would the receiver as a whole.

The second IF would be fixed, enabling BFO operation as well as the use of crystal filters.

DC Receivers

DC stands in this case for direct conversion. This receiver is very similar to the synchrodyne and homodyne described above. The major difference is that the oscillator is not locked onto the received signal. It is therefore best suited for reception of signals that would normally require a BFO, such as SSB, CW, RTTY, fax, etc.

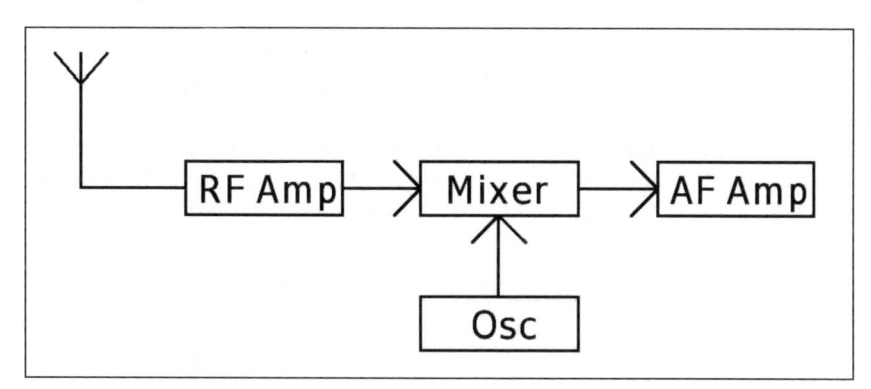

Fig 124: Block diagram of a direct conversion (DC) receiver

The oscillator is tuned to a frequency very close to the signal, so close, in fact, that the difference is still audible. As a carrier is turned on and off on the signal frequency (such as morse code), the information becomes easy to read. A block diagram is shown in **Fig 124**.

A DC receiver is very easy to build. All it needs is an RF amplifier, a mixer, an AF amplifier and an oscillator. Four stages.

Detection is performed by the mixer. When the oscillator signal is mixed with the received signal, one of the resulting signals is audible. Therefore, only a low-pass filter is needed between the mixer and the AF amplifier.

Super-regenerative Receivers

This type of receiver is not recommended. It radiates a lot of radio energy and can easily interfere with other equipment in the neighbourhood. Edwin Armstrong invented it in 1914 whilst still in college, and patented it in 1922.

Its detector has positive feedback which is balanced such that it oscillates in bursts - the oscillations are turned on and off with an inaudible frequency. The result is a sensitive and fairly simple receiver.

Reflex Receivers

A reflex receiver is one where the same valve or chain of valves are used to amplify both RF and AF. The AF is connected back to the beginning of the chain. It is kept separated from the RF with filters, but it is difficult to get right,

VALVES REVISITED

because there is often residual RF in the AF and vice-versa. The reason for its development was the cost of valves. When the valve prices went down, the number of reflex receivers went down too.

The Stenode

The Englishman James Robinson discovered in the mid 1920s that a control crystal could be used to filter out a transmitter. A control crystal can be seen as a tuned circuit with a fixed resonance frequency and a Q of 2,000 or more. The extremely high Q makes the crystal bandwidth extremely narrow.

He filtered out the carrier from the IF and achieved a signal with only very weak residues of the modulated information - a carrier. Connecting a high pass filter after the crystal made it possible for him to restore the information in the modulation. He named the design 'Stenode' and claimed a high quality result.

The usage of crystal filters was later taken over by the manufacturers of communications receivers, and development of crystal filters has evolved ever since.

The 'Stenode Radiostat', shown in **Fig 125**, was manufactured commercially, by a company named Stenode, but it was met by scepticism in the press and never caught on. Fig 125 shows the arrangement of the last IF stage of the Stenode.

Fig 125: Stenode Receiver

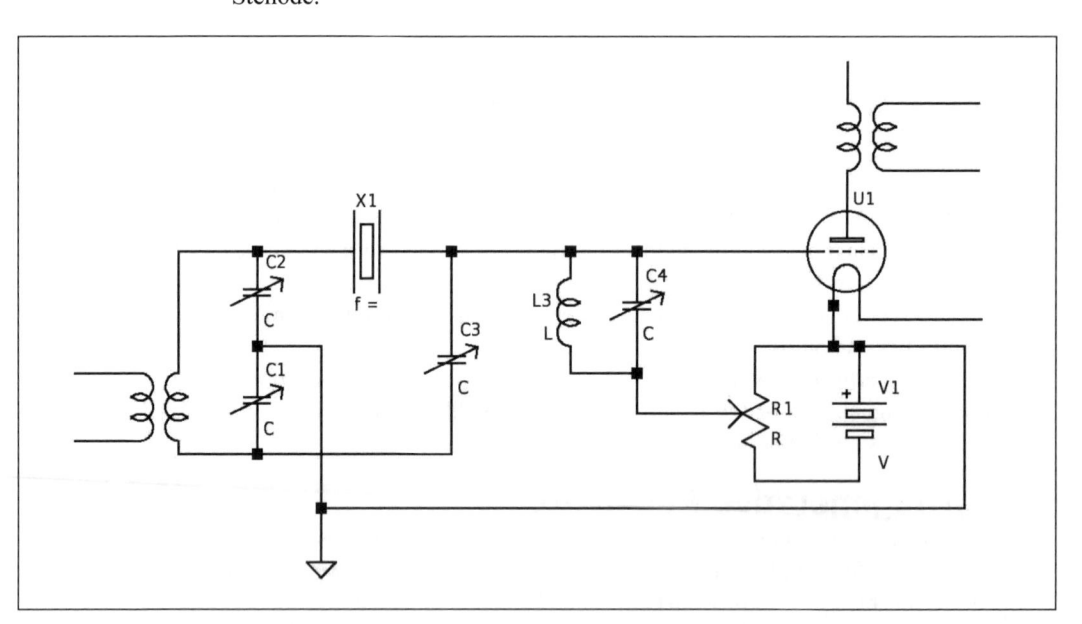

9

The Superhet in Detail

IN ORDER TO understand well how the superhet works, it is necessary to go through the various stages and their alternative designs. Naturally, different manufacturers used different approaches to the various stages to achieve the best results within the economic constraints of the radio, but the differences lie in details only, in many cases.

Frequency Converters

In a superhet, the incoming signal is converted to a lower, fixed frequency. This is done in a frequency converter. A frequency converter consists of an oscillator and a mixer stage. The combination of both stages together is usually referred to as a 'mixer'.

Mixing two frequencies together can be done by virtually any non-linear device. It is the lack of linearity that causes the signals to generate the difference and sum frequencies.

The principle is used not only in superhets, but also in other types of receiver. It is also used in many measurement and laboratory instruments.

An overview of the entire spectrum of receivers for domestic use, as well as communications receivers, shows that a frequency converter can look very different.

In a professional communications receiver, performance is more important than price, so sensitivity and freedom from noise are major factors. In the vast majority of those cases, the frequency converter has a separate valve for the oscillator and one for mixing the signals, whereas a domestic receiver mainly was simpler in design and usually had a combination valve to perform both functions. They were mainly designed for use on MW and LW, even if many of them also had SW bands.

Some advanced commercial receivers have more than one frequency converter. These are dual or triple conversion receivers. The first IF is usually high - up to 9MHz or more - to achieve good image frequency immunity, whereas the last intermediate frequency could be as low as 100kHz or thereabouts to achieve the necessary narrow bandwidth.

The Racal RA117 receiver is an example of thinking outside the box! Its first IF is as high as 50MHz! This eliminates the image frequency problem and allows the receiver to cover 1 - 30MHz continuously, but introduces stability problems instead. The oscillator was tunable between 51 and 80MHz.

Racal solved the problem by phase locking the oscillator to the overtones of a crystal controlled oscillator, with increased risk for introduction of noise. The

Fig 126: Octode mixer stage

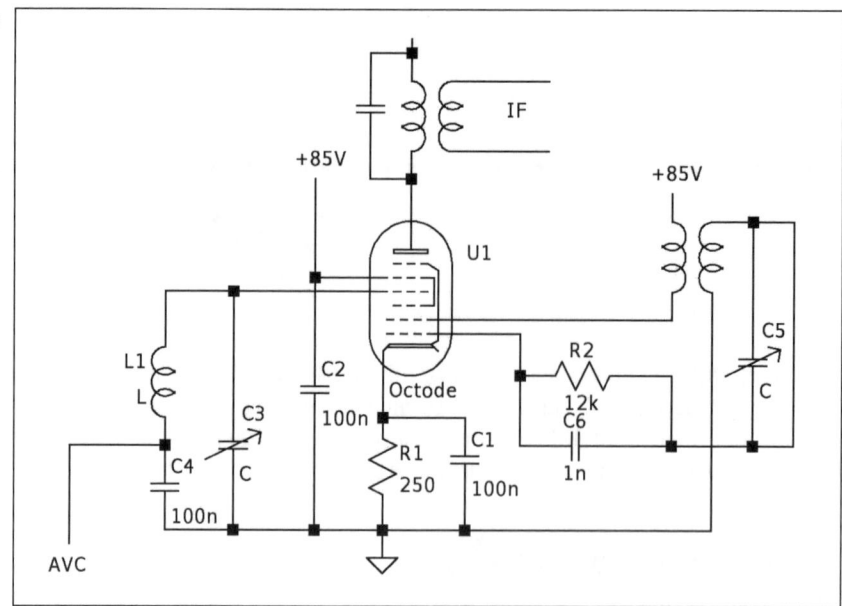

50MHz IF is later converted down to lower frequencies to achieve the necessary bandwidth.

Octode oscillator/mixers

The octode (see Chapter 2) was developed and introduced in the USA in 1939. The DK90 is one such valve intended for portable receivers. There are none with a noval (B9a) socket. However, the principle is interesting, because in some designs, the oscillator and mixer are both built into the circuit as in **Fig 126**, despite the fact that the anode is common to both functions. One such example is shown here.

The oscillator part consists of the two bottom grids, G1 and G2 and associated components, where G2 functions as the oscillator anode. The antenna signal is injected into G4, where it modulates the oscillator signal. G3 and G5 are connected together and act like the screen grid of a pentode.

R1/C1 generate the grid voltage for the mixer portion, whereas the oscillator grid has a bias of 0V, since the grid leak R2 is connected to ground through the coil.

Multi-function valves

In domestic receivers, multifunction valves were used. These are noisy (due to the many grids) and the oscillator parts of the design were neither optimised for stability nor spectral purity. However, they required only one valve for both functions, which is cheaper. It is possible, though, to design a stable oscillator using the triode part of the valve.

Fig 127 shows the converter of a domestic radio. Note the AVC line to the mixer. In general, it is not recommended to connect AVC to the mixer stage.

The oscillator signal is inserted into the mixer's G3, thence modulating the aerial (antenna) signal.

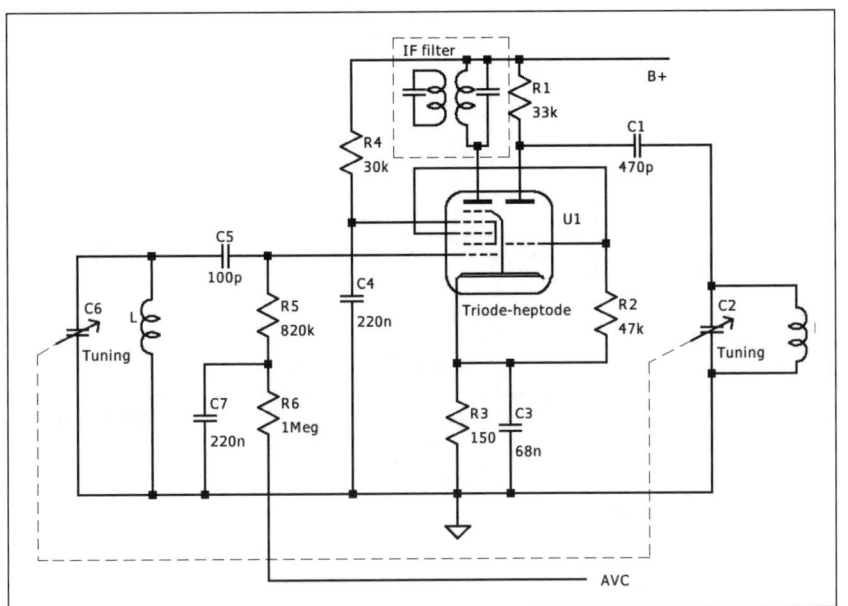

Fig 127: Triode-heptode mixer

Pentode mixers

In a pentode, there are two points into which the oscillator signal can be injected - the cathode and the screen grid. Since the pentode is a single function valve, the oscillator has to be a separate valve.

Varying the screen grid voltage of a pentode will cause the amplification of the stage to vary, analogous with the mixer shown above. **Fig 128** shows the basics of a pentode mixer with oscillator injection on the screen grid. Any small-signal RF pentode will perform the function.

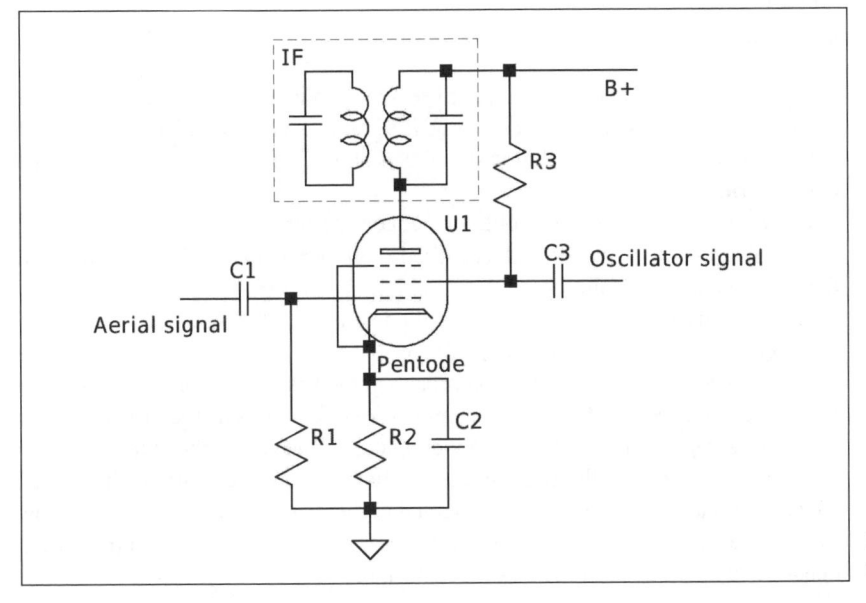

Fig 128: Pentode mixer

113

VALVES REVISITED

Fig 129: Double triode mixer

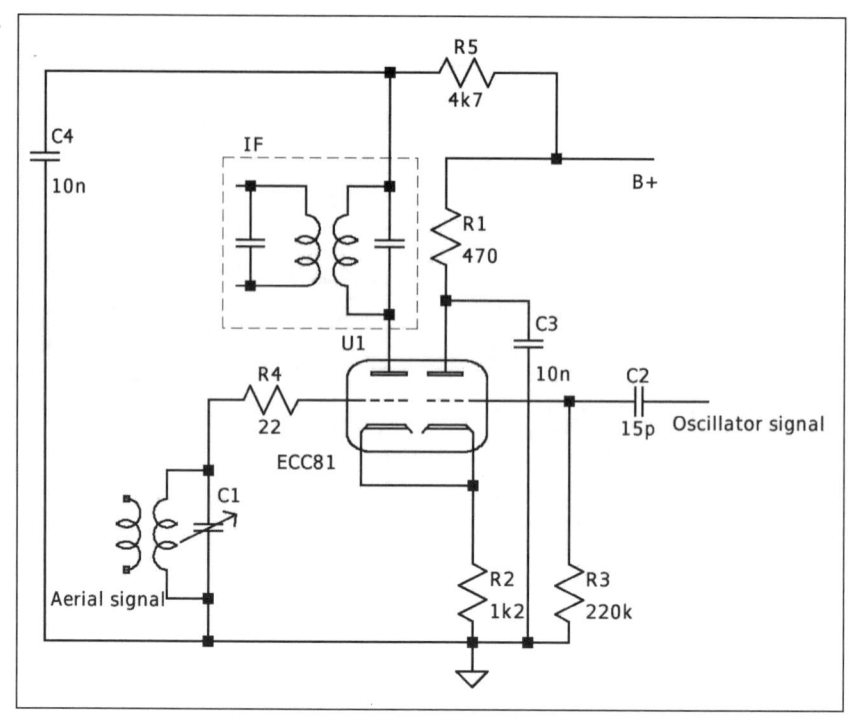

R3 can be replaced by a RFC (radio frequency choke) to obtain increased efficiency.

Due to the higher noise level in pentodes, a pentode mixer as the first mixer is not recommended in high performance receivers. Some pentodes, eg the ECF80, were combined with a triode. The triode was designed to function as the oscillator, whereas the pentode proper was the mixer. This was a usual configuration in television receivers.

Triode mixers

If you want your receiver to be as free as possible from internally generated noise, you should use a triode as in **Fig 129**, or a valve diode balanced mixer.

In a triode mixer, the oscillator signal can be injected into the cathode or the control grid.

To avoid influences on the oscillator frequency from strong signals coming in, the oscillator should be buffered. A cathode follower between the oscillator and the mixer is the best solution.

A double-triode with a common cathode or with the cathodes connected together can be used as a low noise mixer.

Collins uses a slightly different approach in the first and second mixers of the 51-S1 receiver, see **Fig 130**. Here, a single triode functions as the mixer, and the oscillator signal, which is inductively coupled, is injected at the cathode.

The oscillator coil is the point in the oscillator where the signal is the purest, so the oscillator coil is inductively coupled to the mixer cathode. Due to the low impedance of the cathode, the oscillator coil/mixer pick-up coil combination needs to function as an impedance transformer.

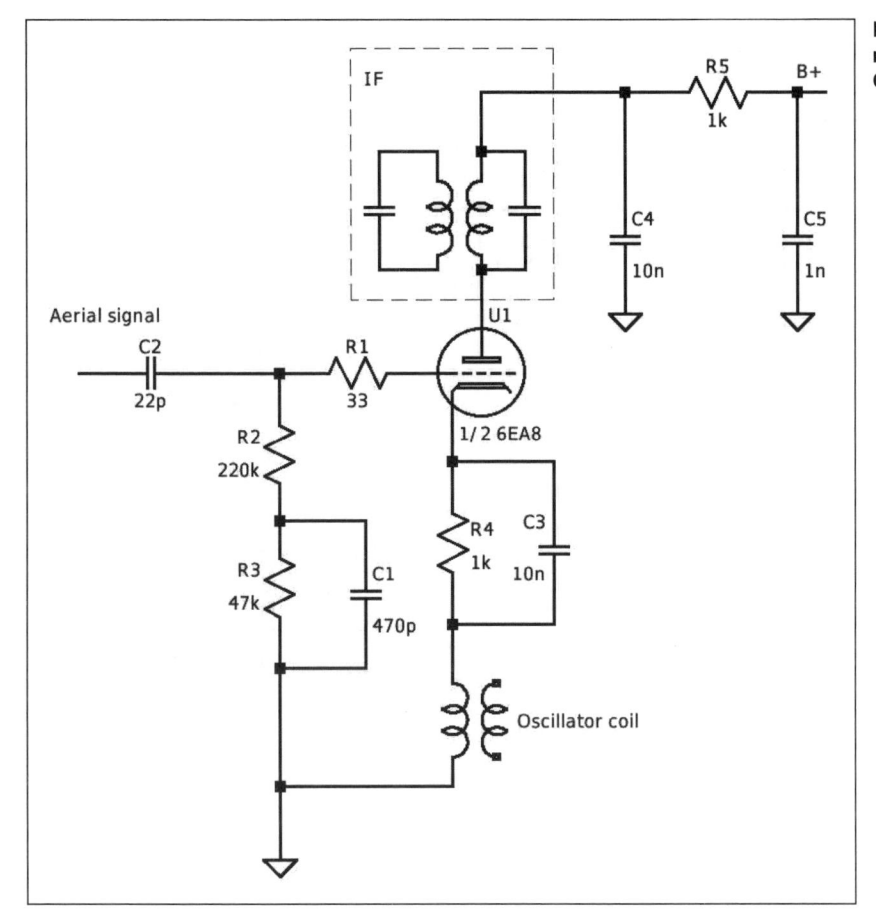

Fig 130: Triode mixer used in Collins 51-S1

Triode balanced mixer

A triode balanced mixer is a useful device. The noise level is low, and since the valve amplifies the signal, the mixer is very sensitive.

The principle is to inject one signal symmetrically and the other asymmetrically. This requires three inputs. The output contains the asymmetric signal plus the sum and difference frequencies between the two. In the example below, the oscillator signal is the symmetric one and the antenna frequency the asymmetric one.

The asymmetric signal from the antenna is connected to the grids via the transformer. The phase relationship between the grids is 180 degrees. The signal is amplified and fed to the output transformer much like a push-pull amplifier. The antenna signal is thus passing straight through the mixer.

The oscillator signal, however, meets the two grids with no phase shift. Therefore, the oscillator currents through the two valves are cancelled out at the output transformer, and if the transformers are wound (and the stage is designed) with some precision, the oscillator signal will not pass through.

The process is shown in **Fig 131**. When both signals are present, a mixing function occurs, and, apart from the antenna signal, the sum and difference signals will

VALVES REVISITED

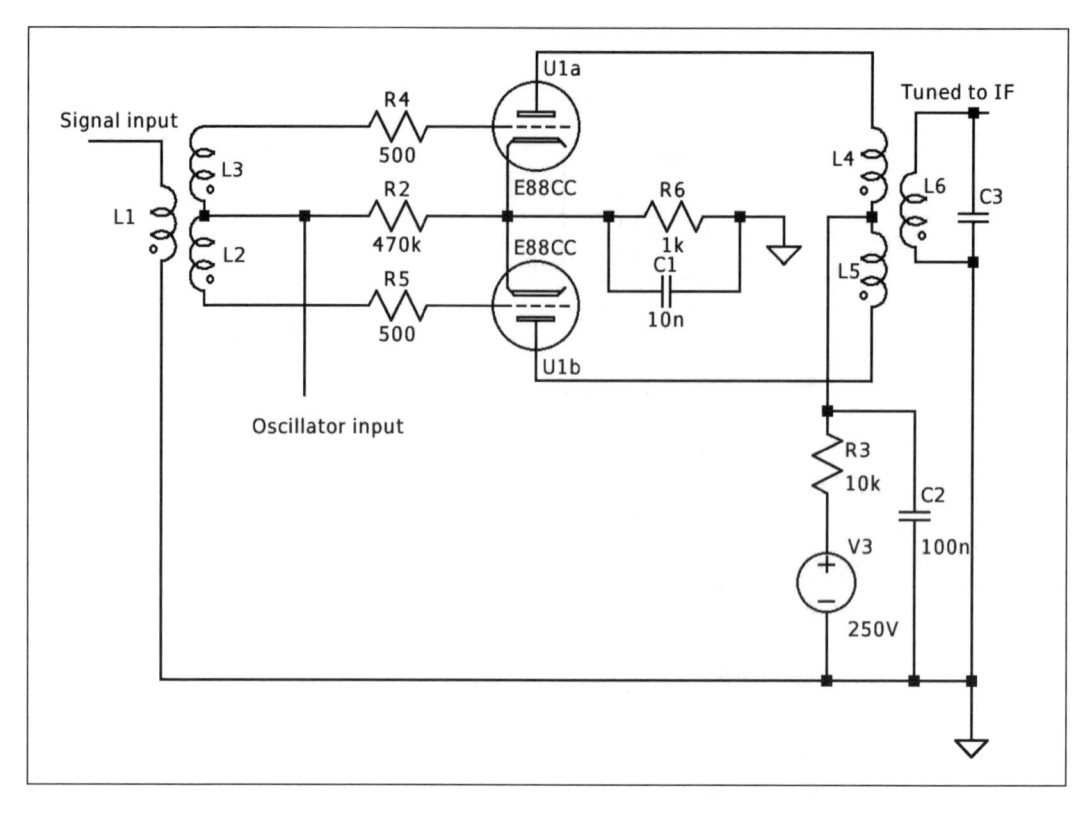

Fig 131: A triode balanced mixer

be present. The IF signal (or AF signal, in case the stage is used as a detector) is filtered out by the tuned circuit at the output. In the detector case, a pi-filter would be appropriate.

In mixer operation, the device is usually referred to as a 'balanced mixer', whereas the detector application usually is referred to as a 'product detector'. It is basically the same device, even though the inputs and output, of course, have to be designed for the kind of signals they handle. A third use for such a device is called a 'balanced modulator'. A balanced modulator is a device which is normally used to modulate an RF signal for SSB usage.

Whatever the application, the principle remains the same - feed one signal unbalanced and the other balanced, and filter out the desired signal. See further below.

The triode balanced mixer is beneficial due to its low noise level and amplification of the signal.

Diode mixers

A diode has no grids, and is therefore the most silent valve of them all. In a balanced modulator, two diodes are connected to the AF and oscillator outputs, and the output of the modulator contains only the two side-bands and AF when perfectly balanced. The same principle can be adapted for RF use, where two signals are mixed to produce a third and fourth. The output needs to be selective and tuned to the IF to suppress the unwanted signals. When used for RF only, the device is called a 'balanced mixer' (see above).

CHAPTER 9: THE SUPERHET IN DETAIL

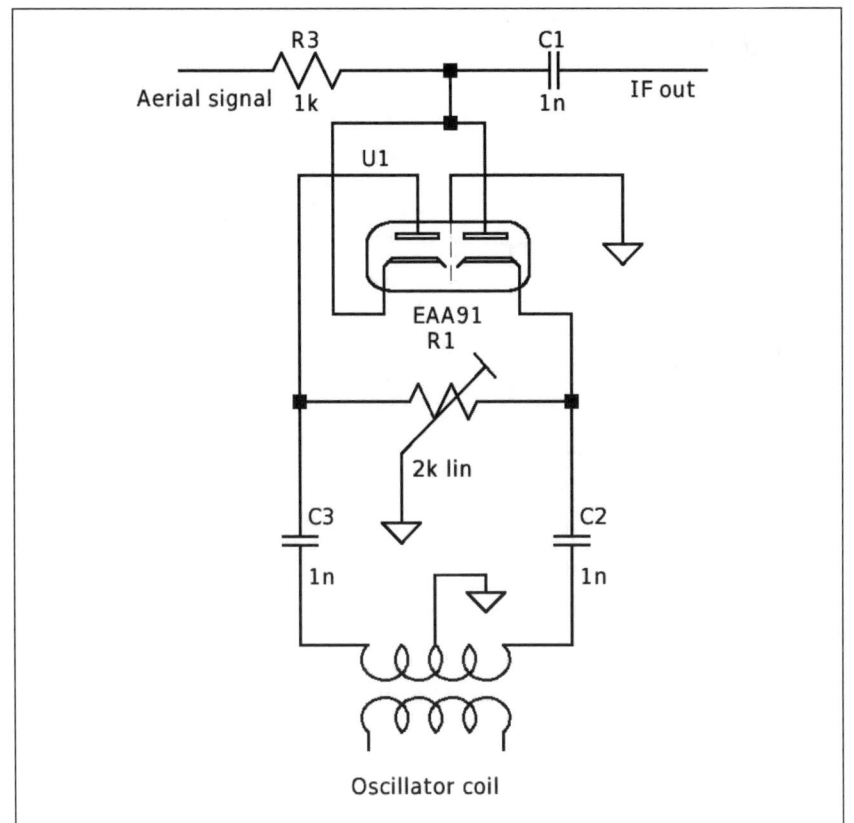

Fig 132: Valve diode balanced mixer

The same design can be used for demodulating SSB/RTTY/CW signals as well, and is then called 'balanced demodulator', 'product detector' or 'heterodyne detector'.

In applications where it is used for detection, the output is an AF signal, rather than an RF signal, and the oscillator input is called a beat frequency oscillator (BFO) or carrier insertion oscillator (CIO).

In **Fig 132**, a dual valve diode is used. The original schematic uses semiconductor diodes, connected as shown in the figure.

Only one signal is symmetrically connected to the device, the oscillator. When there is no signal at the other input, the device is balanced, and generates no output. However, when the other signal is added, the balance is upset and an output is generated.

The trimpot R1 is used to set the balance. With a tuned output, balance is not an issue in this application - the oscillator signal gets balanced out anyway.

This design does not amplify either of the signals and is therefore less sensitive than the triode mixer above.

Balanced mixers and modulators can also be designed so that both input signals are phased out. These are named 'double balanced mixers'. At the output are still the sum and difference signals, of which one has to be filtered out.

Double balanced mixers can be realised using valves or semiconductor diodes.

IF Filters

L/C filters

Fig 133: Ideal IF curve

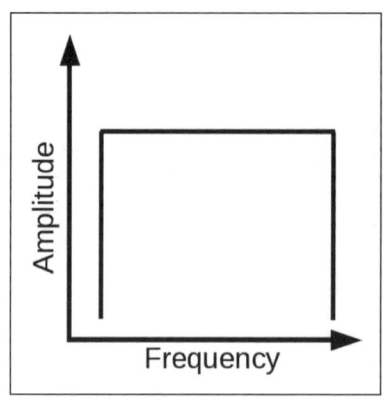

The purpose of an IF filter is to filter out only the station you wish to listen to and reject everything else. It should also pass all frequencies from the transmitter equally.

The ideal filter curve is a square shape with steep sides and sharp corners, as you can see in **Fig 133**. These are, of course, impossible requirements.

Compare this to the practical curve from a single IF L / C-filter shown in the graph of **Fig 134**.

Fig 134: Practical single filter IF curve

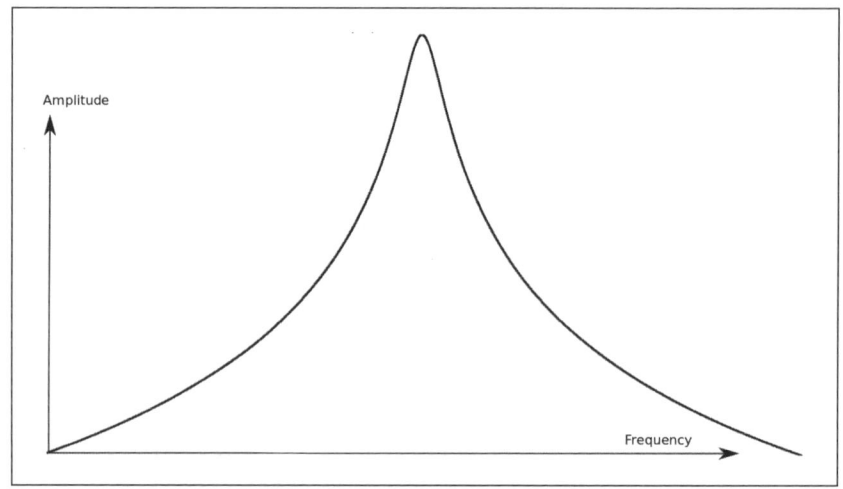

Quite a difference! Generally speaking, the more filters in an IF amplifier, the steeper the sides of the resulting curve. However, if all filters are tuned to the same frequency, not only do the sides get steeper, but, at the same time, the bandwidth gets narrower. **Fig 135** shows a set of curves, the result of up to four circuits, all tuned to the same frequency. The sides get increasingly steeper, and the tops are still rounded at all curves. This means that all modulation frequencies will not be amplified to the same degree, and the output from the detector is going to differ from the original modulation.

If, however, the filters are tuned to slightly different frequencies, the situation will be quite different. In **Fig 136** is another set of curves, showing the outputs from filters number two and four.

The top is now flat, which is what we want. The sides of filter four are, as expected, steeper than that of filter two. Introducing even more filters gives us a final curve that even more resembles our goal. We must remember that each IF transformer actually contains two filters. Six or eight filters is therefore not

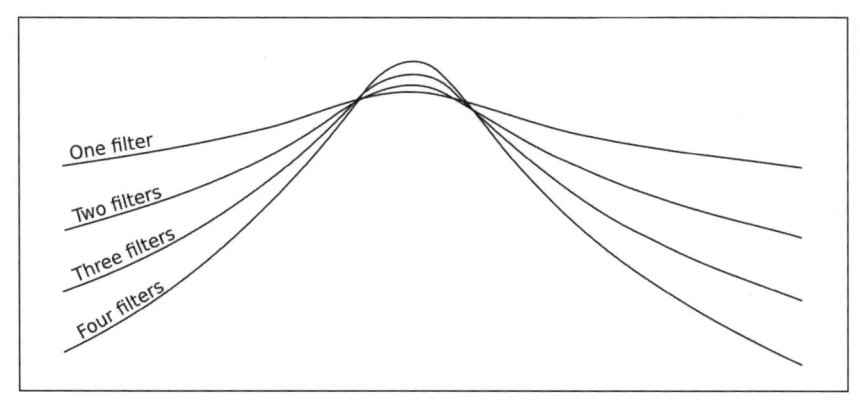

Fig 135: Comparison of curves for one, two, three and four IF filters

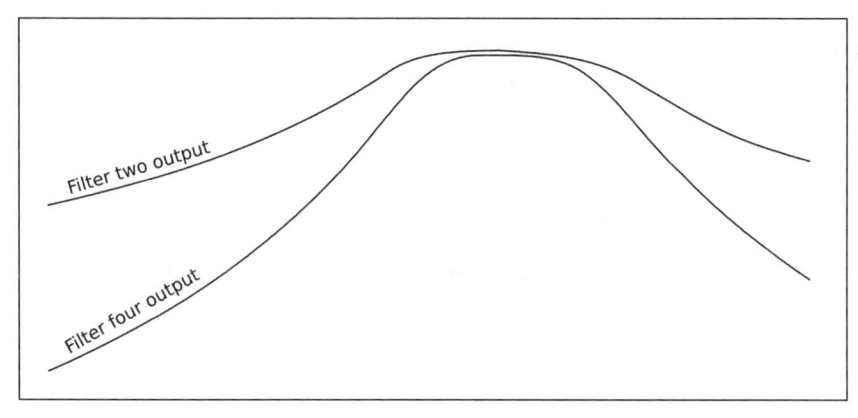

Fig 136: Resultant selectivity from the two and four filters as above, but with staggered tuning

impossible to implement. A word of caution, though. Using three or four IF amplifiers increases the risk of unwanted feedback, resulting in oscillations in the amplifier chain. Reducing the amplification will therefore reduce the risk of oscillation.

Naturally, the lower the IF, the easier it is to achieve a good approximation of the ideal filter using coils and capacitors.

A flat-top bandwidth of about 3kHz (which is suitable for SSB reception) is achievable using four filters with a centre frequency of 200kHz. The -60dB bandwidth (which is also frequently quoted by filter manufacturers) is about 25kHz at 200kHz. The curves shown above are valid for 1.5MHz. This method of achieving a flat top is known as 'staggered tuning'.

A different approach is to increase the coupling between two coils in an IF transformer to beyond critical coupling. The result will be a curve with two peaks as you can see in **Fig 137**. The 'dip' between the peaks can be compensated for by stagger tuning the other two filters.

Fig 138 shows the resulting curve after all four filters. The Y-resolution is 10 dB per division, and the horizontal resolution 1kHz. With this method, the two last filters need to associated with different stages, or their resulting curve is going to be only one single peak, and two are needed to produce a good result.

Of course, the more filters you have in your IF amplifier, the closer to the ideal curve you get, if you tune the filters carefully. This is the reason that some

Fig 137: Hard ('overcritical') coupling between two filters

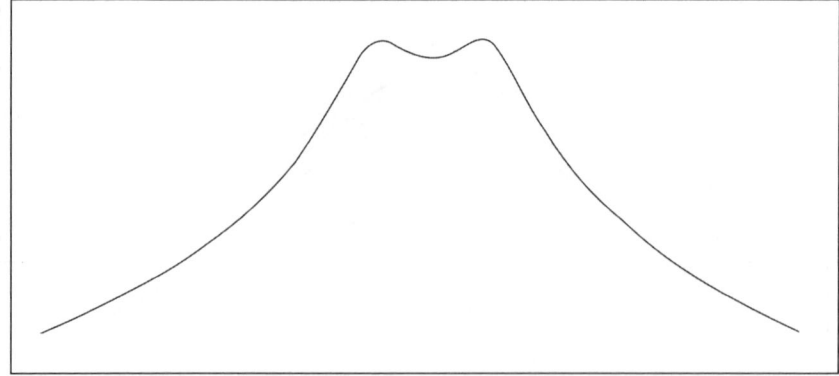

Fig 138: Compensated overcritical coupling

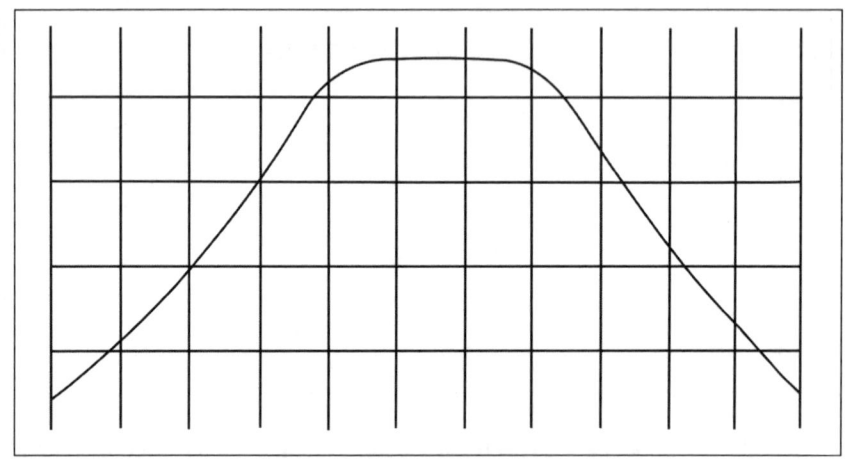

receivers have several IF amplifiers, although the amplification of one stage would be enough. The total amplification of the IF amplifier is not much more than that of a single amplifier, though, to prevent overloading of the detector.

AVC is sometimes generated by a separate detector. The reason for this is that using a separate AVC detector allows for the possibility to amplify the signal additionally. The more amplification, the better control of the AVC voltage and hence the less difference between different stations is achieved. Also, fading on a signal is reduced.

Crystal filters

Crystal filters are quite different from L/C filters. Readily manufactured filters, of course, yield the best results.

A crystal is a device with a very high Q, up to 2000 or more. It can be regarded as if it consists of inductors and capacitors with very small losses. Therefore, connecting capacitors and/or inductors to a crystal will affect the crystal's resonant frequencies.

It is possible to combine a set of crystals to achieve a flat-topped response and very steep sides at fairly high frequencies. Most professional receivers and some amateur receivers contain a number of crystal filters with different bandwidths to accommodate the requirements of different types of transmission.

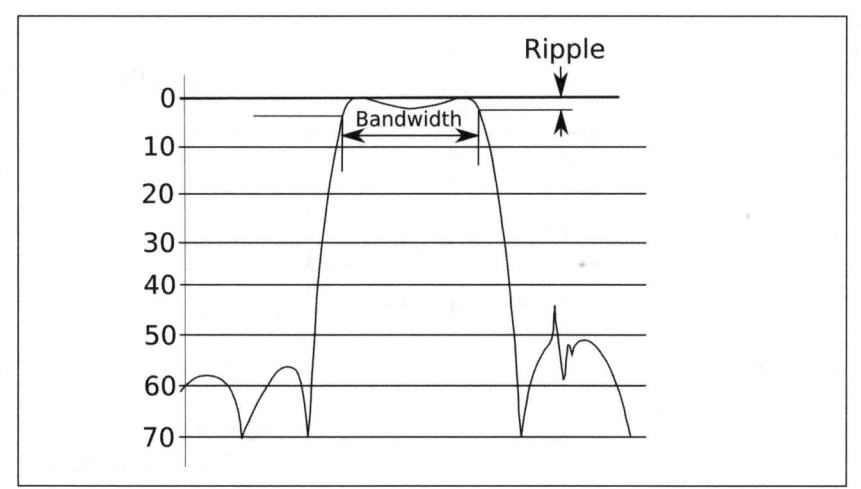

Fig 139: Typical (theoretical) curve for crystal filter

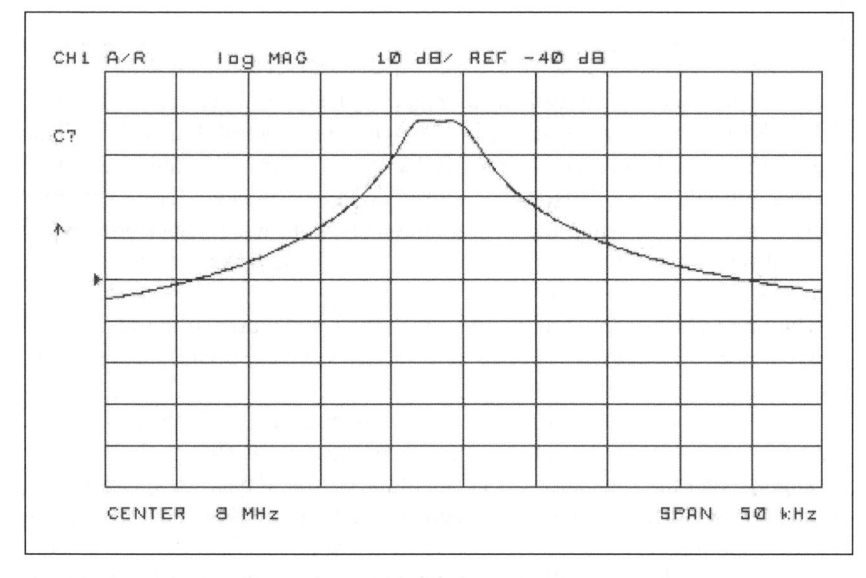

Fig 140: A measured curve of a home built crystal filter for SSB

Crystal filters are low impedance devices. The impedance is generally 50 ohms. If you build a crystal filter, it is important that the input and output impedances are correct, otherwise it will not work as predicted.

In **Fig 139** you can see a principal curve for a commercial crystal filter whereas **Fig 140** shows the curve for a practical home-made filter.

As you can see, the curve is still not ideal, but much closer to the ideal than L/C filters. It shows that it is quite possible to make a good filter for little expense.

A crystal filter may be as simple as a crystal in series with the signal path. However, such a filter would be too narrow for most applications. Therefore, if the bandwidth could be varied in some way, much would be gained. **Fig 141** shows such a filter.

The crystal resonates at the IF. C1 and C2/C3 tune the IF transformer to the crystal frequency. C5/L1 are tuned to the same frequency. Bandwidth is controlled

Fig 141: IF amplifier with a simple crystal filter

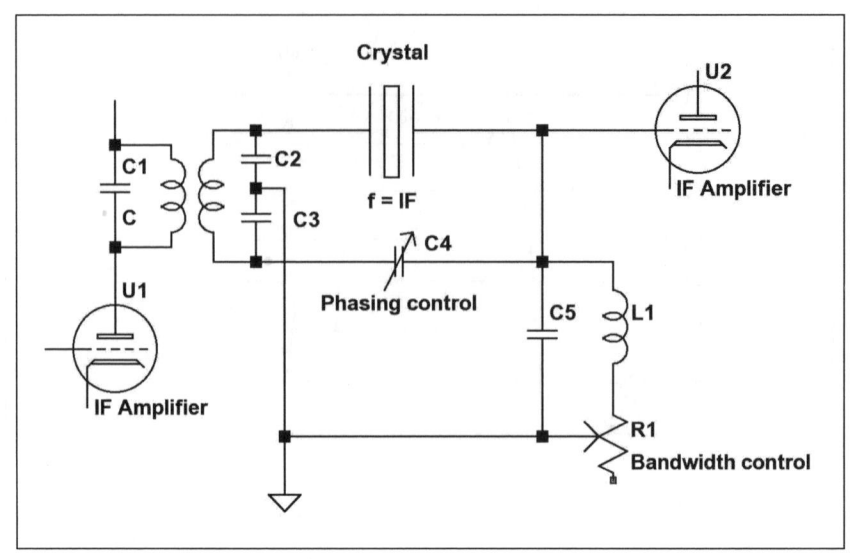

by R1. The filter selectivity is at its highest when the impedance of the filter L1/C5/R1 is at its highest. Capacitor C4 is a 'phasing' control. It takes a tiny portion of the signal before the filter and feeds it to the other end of the filter in such a phase that the stray capacitance across the crystal is neutralised. This minimises the coupling across the filter additionally, improving its performance.

As always when filters are involved, it is important to make sure that no signal can leak from input to output.

The more crystals that are involved in a filter, the closer to the ideal the IF curve comes. The crystals have to be carefully selected, because a crystal has both series and parallel resonance.

Old style crystals are housed in a holder which could be opened and closed again, allowing access to and modification of the crystal. This made it possible to slightly alter the resonance frequency of the crystal, if necessary. Using a soft pencil to shadow the crystal would decrease the frequency, and etching it would increase the frequency.

Today's crystals do not allow such manipulation. On the other hand, modern crystals can be custom ordered to fit requirements, but at an additional cost, of course.

A host of different crystal filters can be home built, and a couple of them will be shown here.

Filters with Multiple Crystals

The lattice filter

Lattice filters are filters in which the resonant frequency of each crystal is different. They complement each other, much like a staggered L/C filter bank. The resulting filter curve is reasonably flat topped, and the sides are steep. Two kinds exist: Full lattice and half lattice.

The lattice filter is fed through a phase shifter, usually a transformer, with a ground referenced input and a floating, phase shifted output. Consequently,

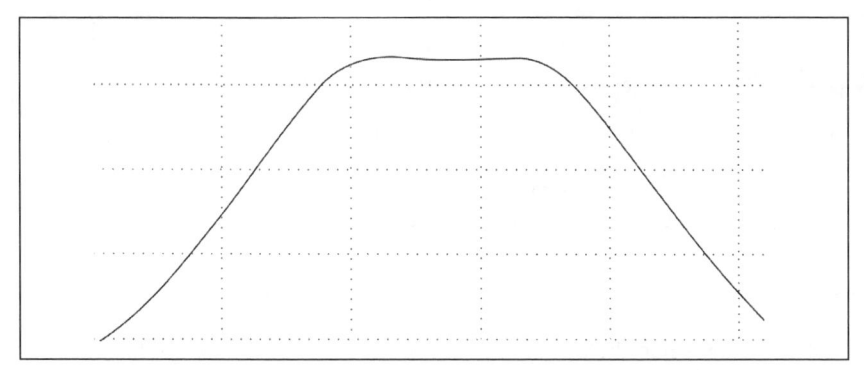

Fig 142: Curve for 6-pole half lattice filter

another transformer is needed at the output to restore the ground referenced nature of the signal for the next stage. There is one crystal in series with each branch of the output of the input transformer, and two or more crystals across the branches. Since the crystals need to be on different frequencies, they have to be very carefully selected in respect to their series as well as parallel frequencies.

The (theoretical) resonance curve of a 6-pole (six crystals) half lattice filter is shown in **Fig 142**.

Ladder filter

A filter that is easier to realise is a ladder filter. It is built of a number of crystals in series, each crystal also connected to ground through a capacitor. All crystals resonate on the same frequency, and the internal parameters are identical (or nearly so).

What gives the filter a usable curve are the capacitor values, and capacitors are a lot easier to change than crystal resonance frequencies. If the capacitors values are off nominal, the curve may change, sometimes drastically. The resulting (theoretical) curve of a 6-pole ladder filter may look like **Fig 143**.

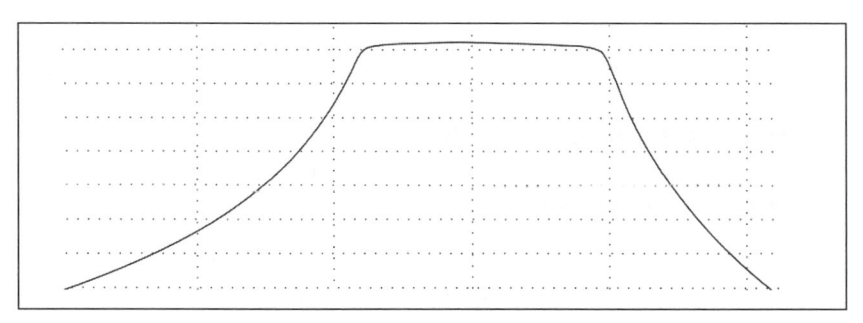

Fig 143: Curve for ladder filter

The top of the filter is flat and the sides are steep, an ideal curve for a IF filter. Minor deviations from perfect match between the crystals will affect the curve. The steeper (right hand) side of the curve leans towards the parallel resonance point. A practical implementation of a ladder filter is shown in **Fig 144**.

The distance between the serial and parallel resonance frequencies determine the bandwidth of the filter.

The principle of staggered tuning is applied to ladder filters too. They are tuned by the capacitors.

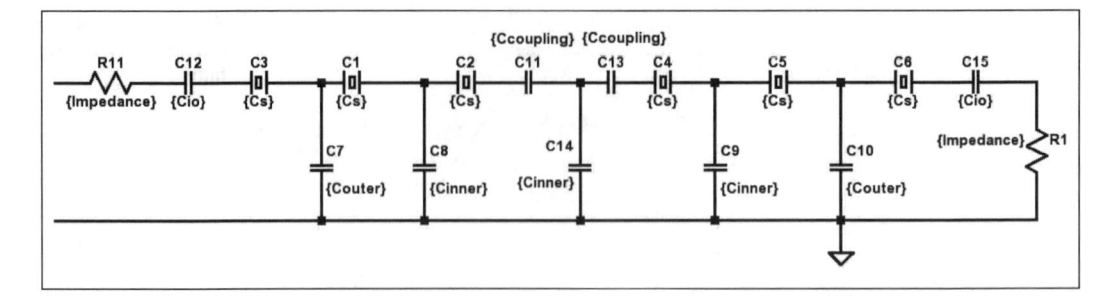

Fig 144: Implementation of 6-pole ladder filter

The typical values below are intended for a narrow CW filter. Selecting other component values will result in wider bandwidths. The crystals' serial resonant frequency is in this case 6400kHz. The filter can, of course, be designed for other frequencies by selecting other crystals.

Impedance:	250Ω	(R11 and R1)
Cio:	100pF	(C12 and C15)
Couter:	82pF	(C7 and C10)
Cinner:	100pF	(C8, C9 and C14)
Ccoupling:	470pF	(C11 and C13)

Commercial filters

Since the invention of the crystal filter, commercial filters have been available. Today's filters are monolithic filters and enclosed in cans. They can be manufactured with a very high precision and very close tolerances, using laser and etching techniques.

The most common use for crystal filters today is in IF amplifiers of FM receivers. The centre frequency of such filters is 10.7MHz. If it weren't for the wide bandwidth requirements of public FM signals, those filters could be used for short wave reception too.

The best known manufacturer of high quality crystal filters for short wave reception is perhaps Collins, but many others make them too.

Other Narrow Band Filters

Different techniques are applied to the design of filters. Many of them are intended for use on microwave frequencies, and therefore fall outside the scope of this book.

However, two filter types will be mentioned here for the sake of completeness, both of which may be designed into short wave receivers.

Mechanical filters

It is a known fact that a crystal vibrates mechanically when stimulated electrically. This phenomenon is used in mechanical filters, where the input and output devices are crystals. The vibrations from the input crystal are propagated to tiny bars or rods, sized to match the mechanical vibrations of the crystal. A number of such rods or bars can be mechanically connected together, whereby the bandwidth of the filter decreases. The output crystal senses the vibrations and converts them back to electrical signals.

Mechanical filters are mostly manufactured for 455kHz and different bandwidths. Input and output impedance is typically 600 ohms.

Ceramic filters

Also known as ceramic resonators. These filters or resonators are not widely used in precision applications, since the precision of a ceramic resonator is limited. Neither is their filter curve particularly attractive for use in radios.

They find their use mainly in fixed oscillators, where the frequency and stability are uncritical.

IF Amplifier

The function of an IF stage is to achieve selectivity. Pentodes are used in many cases for IF amplifiers, due to their high amplification. At this stage in the receiver, noise is a lesser problem, in particular if an RF stage precedes the mixer.

However, if there is no RF stage and the mixer output amplitude is low, a triode IF stage is still recommended. It could be built as a cascode stage (see the chapter on compound amplifiers) to achieve the desired degree of amplification with a low noise level.

Adjacent channel selectivity

If you are trying to receive a station on one frequency and there is a strong station on a nearby frequency, there is a risk that the strong station also reaches the detector. Consequently, the strong station will also be heard, even though the receiver is tuned to another frequency.

The cause of this problem is that the input filter is not narrow enough to dampen the disturbing station enough. However, the input filter is not the only point of interest here. The quality of the IF amplifier and its associated filters are also important.

If you wish to improve adjacent channel selectivity in your IF amplifier, a solution is to implement yet another filter. Two stages of moderate amplification yield a narrower filter with steeper sides than a single stage would. A typical pentode IF stage is shown in **Fig 145**.

IF amplifiers are frequently volume controlled by the AVC chain. The AVC voltage is disconnected from the low end of the input coil from ground and instead connected to the AVC chain. A pentode connected to AVC needs to be of the 'variable μ' type. In some cases, the cathode resistor is a potentiometer to enable manual sensitivity control. **Fig 146** shows the curves for such a valve, an EF83. The curve shows the relationship between the grid bias (VR) and amplification (Vo/Vi). With the given parameters, it can be seen from the curve that the amplification can be varied between about 14 times and over 100 times.

In case you decide to add another IF amplifier for selectivity reasons, you should try to keep amplification levels reasonable. A pentode has inherently a very high level of amplification, and adding yet another stage is bound to create problems. One way of doing this is to disconnect the decoupling capacitor (C2) from the cathode, resulting in some negative feedback.

In AF applications, though, variable μ valves would give a considerable distortion, in particular at high signal levels.

Fig 145: Example of pentode IF amplifier

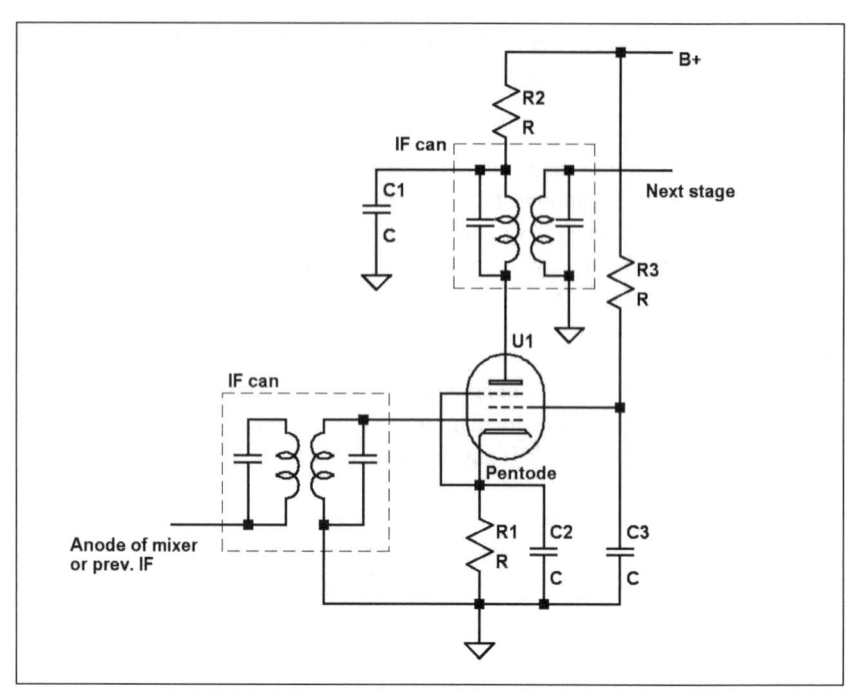

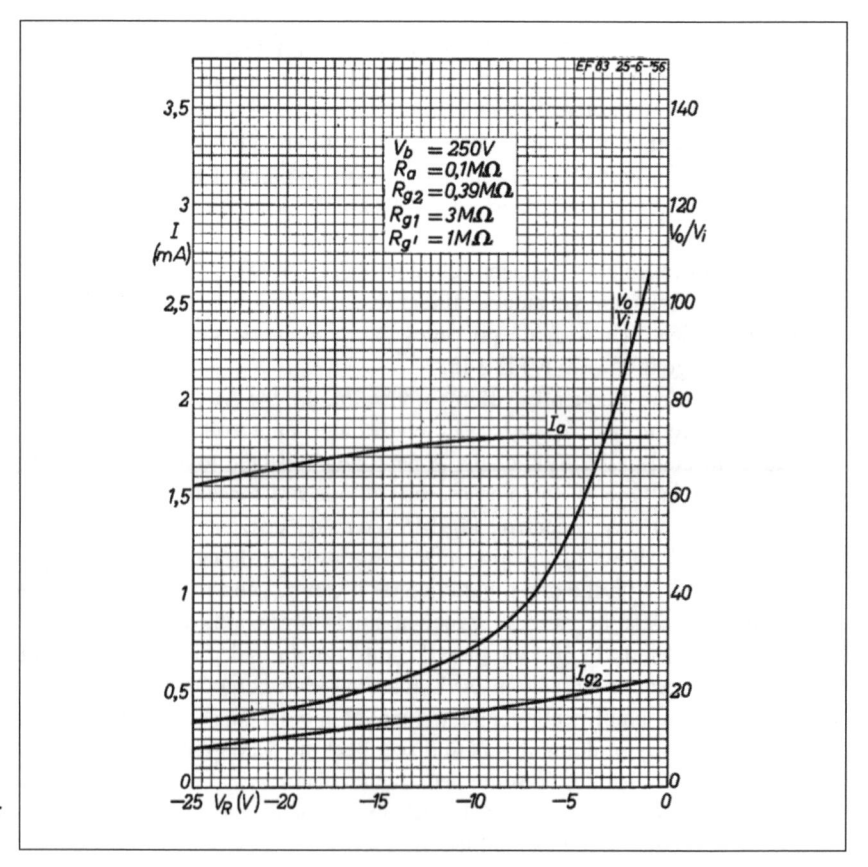

Fig 146: Curves for EF83

The Detector Stage

One stage that has evolved over the years to a degree almost as high as that of the oscillator, is the detector.

In the very beginning there was only the crystal detector. Amplifier stages were added before and after the detector in order to increase the sensitivity of the receiver.

However, various problems turned up, which caused the designers to take a closer look at the detector in an effort to improve it. The detector was considered to be the bottleneck of a sensitive, high quality receiver.

There are three major types of AM detector:

- Rectifiers
- Synchronous detectors
- Heterodyne detectors

The first four detectors we shall deal with are of the rectifier type:

The diode detector

The idea of a detector is to rectify the signal and filter the resulting signal so that only the AF remains. Rectification is exactly what a diode is made for. A diode detector can be made very simple.

A typical stage is shown in **Fig 147**. Note that the cathode is connected directly to ground. This gives a negative going DC signal, proportional to the strength of the incoming signal, upon which the AF is overlaid. The filtering capacitor, C2, needs to have a reactance for the given RF which is small compared to the resistance of R1. If C is too large, some of the AF will be lost. 250pF is a typical value.

Connecting a low pass filter with a very low cut-off frequency (a couple of Hz) to the output will create a DC suitable for use as AVC signal.

Since the detector draws current from the input circuit, the circuit is loaded, and its Q will decrease.

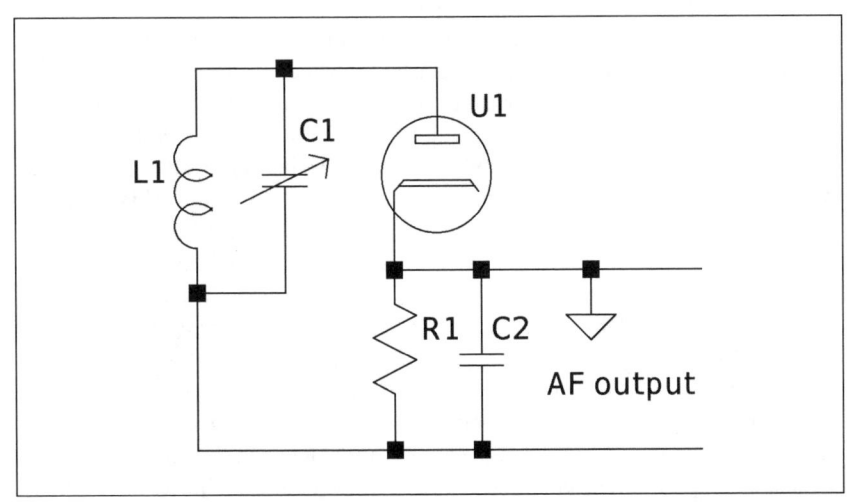

Fig 147: Example of diode detector

VALVES REVISITED

Fig 148: Example of diode detector with double diode

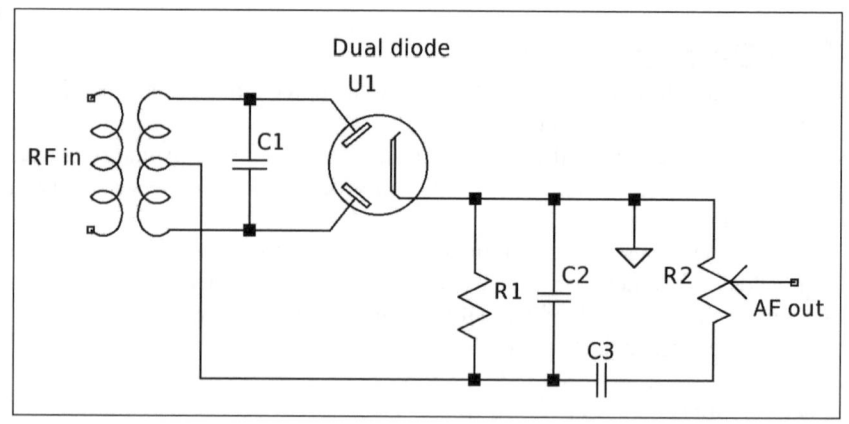

A diode detector can also be a full-wave detector, similar to that of a power supply. It might look like **Fig 148**.

Note again that the ground is directly connected to the cathode. C1 tunes the circuit to the desired frequency. C3 blocks the negative DC, so AVC should be taken at the junction of C2 / C3. The same points regarding the filter capacitor C2 apply as above. C3 should be large enough to pass the lowest AF frequencies.

Diode detectors caused problems in early receivers because they caused the selectivity and sensitivity to decrease due to the fact that they loaded the previous tuned circuit. So a race started to develop a detector that didn't load the circuit, or that could amplify the signal to compensate for the loss.

Nonetheless, more modern valve receivers and domestic radios use diode detectors because of their good linearity and simplicity. The losses caused by the diodes could be compensated for by additional amplifier stages as valve and component prices went down.

There are three major types of rectifying detectors that do not involve diodes. They are described below.

Grid leak detector

The grid leak detector, shown in **Fig 149**, amplifies the signal in addition to rectifying it. This overcomes the problem with low sensitivity detectors.

It is an amplifier where the grid bias is automatically set through rectification of the carrier at the grid. The grid leak and the coupling capacitor are so dimensioned that the grid bias follows the amplitude variations of the modulation. This means that the grid acts as an anode in a diode during the positive parts of the signal and as a grid during the negative going parts of the signal. This lowers the grid bias to a level determined by the amplitude of the carrier.

The negative portions of the signal are amplified in the valve in the normal way and are found at the anode of the valve. A resistance/capacitance filter carries most of the RF to ground, which leaves the AF signal to be amplified in the following stages. Since the grid leak detector acts as a diode, it also loads the previous stage, reducing its Q, thence widening its bandwidth.

As you can see, there is no cathode resistor. Grid bias is generated at the grid itself by the combination C1 / R1. The grid leak (R1) can be very high, up to a couple of megohms. The coupling capacitor value must be adjusted accordingly.

CHAPTER 9: THE SUPERHET IN DETAIL

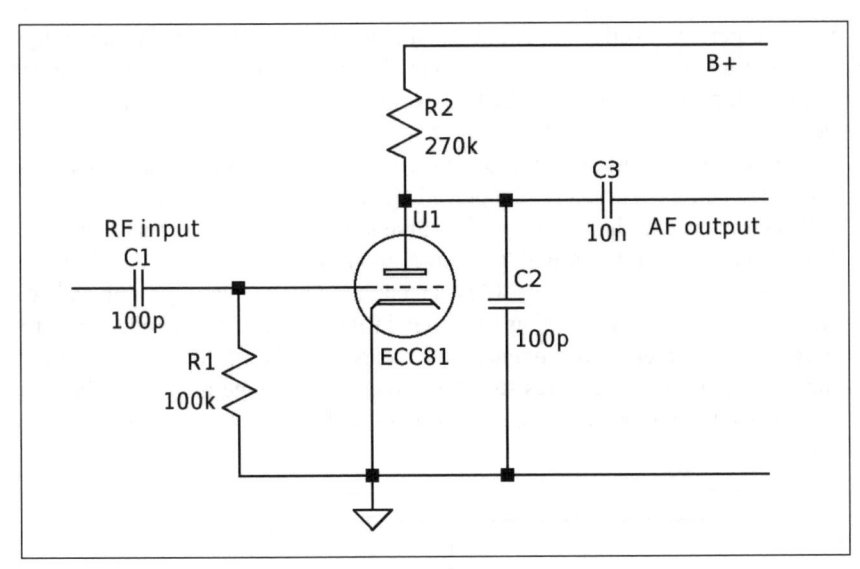

Fig 149: Grid leak detector

The grid leak detector is recognised by its lack of cathode resistor and the RF decoupling capacitor in the anode (C2). Due to the non-linear nature of a triode, this detector introduces some distortion on the resulting signal.

Anode bend detector

Without signal, the grid leak detector is biased at zero volts. The anode bend detector works at the other end of the curve, near the cut-off point. The curve of **Fig 150** is the dynamic curve of ECC81, an RF valve. At the bottom you can see how the curve bends in a smooth fashion towards the 150V-line. The anode bend detector takes advantage of the fact that the amplification is higher towards the centre of the curve, and lower towards the bottom end. This causes an unbalance in amplification as the signal moves up and down the curve, and a certain amount of rectification occurs.

Not all valves are suitable for anode bend detection, though. Compare the curves of Fig 150 and **Fig 151**. The second curve, which is a dynamic curve of an E88CC, is considerably straighter towards the bottom. For this

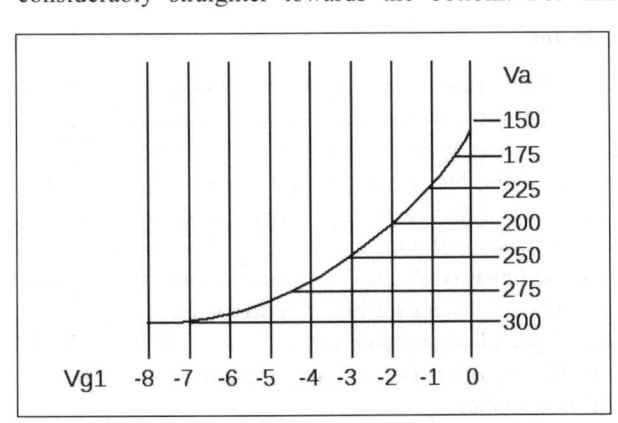

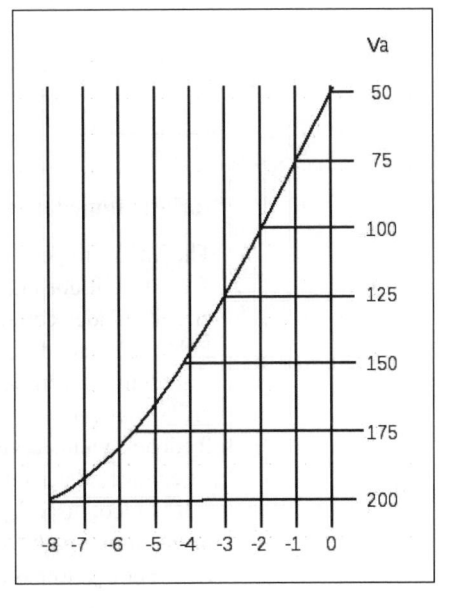

(below left) Fig 150: Dynamic curve for ECC81

(below) Fig 151: Dynamic curve for E88CC

129

valve to perform well as an anode bend detector, the bias needs to be so low that portions of the signal fall beyond the cut-off point, class B, in other words. This suggests that the ECC81 would be a better valve for anode bend detection.

The grid is at no point going positive, which means that the valve does not draw grid current. Therefore, the preceding circuits will not be loaded.

Since the bend at the bottom of the curve is inevitable, fidelity is not perfect. Due to the low current through the valve, anode voltage is going to be high.

The anode bend detector (**Fig 152**) is recognised by its relatively high cathode resistor and the RF decoupling capacitor in the anode. The cathode capacitor needs to be big enough to avoid negative feedback of the AF frequencies. Due to the high value screen grid resistor, the screen grid voltage is going to be very low. It is set low to move the control grid bias closer to the zero anode current point.

Fig 152: Anode bend detector

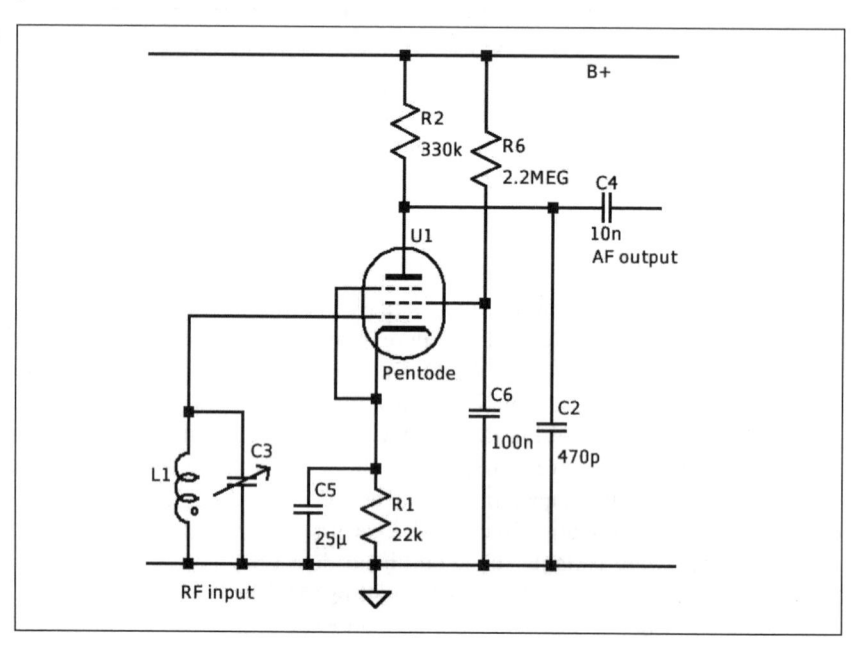

Infinite impedance detector

The input impedance of this detector is very high, although not infinite. Therefore it does not load the previous stage, which very well may be a tuned circuit. Under certain circumstances, the impedance can even be negative. It falls under anode bend detectors, but is somewhat differently designed. It is essentially a cathode follower with a very high value cathode resistor. The grid's bias point is close to the cut-off point on the curve. The 100% negative feedback yields a very good linearity. A small capacitor, about 100pF, should be connected across the cathode resistor for better performance.

It is recognised by the high cathode resistor and filter network C5 / R3 / C2 in the signal path to the AF amplifier. The coupling capacitor C4 needs to be large enough to pass AF frequencies.

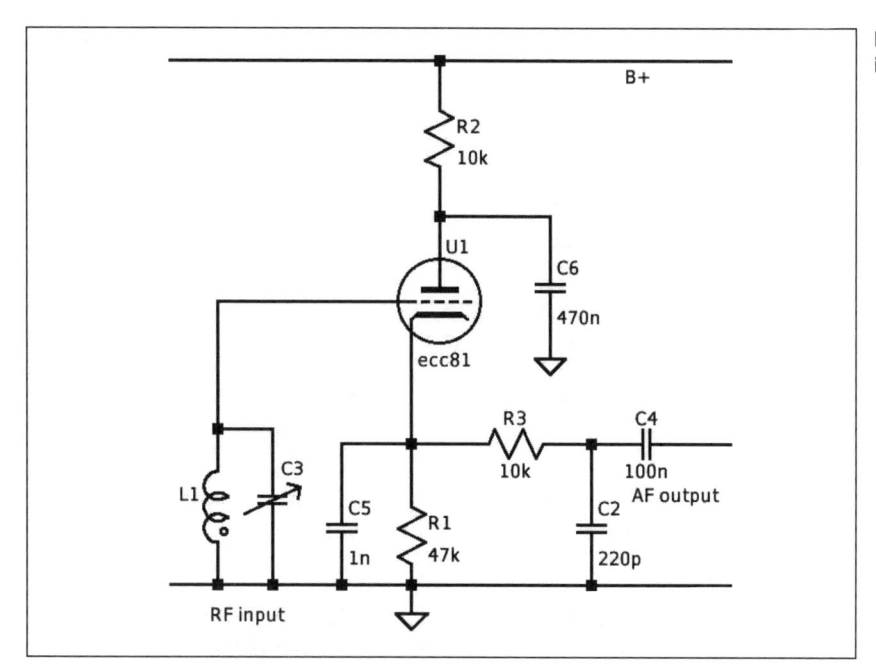

Fig 153: Infinite impedance detector

The infinite impedance detector, shown in **Fig 153**, does not generate a negative going DC signal suitable for AVC. The output signal at the cathode is a DC that goes positive as the input signal increases. However, a negative DC signal could be generated using a separate AVC detector.

Heterodyne detectors

A heterodyne detector is where an oscillator signal (BFO) is mixed with the IF to yield an audible signal. As you tune past an AM transmitter or just a carrier with the BFO turned on, you will reach a point where the BFO frequency and the IF frequency are the same. At this point, there is no beat between the two frequencies. However, the quality of the detection has improved, and the receiver is more sensitive. Furthermore, selecting just one side-band improves reception additionally because possible interference at the other side-band disappears. This is in essence the principle of a heterodyne detector.

Synchronous detectors can be considered a special type of the heterodyne detector. In a synchronous detector, the oscillator is tuned and locked to the incoming frequency. The result is a good quality AF signal, where the phase distortion that normally accompany selective fading is reduced or eliminated.

It is rather complicated since the oscillator needs to be locked to the incoming signal, whose important parameter, the amplitude, is beyond the designer's control. Nonetheless, it is used in more advanced modern 'world radios'. The detector is arranged such that it switches to a normal AM detector if the incoming signal's amplitude becomes so low that the oscillator can't lock. This principle also prevents the 'squeak' that normally appears before the oscillator locks.

There are two major types of synchronous detector, the homodyne and the synchrodyne.

Fig 154: Homodyne detector

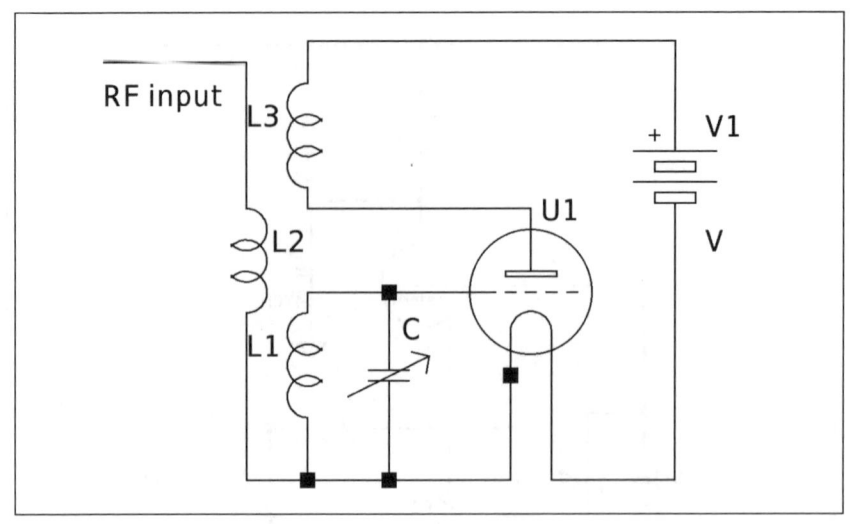

The homodyne

The original homodyne receiver uses an amplifier with positive feedback into which the signal is fed. When the amplifier begins to oscillate on the signal frequency, there will be no beat note, and the station is heard clearly. There are drawbacks with this system, though - the narrow bandwidth. Since the oscillator is in the path of reception, the amplifier operates like a Q-multiplier, and the bandwidth is therefore severely limited.

One homodyne detector is illustrated in **Fig 154**. The signal is brought in through L2, and the amplifier is made up of the valve, L1, L3 and C. If you look at the amplifier alone, you shall find that it in reality is an Armstrong oscillator. The tuning is done with C. The amount of feedback is controlled by the coupling between L1 and L3.

The detector is very simple, but does not produce much in terms of hi-fi. The homodyne was improved upon by filtering out the signal carrier to use instead of the oscillator. This principle allows for the entire spectrum of the modulation be transferred through the receiver, and gives a much better result. The problem in this case is that the design requires a high-Q circuit to separate the carrier from the complete signal, which makes tuning a bit more difficult to design.

The synchrodyne

The synchrodyne is a development of the homodyne. It has an oscillator separate from the signal path, and a locking circuit, like a phase-locked loop (PLL), in modern terms. A block diagram is shown in **Fig 155**. This implies the necessity for a phase detector, correctional circuits for the oscillator, and the oscillator itself.

When the detector is not locked to the signal, for instance during tuning, a whistle occurs as the result of the oscillator beating with the signal. So, what remains is a device that is capable of eliminating those whistles. If the signal is weak, the oscillator may temporarily fall out of synchronisation, and on each occasion the whistle starts. The whistle suppressor is placed after the mixer stage.

CHAPTER 9: THE SUPERHET IN DETAIL

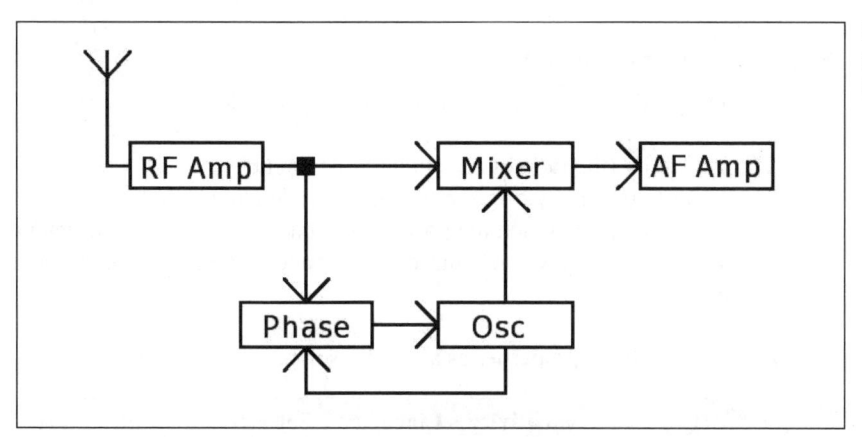

Fig 155: Block diagram of Synchrodyne detector

The design is complicated, in other words, and therefore expensive to implement in valve technology. These days the whole process is performed by an integrated circuit.

The RF stage of a synchrodyne receiver is shown in **Fig 156**. It consists of two pentodes. Note the potentiometer across the tuned circuit. It controls the amount of signal received by the first valve, but, at the same time, dampens the circuit, increasing its bandwidth.

The signal is further amplified by the following stage and a portion of it brought back again to the cathode of the first stage, providing negative feedback, which increases the stability of the stage. The amplified signal is also brought to the mixer and phase comparator.

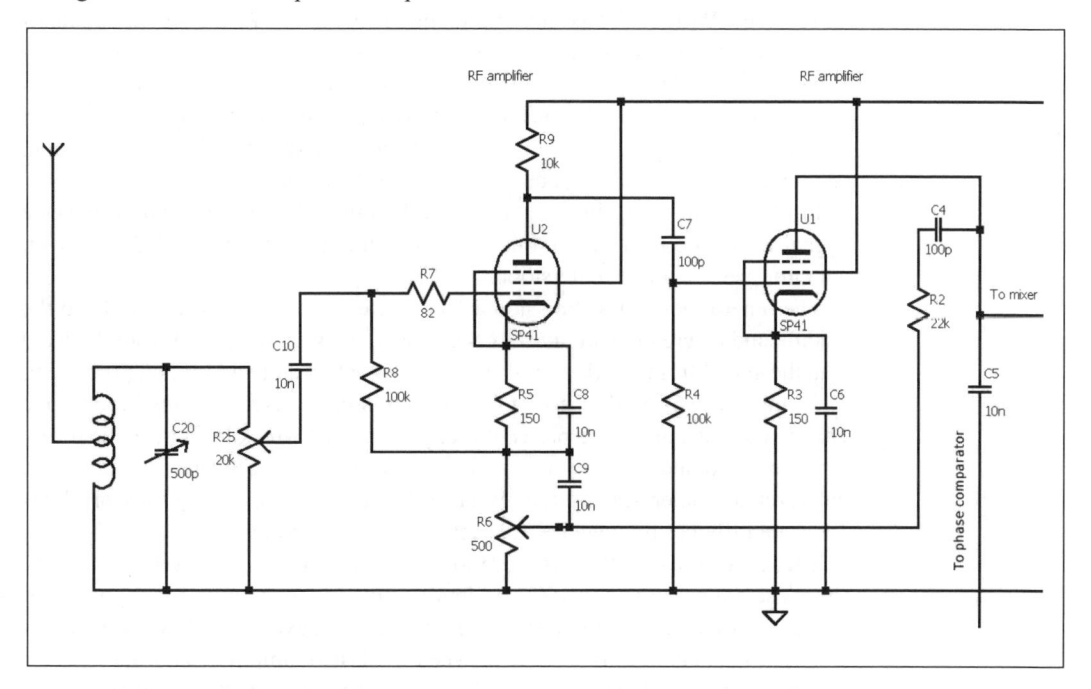

Fig 156: Implementation of RF stage for synchrodyne receiver

133

Selecting an appropriate detector

When selecting a detector, there are many factors to consider:

- The sensitivity (the ratio of AF output to RF input)
- The linearity (accuracy of reproducing the modulation signal)
- The selectivity (how much the previous circuit is loaded by the detector)
- The input ability (how well the detector handles large signals without overloading)

Table 15, shows these properties as a comparison between different detectors:

Detector type	Sensitivity	Linearity	Selectivity	Input ability
Diode	Low	Good	Poor	High
Anode bend	Medium	Good	Excellent	Medium
Grid leak	High	Poor	Poor	Limited
Regenerative	Higher	Poor	Excellent	Poor
Infinite impedance	Low	Excellent	Excellent	High
Synchronous	High	Excellent	Excellent	High

Table 15. Comparison of different detectors

The BFO

So far, we have talked about detectors for amplitude modulated signals. In order to receive CW, RTTY, Fax and many other signals on SW, the signal is inaudible when listened to with an AM receiver. All these signals are frequency modulated or on-off keyed, and an AM detector can not cope with them. SSB is slightly different, though, because SSB is basically an AM signal. It can be heard, but is not intelligible, since the carrier is missing from the transmission. With a BFO, however, all such signals can be received.

In the case of SSB, the BFO replaces the missing carrier and so produces a signal which looks like an ordinary AM signal to the remainder of the receiver, and the speech can be understood.

The other types of transmission are converted by the BFO from the IF to the audio band by creating an interference, a beat, between the signal from the BFO and the signal from the IF amplifier. These signals require additional processing to be made use of. In the case of CW, which in most cases contains Morse code information, it can be interpreted directly by anyone who knows the code (or by a piece of computer software), and there is plenty of software that is able to interpret the other signals too. What is heard is a beep that jumps or slides between different frequencies at different rates, depending on the type of signal.

The frequency of a BFO is close to the IF, eg 455kHz or 100kHz. By far the simplest way of adding a BFO and its functionality to an existing receiver is to build an oscillator (a number of different types have been described in this book), and connect it to a point between the last IF amplifier and the detector using a small capacitor (5 - 10pF). BFO reception requires the receiver (and, of course, the BFO itself) to be very stable, though. If not, the tuning needs to be adjusted from time to time to keep the intelligibility of the signal. Also, the signal from the BFO must be the right amplitude for good reception; too high or low, and the amplitude of the resulting signal will be too low.

If you are building or heavily modifying a receiver, a BFO can preferably be connected to a separate detector, eg a product detector, with its output low pass filtered.

A BFO is frequently crystal controlled, but as an add-on it might just as well be adjustable. Its tuning range needs not exceed the passband of the IF amplifier by much - the whole idea is to keep the signal audible.

A crystal controlled BFO is often provided with two crystals to enable easy switching from one sideband to the other. On the amateur bands, 160, 80 and 40 meter (1.8, 3.5 and 7MHz) SSB is standardised to the lower sideband, whereas the other bands use the upper sideband. Commercial SSB stations (such as weather stations, maritime communications, etc.) use the upper sideband almost without exception.

The IF filter of a receiver for communications or amateur use is often a set of crystal filters, one for each sideband, with very steep sides, and with one side steeper than the other. The steeper side of the filter is is placed on the frequency of the BFO. This prevents the BFO signal from actually entering the receiver and upsetting, for instance, the AVC control path.

The AF Stage

In order for the audio signal to become audible, an AF amplifier is necessary. It amplifies the signal from the detector and provides enough power to drive a loudspeaker or a headset.

The AF stage is usually very simple in communications receivers or radios designed solely for AM reception. The reason for this is that the signal from an AM station is very limited in bandwidth; voice communication only requires some 2.5 - 3kHz (corresponding to an AM transmitter bandwidth of 5kHz, SSB 2.5kHz). Morse code transmissions require even less, some 50 or 100Hz.

This makes the need for a high quality amplifier completely redundant in AM receivers. A simple cathode coupled stage followed by a simple, low power loudspeaker driver is quite sufficient.

One valve can take care of both tasks. A triode-pentode such as ECL82 does the job quite well. A suitable circuit is described in the chapter about hi-fi amplifiers.

Additions

A domestic radio can be transformed into a decent communications receiver by the addition of a few units. The first thing that needs to be done is, of course, make sure the VFO is stable. There are plenty of pointers about how to build stable valve VFOs in this book.

Q multiplier

We have learned that the Q of a tuned circuit depends on losses in the circuit, mainly in the coil. However, if those losses were compensated for, as is done in an oscillator, the circuit bandwidth would be greatly reduced, indeed, to such a degree that an AM transmitter would become unintelligible. This is a useful property if you are listening to morse code or other narrow-band transmissions.

Fig 157: Q-multiplier

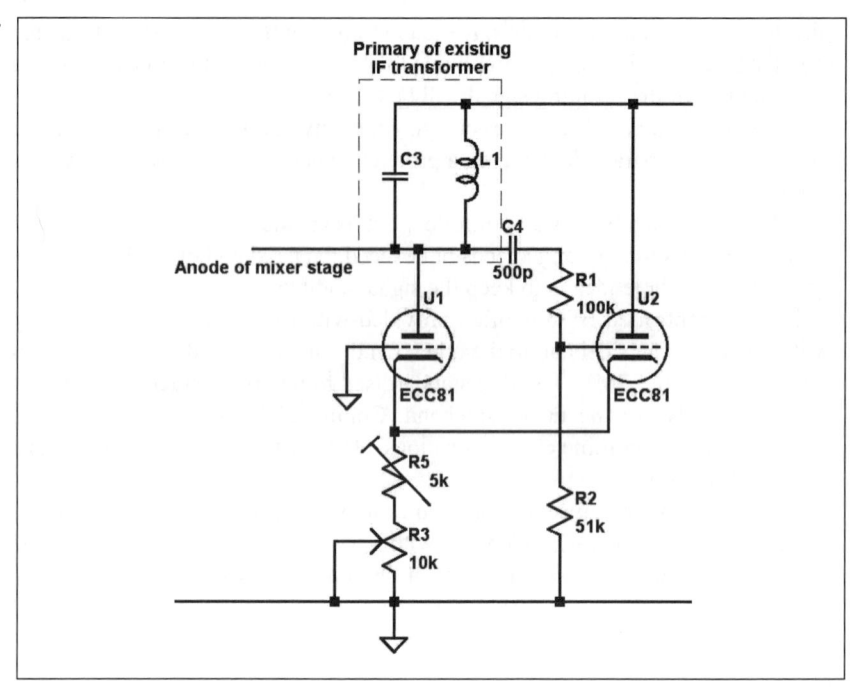

An aid to reach that goal, and also to improve selectivity in general in your radio, is the Q multiplier, shown in **Fig 157**.

A Q-multiplier is essentially an oscillator which is not driven into oscillation. The signal can be taken from the IF transformer and a tiny portion fed back into the transformer, just enough to compensate for the losses. This increases Q and amplification, and reduces the bandwidth. Fig 157 shows an example of a cathode coupled configuration.

The stage is connected to the primary of the mixer stage IF transformer as shown. A portion of the signal is fed to the grid of the first triode in the chain, U2. As the grid voltage varies, the cathode current also varies, and the voltage across the resistors in the cathodes, R5 and R3, varies accordingly.

Since the grid of the left hand stage is grounded, the anode current of U1 varies too. A cathode coupled stage has no phase shift between input and output, so the signal fed back into the tuned circuit will compensate for the losses of the circuit.

The amount of feedback is controlled by the potentiometers at the cathodes but also by the ratio of the resistors R1 and R2. There is a possibility that you might have to experiment with these as well.

The stage contains no selective components of its own, which makes it simple to build. Due to the added capacitances in the IF transformer, the primary and secondary are likely to need re-adjustment.

The stage gives amplification, and there is a risk that the following IF amplifier goes into oscillation. This can be remedied by removing its cathode decoupling capacitor.

R5 is the coarse control of the amplification and R3 is used as the selectivity panel control.

CHAPTER 9: THE SUPERHET IN DETAIL

To adjust the stage:

- Set R3 to lowest resistance
- Adjust R5 to the point where the stage just begins to oscillate. This is the point where the stage can be used as a BFO to enable reception of CW and other such signals.
- Increase R5 slightly to cause the oscillations to seize, amplification of the stage should be high and the bandwidth narrow. As R5 is increased, the bandwidth should increase.

Magic eye tuning indicator

The magic eye was invented by Dr Allen B DuMont as a means of assisting the user when tuning a domestic radio. It is basically a cathode ray tube with a tiny screen upon which was projected a pattern which changed as a station was tuned in. There were several designs. Some had a circular metal screen with four segments that grew when the station was correctly tuned in. Others had horizontal or vertical bands that grew. Yet others displayed a V-shaped structure that closed, etc.

The magic eye was connected to the AVC voltage - the less negative the control electrode, the bigger the indication in some valves - the other way around in others. The data sheet illustrates the relationship between the grid voltage and the deflection. Below, in **Fig 158**, are two symbols, to an extent implying the practical implementation of the valve:

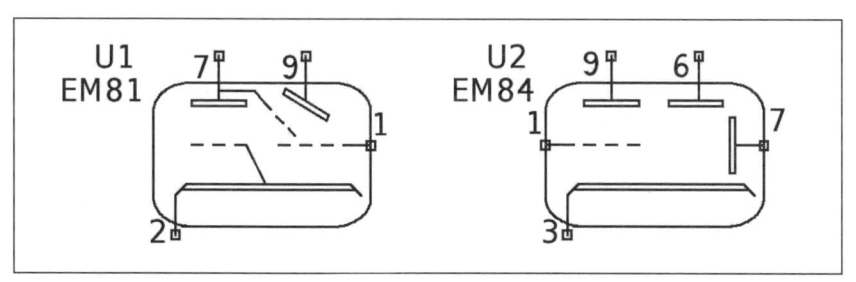

Fig 158: 'Magic eye' symbols

The symbols could vary quite a bit, as you can see. Besides, different manufacturers used different symbol designs. However, the function is basically the same. There is frequently a triode built-in to increase sensitivity. To get a hint of the grid voltage requirements for various indicators in the E-series (heater voltage 6.3V), see **Table 16**. The lowest grid voltage is generally 0V.

Type	Grid voltage
EM80	-14V
EM800	-10V
EM81	-10.5V
EM84	-22V
EM840	-21V
EM84a	-10V
EM85	-5.8V
EM87	-10V

Table 16: A comparison of grid voltage between different indicator valves

137

Magic eyes were fitted in domestic radios from about 1936. The benefit of using a tuning indicator for indication is that it is fast - far faster than any panel meter, far faster even than the eye. The only moving parts are electrons.

10

Designing a Receiver

DESIGNING AND BUILDING a receiver is great fun! If you are into valves, you should try it. Hearing a station coming through in your own home-built receiver gives you great satisfaction.

Be prepared for a lot of planning, though, in particular if your receiver is going to be a dual or triple conversion superhet. It is impossible to build an oscillator, be it a VFO or a crystal controlled one (aka XFO or XO), without it also generating overtones (harmonics). Overtones give rise to unwanted beeps and squeaks which are extremely annoying if you are trying to receive a weak station.

Super, TRF or DC?

If you decide to design and build your own receiver, the first thing you need to decide is what you want to listen to, as well as your experience with building radios.

There are four basic types of receiver to choose from as described earlier. To recap:

Crystal receiver:

This is the very simplest of all receivers. It doesn't (normally) contain any amplifying elements, and therefore depends very heavily on a good aerial and ground connections, and very high Q coils. Like the TRF, it is only suitable for AM signals. There is no way it can be made to oscillate, so other types of signals cannot be received.

Superheterodyne

This is the most complex type of receiver, and not suitable for beginners. On the other hand, it is the type of receiver that gained popularity towards the end of the 1930s and has been prevalent ever since. With suitable additions, it could be used for any type of signals.

TRF

This is a receiver that can be very simple to build but somewhat more complicated to use. It is the first type of receiver that contained valves and thus had the ability to amplify the signal before it could be heard in a loudspeaker. It dominated the radio market during the 1920s and far into the 1930s.

It is suitable for AM stations. It is possible to adjust it to receive signals that would require a BFO in a superhet, but not stable enough when oscillating to receive SSB signals. The TRF is, when not oscillating, inherently very frequency stable.

Direct conversion

DC receivers fall somewhere between TRF receivers and superheterodynes in complexity. They are built on the principle that an oscillator is brought to interfere with the incoming signal very near the original frequency. Thereby, an audible tone is heard when the station is tuned in. It can be seen as a receiver with an inherent BFO. The BFO can not be turned off, so the receiver is not suitable for reception of AM. It is, consequently, suitable for reception of CW, RTTY, SSB, fax, and other similar signals. The stability of the oscillator determines the stability of the entire receiver.

Size

The final decision you need to make is how big to make it. A tip: Don't decide to make an enormous chassis with lots of room for future extensions. You are going to end up with a big, nearly empty chassis with a powerful power pack and a couple of valves tucked away in a corner. Set a goal and stick to it.

Frequency Considerations

If we take a look at the frequencies that may occur in a superhet, we find that it is very easy for an overtone to find its way into an IF stage or, even worse, an RF stage. The distribution of various frequencies in a standard superhet receiver is very important.

The VFO inevitably radiates overtones as well as the desired signal, and with bad planning, those overtones may find their way into parts of the receiver, where they can be heard. Therefore, they must be prevented from entering, for example, the IF amplifier or the RF end of the receiver. This is done in a standard superhet by placing the VFO above the IF, rather than below it. No signals are (normally) generated below the fundamental frequency of the VFO.

Image Rejection

If there were no tuned circuits at the RF end of the receiver, unwanted stations on the 'image frequency' would be received at the same time as those on the intended frequency.

At the output of the frequency converter (mixer), the sum of the VFO frequency and the IF, as well as the difference frequency between the two, appear. We are only interested in one of those two, and, since we placed the VFO above the IF, the frequency of interest is the difference. The other is the image frequency. Let us look at an example to illustrate the point:

If we wish to receive a station on 1,500kHz, and the IF is 500kHz, then the VFO must tune to 1,500 + 500kHz = 2,000kHz. The mixer will produce the wanted signal on the difference frequency (2,000 - 1,500 = 500kHz), which is the IF. However, a signal on 2,500kHz will also produce a signal at the IF (2,500 + 2,000 = 500kHz). 2,500kHz is the image frequency.

Now, another station has caught our attention. Its frequency is 1.600kHz. So, to achieve the IF of 500kHz, we tune the VFO to 1,600 + 500 = 2,100kHz. The image frequency will then be at 2,100 + 500 = 2,600kHz.

If our receiver has no filters at the RF input, both our desired station on 1,600kHz and a possible station on 2,600kHz will be heard at the same time, which is not what we want. We must make sure that the input filter in our receiver is narrow enough to reject the image frequency, and that it follows the VFO tuning at a fixed distance in frequency (in the example 500kHz) at all times.

What has been described here is the behaviour of a standard superhet receiver. It is inherent in the design, and there is no way around it.

Choice of IF

In the above example, we selected 500kHz as our IF. This is all very well on MW and LW. However, it is not such a good choice on SW. The reasons are these:

An L/C filter will increase its bandwidth the higher frequency it is tuned to (with Q preserved). This means that the bandwidth at Q = 100 when tuned to, say, 1,500kHz will be 15kHz, and that is more than enough to select only one of the frequencies. Suppose we managed to preserve Q (which is unlikely, but let's toy with the thought) and tune the filter to, say, 30,000kHz instead. That would give us a bandwidth of 300kHz. Now, as frequency increases, Q decreases. On 30MHz, Q might, due to the skin effect and other factors, have gone down to, say, 10, which yields a bandwidth of 3MHz. So, on our high frequency, the bandwidth has increased to such a degree that the filter would cover both the desired signal and the image frequency.

European domestic receivers mostly had an IF at 455kHz. UK-made radios had various IFs, but most of them were around 475kHz.

There are ways of getting around the image problem, but they may create others instead.

Higher IF

Since the image frequency is at a distance from the wanted frequency equalling twice the IF (as we have seen above), one possible solution to the image problem lies in using a higher IF. An IF of, say, 5MHz would place the image at a distance of 10MHz from the desired signal, which would allow for a reasonable Q of the input filter, but still eliminate the image response, or, at least reduce it to an acceptable level.

Doing so, however, will introduce another problem. We need an IF filter which is narrow enough to make the receiver immune to 'adjacent channel interference', ie an unwanted station near the wanted frequency. If we are listening to speech, the IF filter should ideally have a bandwidth of some 3kHz. It would need to be wider, of course, for music reception. Regulations state that the distance between stations on SW be 10kHz, which means that the bandwidth necessary for listening to music on SW (AM) would be 5kHz. For a frequency of 5MHz and a bandwidth of 5kHz a filter with a Q of 1000 would be required. This is not realistic.

The words 'crystal filter' spring to mind. A crystal filter does introduce problems, though. One is practical, the other is economical. Quite a few manufacturers offer crystal filters, but most of them have a centre frequency well above our desired 5MHz, and those that do exist for SSB reception at this frequency are not particularly cheap. The solution would be building your own crystal filter, which is not really easy, but achievable.

Multiple conversion

Another solution to the problem is to build a slightly more complex receiver - a dual conversion one. In such a receiver, the incoming signal is first converted to a high first IF for good image reduction, then down again to a low second IF for good selectivity. Many valve ham receivers in the 1960s were built like this.

Using the Drake 4A as a (simplified) example: The first converter mixes the signal from the VFO, which tunes between 4955 and 5455kHz, and the incoming signal to a first IF of 5645kHz. This signal is then mixed down by a crystal oscillator to the second IF, which is as low as 50kHz. The first IF provides good image reduction, and the second IF opens possibilities for narrow bandwidth and adjacent channel reduction.

The method does have implications, though. If you wish to listen to frequencies within the tuning range of the VFO, you would have to re-design your receiver to prevent the VFO interfering with the desired signal. In fact, many receiver manufacturers did just that! One of the ultimate examples was the Collins 51-S1, a receiver that covers 0 - 30MHz continuously. On some bands it is a dual conversion receiver, on others a triple conversion receiver.

Yet another receiver could have a crystal controlled first mixer, a tunable first IF, and a VFO controlled second mixer. The result would be virtually the same, regardless which way you choose to go. Your personal needs have to determine which approach you wish to take.

Front End Considerations

There are two main considerations to make when designing the front end:

- Connecting the aerial (antenna) to the receiver
- Image selectivity

Aerial connection

It is common practice to use an ideal half-wave dipole as a reference for aerial assessments. If fed at the centre, this has a theoretical low impedance, in the order of 50 or 72Ω.

In fact, this is true only when the aerial is receiving signals on a wavelength for which the dipole is designed. At all other frequencies, the impedance is different, and it may vary from the ideal 50Ω up to several kΩ. Additionally, the environment in which the dipole hangs affects the impedance. Properties such as the aerial height, trees nearby, the properties of the ground below the aerial, air moisture, all affect the impedance. Therefore it is virtually impossible to predict the actual impedance of the aerial.

A great variety of aerial tuning units have been developed over the years. They are in fact mis-named as they do not tune the aerial at all, instead they match the aerial's impedance with that of the receiver's input. They work well and are frequently invaluable when the aerial is used as a transmitting antenna. Without it, the transmitter may be seriously damaged.

CHAPTER 10: DESIGNING A RECEIVER

For receivers, though, in particular domestic radios, an aerial tuning unit is too expensive and fiddly to adjust. So as a compromise the manufacturers added a coupling coil to the input tuning circuit. This input coil has a small number of turns to transform the high impedance demands of the tuning circuit into the low impedance feed of the aerial.

There was another problem involved with domestic receivers. Few people actually paid due attention to the aerial, and, besides, a domestic receiver tunes a wide range of frequencies, not just the one for which the aerial was designed. Consequently, the aerial could have virtually any impedance, and the inevitable mismatch between the aerial and the input of the radio caused loss of sensitivity. The average radio listener didn't hear the difference, though, so the compromise stayed.

A device which can transform any impedance into a fixed known impedance would be ideal. Even better would be if it could have a high impedance output to which a tuned circuit could be connected. Some degree of amplification would be nice.

This sounds like a job for a cathode-coupled amplifier (again!). The input stage is a cathode follower (high input impedance), and the output stage is a grounded grid amplifier (amplification, high output impedance). The cathode follower would transform the varying aerial impedance to the low input impedance of the grounded grid stage.

If you are re-building an old radio, you need to identify and by-pass the low-impedance aerial input, and connect the new front-end directly to the tuned circuit. As mentioned in connection with the Q-multiplier, the tuned circuit would probably need to be re-tuned after the operation.

Fig 159 shows an example of such a front end. The aerial is capacitively connected to the input of the cathode follower. The common cathode resistor carries the current from both halves of the valve. R5 / C5 and R6 / C4 are filters

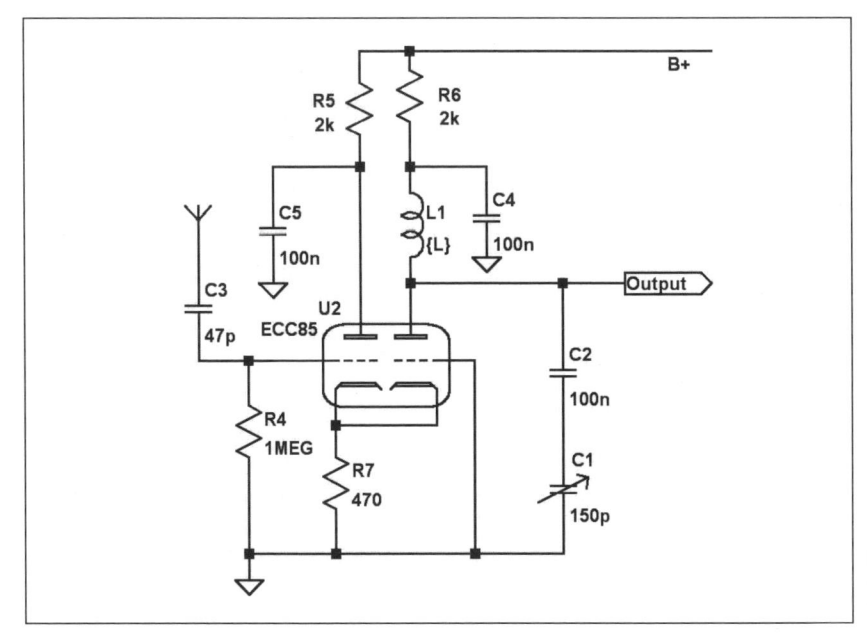

Fig 159: Receiver front end, employing dual triode

to prevent RF from spreading through the power supply. L1 and C1 constitute the tuned circuit, and C1 tunes the stage. C2 prevents the high voltage from reaching the tuning capacitor. At the same time, it decreases the value of C1 somewhat.

The input of an amplifier should in theory be tuned. However, the cathode follower is a wideband device and the danger is that a strong station on a nearby frequency may find its way into the filter through the amplifier.

Hi-fi Amplifiers

PERHAPS THE MOST popular device to build today with valves is the AF amplifier. This is not surprising, since a valve amplifier does sound 'better' than one built with semiconductors. The cause for this is that the distortion that emanates from a valve is different from that of a semiconductor, and more pleasant to the human ear.

Development of amplifier stages and complete amplifiers for hi-fi use is constantly going on all over the world. Excellent sources of information on the current state-of-the-art can be found on the Internet. Development is very intense, and new designs evolve almost daily. The price of a valve amplifier for domestic use on the market today can range from a couple of hundred pounds to some tens of thousands of pounds. This gives an idea of the popularity of valve amplifiers today.

There are excellent sources on the Internet that specialise in the ever changing world of valve hi-fi amplifiers. The information in this book is intended as a stepping stone into the fascinating world of thermionics, so it can contain but a few examples of what has been going on in the past.

In domestic radio receivers, the quality requirements of the AF amplifier are very low. The channel width available to an AM broadcast transmitter is in the order of 10kHz, since the channel separation on the AM bands is 9kHz (LW, MW) or 10kHz (SW). This means that the bandwidth of the information transferred is half of that, at the best. Point-to-point communication is even narrower, in the order of a couple of kilohertz.

Digital transmissions (DRM, Digital Radio Mondiale) have begun on SW, but the available bandwidth still remains no more than 10kHz. Consequently, in order to achieve higher reception quality within the bandwidth constraints, the modulation must be very complicated. The problems are similar to those of high-speed telephone modems, where as much information per unit time as possible has to be transmitted within the bandwidth constraints of a standard telephone line.

The situation is different if music is to be reproduced from a CD, an analogue vinyl disk, a tape recorder or a microphone. The reproduction chain of such a system has wide bandwidth limits, which means that the limiting factor is technology alone.

However, even the simplest AF valve amplifier yields a noticeable improvement in sound quality. Distortion figures may be lower for a semiconductor amplifier, but a valve amplifier does sound better. Many an analysis has been performed in order to find out exactly where the difference lies, and many a book and article has been written about it, but it is extremely difficult to point a finger at any specific reason (even though some authors strongly claim to have found the answers).

Researching the subject is difficult too, because what once was considered hi-fi (10kHz bandwidth) is no longer even acceptable quality. Also, according to some enthusiasts, one make of a valve differs in quality from the same type of another make. The bottom line is that the subjective perception has to be the guideline, whereas reality (the actual live musical experience) remains the standard.

A couple of designs will be presented here. The collection can never be complete, and must be considered pointers only.

Tone Controls

A word of warning. Other sources may express different opinions than those stated here!

In the golden years of valve technology, most domestic radio receivers and quite a few hi-fi amplifiers had front panel controls to set the levels of bass and treble. In the beginning, this was quite understandable, since loudspeaker technology seriously lagged that of electronics. During the mid and late 1960s, however, loudspeaker technology was beginning to catch up and tone controls disappeared gradually from high end amplifiers. These days it is hard, if not impossible, to find anyone who designs an amplifier with tone controls.

The background is this. The brain is designed to assess from where a sound comes by comparing not only the amplitude but also the phase of the sounds that reach the ears. One can even assess whether a sound comes from above the head or from behind. Additionally, a portion of the sound from, say, the left, also reaches the right ear. All this is information that the brain processes to determine the direction of a sound source. In addition, echoes from the environment, such as walls, furniture, trees, and whatever have you, reach the ears (which, by the way, is extremely useful information that vision impaired people use extensively).

Introducing tone controls into the chain of amplification distorts the information that the brain needs, because they destroy the original phase information. Each tone control is a filter, and filters are not very kind to phase information (by design, actually). This is the reason tone controls have disappeared from high quality amplifiers of today. There must be only one reference in hi-fi - reality.

Another story is what kind of music you prefer to listen to. Synth music and other artificially generated sounds may be interesting as such, but since there is no reality behind them, they are not suitable to assess the quality of amplifiers. Unfortunately, the sounds on CDs and other fabricated media are heavily processed in most cases, which means that what you listen to is somebody else's opinion of how a particular sound should be reproduced. This is sadly true, however expensive and advanced and sophisticated an equipment you use for your listening pleasures, all the way from your vinyl or CD player to your brain. Also, a CD player has a low pass filter built-in on each channel.

Besides, the CD and other digital media display a poor low level signal quality, which has to do with the sampling technology. The signal (in this case the electrical representation of sound) is sampled at a fixed rate, around 22kHz per channel. Let us assume that your sound has a 1V peak value at full blast. Using 16 bit sampling gives you 65,536 discrete signal levels. Each level is then

1/65,536V or about 15.2µV per level. Now, assume that the music level drops by 30 dB. The smallest step is still 15.2µV, but now the step has to be compared to 1µV rather than 1V. Our 15.2µV suddenly grew by a factor 1000! Food for thought.

Stereo

As explained above, the ability to distinguish from which direction a sound comes is built into our ears, and our brains are mainly using the phase relationship for that purpose.

Stereo recordings and reproduction systems use two separate channels and loudspeakers to trick the brain into believing that the listener is in a real environment. This is a stereophonic or stereo system.

The illusion is far from perfect, though. Other factors are involved, like walls and furniture in your listening room, even your own body structure. Experiments have been performed, where dummies, complete with the texture and consistency of a human head, including ears and shoulders, have been used for recording sound. The system has two microphones, one inside each ear (in place of the ear drums), and the result has been played back through earphones. The method is called 'Dummy Head' recording. In fact, you can make recordings using your own head. A couple of tiny microphones fixed inside your ears, but pointing outwards yield an excellent result. The drawback is that, unless you stay completely still, the 'sound image' will move around. Occasionally, some radio stations make dummy head recordings, mostly for the fun of it.

The result of a dummy head set-up is astonishing, though! The brain is completely fooled, and has no problems whatsoever distinguishing directions. A conversation between a number of people, say, in a meeting, recorded through a dummy head, turns out to be very useful. It is easy to distinguish individual speakers when several people speak simultaneously. The problem is that if loudspeakers are used in the playback system, much of the information required by the ear and brain gets lost. Another problem is that if your body constitution is different from that of the dummy head, you will experience the sound world through unfamiliar ears.

These days, in the era of digital signal processing, very few recordings are true stereo. Oft-times each instrument has its own microphone, the soloist, if any, has its own, etc. The result is a multi-channel recording, which, during a mixing process, is compressed into two channels.

Undoubtedly, a true stereo recording is best enjoyed through earphones. Your ears pick up tiny details, suck as the tiny noise made by the bow against the violin string, or someone in a big symphony orchestra gets carried away and stomps his foot, or the tiny sound made by the valve of a wind instrument when released. This gives an entirely different experience from the same record listened to through loudspeakers. All this information gets lost through loudspeakers. It all boils down to the preferences of the listener, though.

Stray Capacitances

Some of the most vicious enemies when building your amplifier is stray capacitances and stray couplings. The tiniest coupling in the wrong place may ruin the entire design!

The most sensitive point is, of course, the input of the amplifier. This is usually a high impedance point, where all kinds of disturbances can be picked up. Particularly important are stray capacitances between input and output. An additional 5pF between the grid and anode can cause the rise time of a signal to increase by a factor of three or four, or even more. In severe cases, they may actually cause the amplifier to oscillate, which would render it completely useless.

Another point of interest, albeit not quite as serious, is capacitance between the grid and ground.

The bottom line has to be: Don't strive too much for neatness when you build your amplifier. Keep components well away from each other and from ground. Above all, keep inputs well away from outputs. Observe the guidelines for RF stages.

RIAA Compensation

The RIAA, Recording Industry Association of America, created a *de facto* industry standard for vinyl records during the 1960s. The idea behind it is to compensate for the poor reproduction of higher frequencies in vinyl.

When the vinyl records are manufactured, the sound from the original material is processed such that higher frequencies are boosted according to a specification.

When playing back such a record, it sounds very sharp and unpleasant, unless the amplifier is provided with a RIAA compensation network. A side effect is that the noise level from the grooves, which is most prominent at high frequencies, is reduced. Other media are not RIAA-compensated. **Fig 160** shows a RIAA curve.

Fig 160: Typical RIAA-curve.

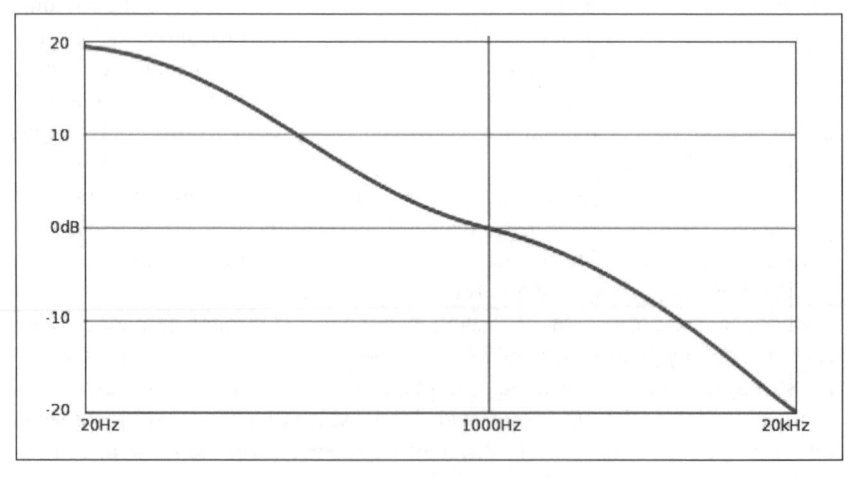

The amplifier should have this curve to reproduce a vinyl record correctly. If it does, the resulting curve from the record will be given a flat frequency characteristic during reproduction.

A RIAA filter should contain only passive components. There are several ways of implementing the RIAA curve. One filter that matches the RIAA specification is shown in **Fig 161.**

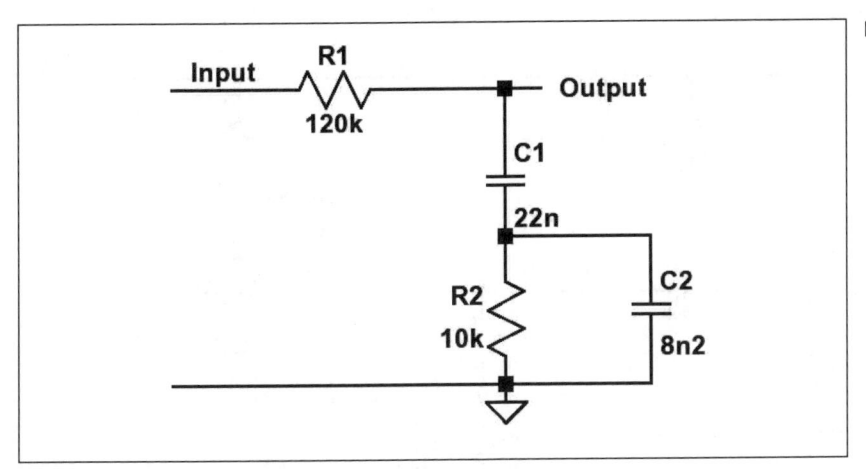

Fig 161: RIAA filter

Four components of standard values do the trick. The closer the match, the better the result. The resulting curve in **Fig 162** is a simulation that shows a decent match to the RIAA curve above:

However, note how the phase (the thinner curve on Fig 162) is seriously distorted, as always when filters are involved (the phase curve's resolution is nine degrees per division). The filter can be connected at the input of the amplifier. If you do, be careful with the coupling capacitor and grid leak of the amplifier input! If you are unlucky, they might distort the curve.

Best result will, of course, be achieved if the input stage of the amplifier is a cathode follower. That would increase input impedance to a level where the filter is not disturbed, and decrease the Miller effect to a negligible level.

Alternatively, two cathode followers, one before the filter and one after, isolates the filter from the output impedance of the record player and cable capacitances in the coaxial cable from the player, as well as the input of the amplifier proper.

This means that the RIAA filter can be built as a separate unit and connected to an existing amplifier without problems. If your amplifier already contains

Fig 162: Curves resulting from the filter in Fig 168

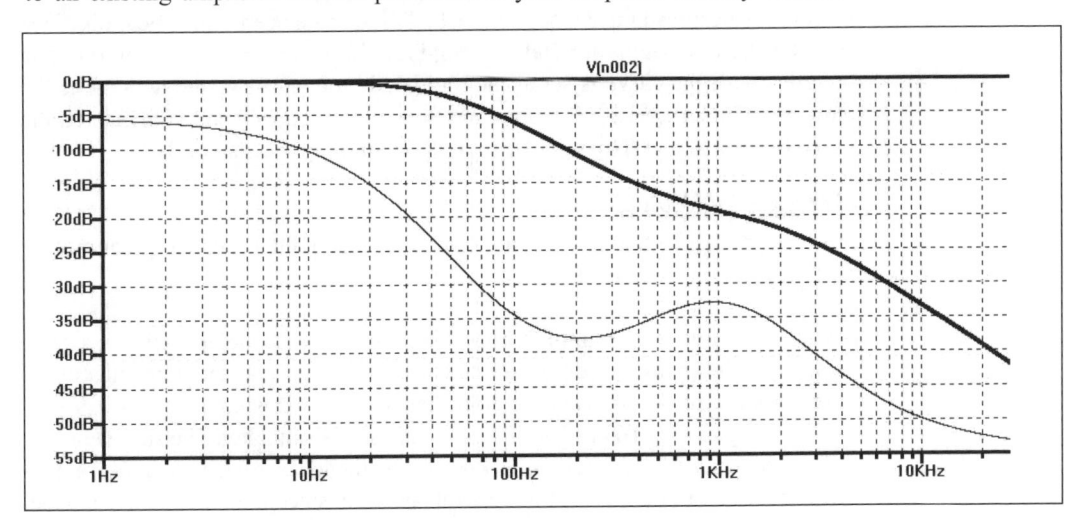

VALVES REVISITED

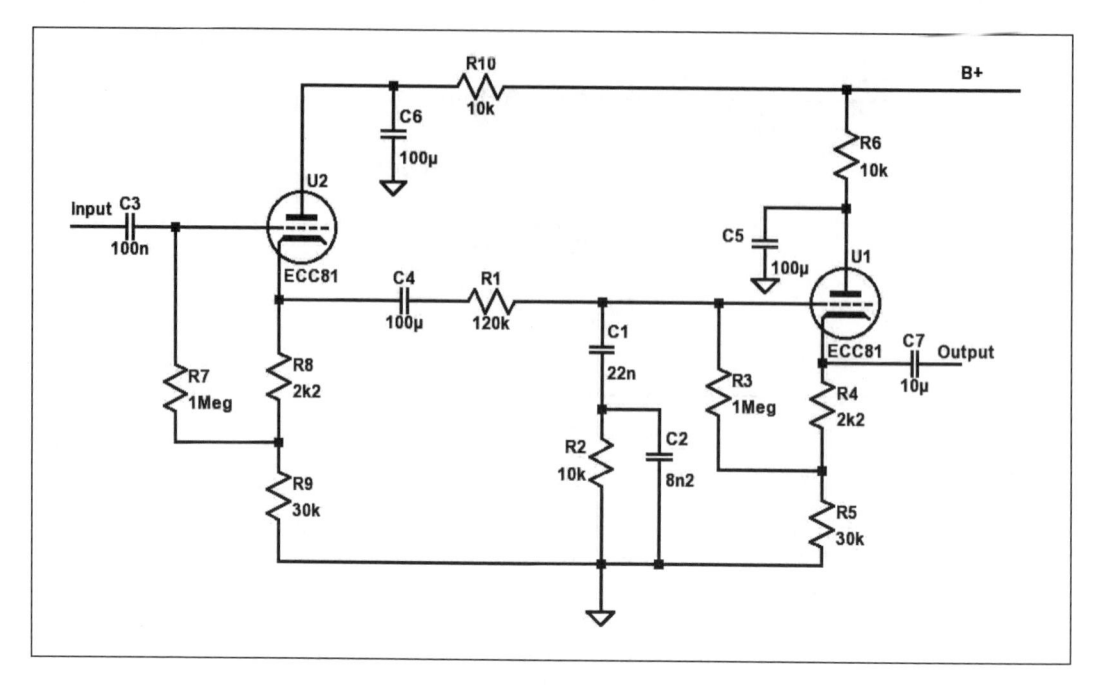

Fig 163: Amplifier with RIAA compensation

valves, then you can steal some heater voltage and high voltage from there, thus reducing the cost and size of the added filter box.

Fig 163 shows a suggested design. U1 and U2 above are two halves of one and the same valve and therefore well matched. If you are lucky, the two valves required for two channels may fit into the turntable casing.

R1, R2, C1, and C2 are the RIAA components. C4 is chosen high so as not to influence the filter. R10 / C6 and R6 / C5 are filter components to prevent the signals from propagating through the power supply. Keep inputs and outputs well away from each other. Also, keep the channels well separated.

You should be able to disconnect or bypass the filter when playing 78RPM records; it is only needed for LPs and EPs.

It is a widespread misconception that a RIAA stage needs amplification, since the RIAA curve begins at +20dB and ends at -20dB. This is simply not true. The reference for the curve is set at the mid bend around 1kHz, and as long as the low frequency end is 20dB higher and the lower frequency end 20dB lower, the specification is met. No amplification is required by the standard.

Preamplifiers

More elaborate hi-fi amplifiers are often provided with a separate preamplifier.

Cascode

When a high degree of amplification is required, a cascode amplifier is a good solution. The input and output impedances are high, and the amplification is high (see the section on cascode amplifiers earlier in this book.

The design in **Fig 164** has been developed using a simulator. The simulator output is shown in **Fig 165**. C1 and C4 determine the lower frequency limit - in particular C1. C4 needs to be of a high voltage type, since the voltage of the upper

CHAPTER 11: HI-FI AMPLIFIERS

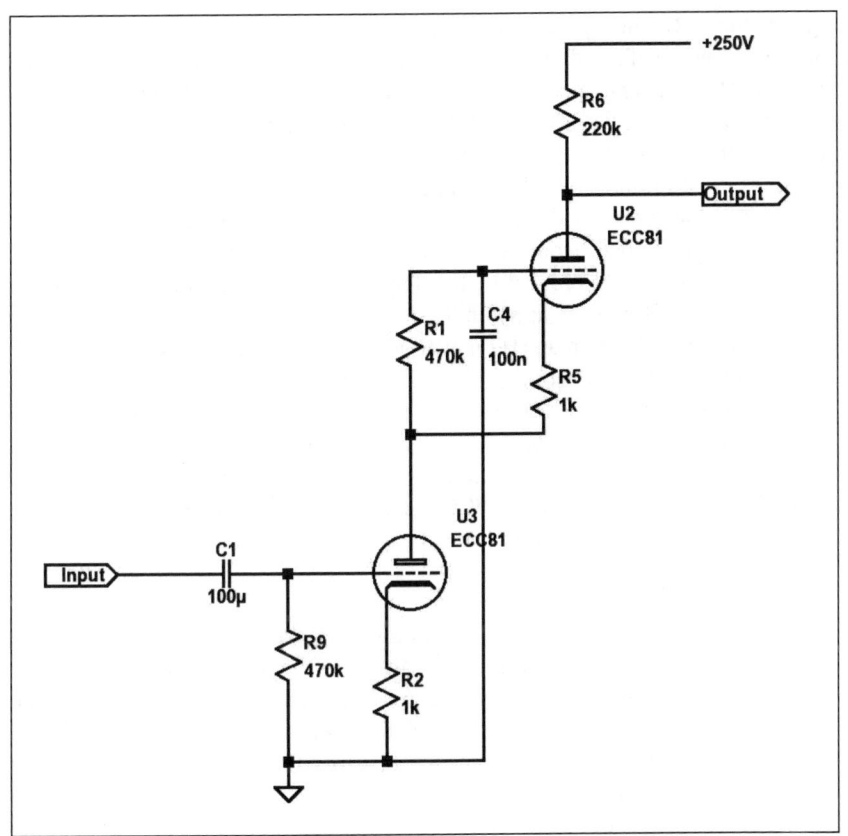

Fig 164: Cascode AF amplifier

(below) **Fig 165: Simulated curves for the cascode above**

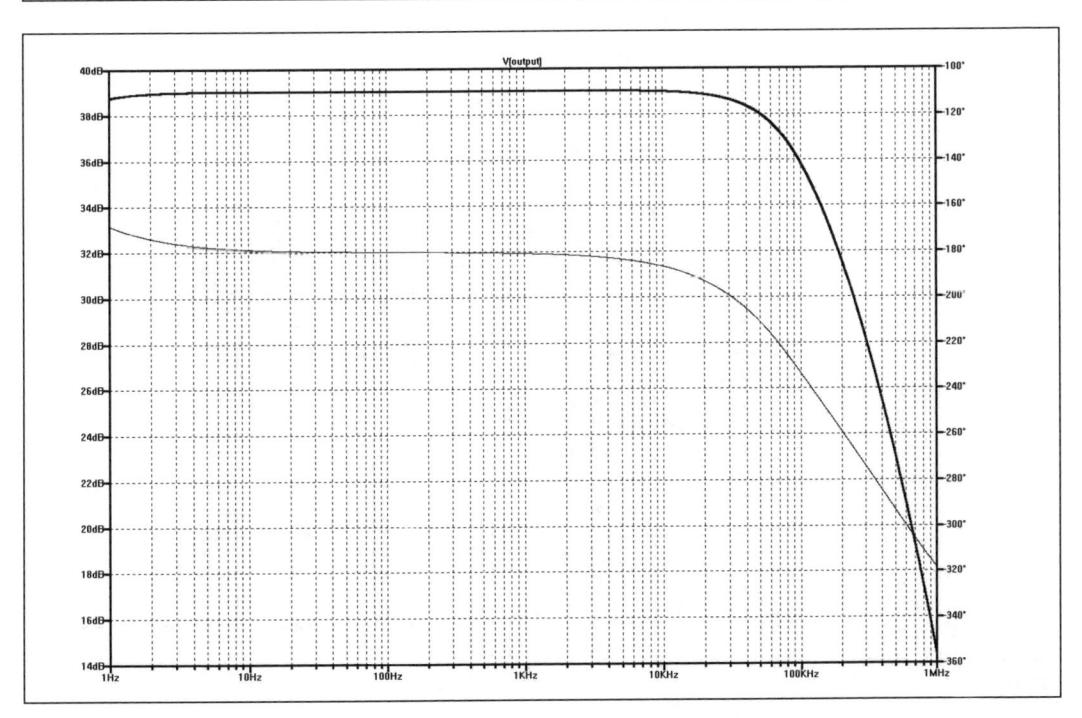

grid is high. R2 and R5 determine the bias voltages. They are not decoupled to achieve a degree of negative feedback, which is beneficial for both bandwidth and fidelity. This is a simple design, but it can be very rewarding if well built.

The simulated curves show that the bandwidth is wide and the amplification even over most of the bandwidth. The thick line denotes the amplification, whereas the thin line shows the output phase in relation to the input.

If the amplifier is to drive a low impedance input, a cathode follower is (again) recommended. Don't forget the coupling capacitor at the anode of U2, though. It is needed to prevent high voltage reaching the grid of the next stage.

The upper 3dB point is at about 95kHz, which would be sufficient for most needs. The amplification at 1Hz is well above the 3dB limit. In fact, the lower 3dB point is at about 0.3Hz.

When evaluating this design, we must remember that the curves are simulated. The quality of the curves depend strongly on the valve models used and not so much on which simulator has been used.

Replacing the ECC81 with an ECC82 (which is an AF triode) yields a slightly narrower bandwidth and a couple of dB lower amplification.

Grounded cathode

A simple grounded cathode stage gives much less amplification and narrower bandwidth. The bandwidth problem is caused by the Miller effect.

However, **Fig 166** shows a well designed pre-amplifier with a ECC83. Note in particular R8, which is a DC feedback channel.

The gain can be set by changing the resistors R8 and / or R4, and can be approximated as: gain = 1 + R8 / (R4 + 1100).

The amplifier has low distortion (below 1%) due to the feedback resistor and the non-decoupled cathode of the first stage (which also contributes to negative feedback). See **Fig 167** for the resulting curves.

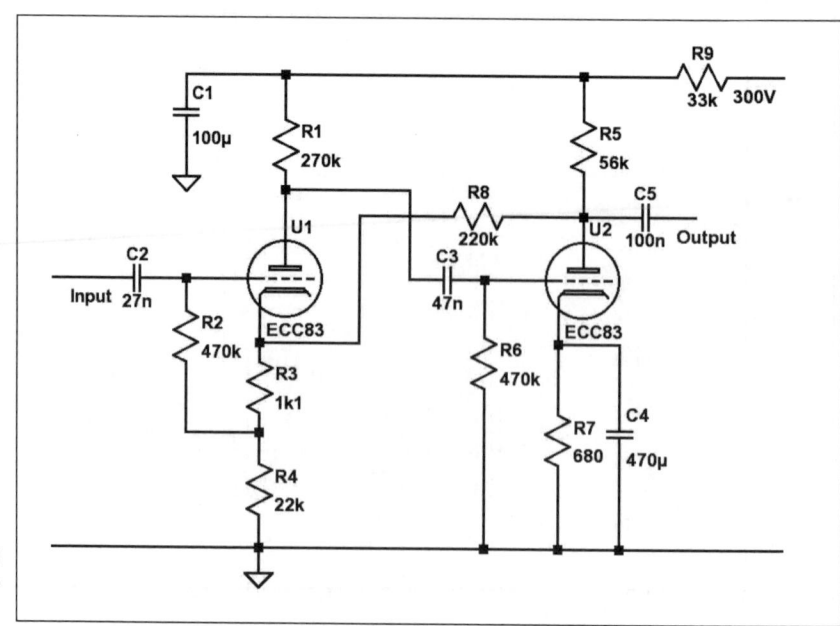

Fig 166: Grounded cathode AF pre-amplifier

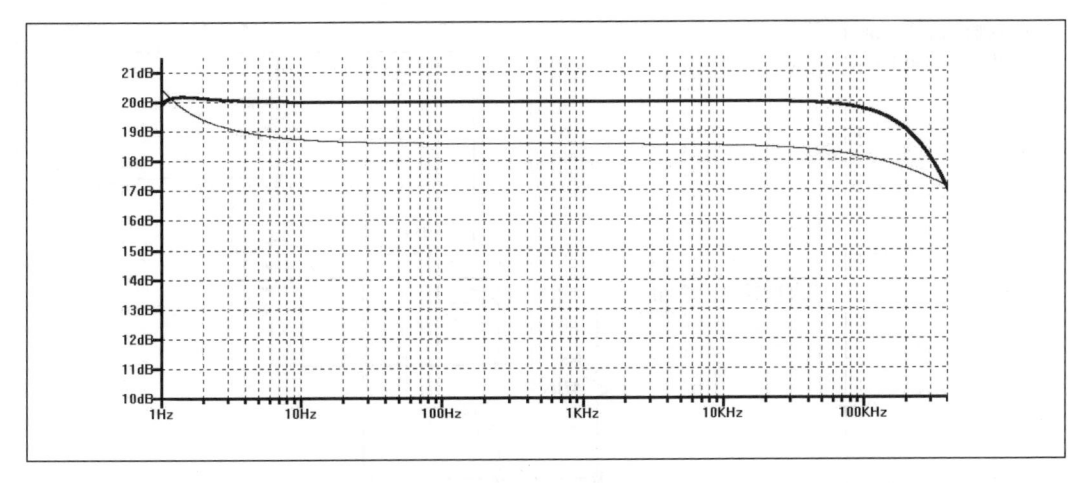

Fig 167: Curves for grounded cathode amplifier above

Bandwidth is from about 1Hz to about 200kHz within 1 dB. Also note the phase curve (thin line), which is reasonably even up to about 100kHz! Note, however, that the curves are the result of simulations with ideal components (as are all curves in this book). Stray capacitances are not taken into account.

Ref	Description	Ref	Description
C1	capacitor, 100µF	R4	resistor, 22k
C2	capacitor, 27nF	R5	resistor, 56k
C3	capacitor, 47nF	R6	resistor, 470k
C4	capacitor, 470µF 25V	R7	resistor, 680
C5	capacitor, 100nF	R8	resistor, 220k
R1	resistor, 270k	R9	resistor, 33k
R2	resistor, 470k	U1	ECC83, first half
R3	resistor, 1.1k	U2	ECC83, second half

Components list for the grounded cathode amplifier

Cathode coupled

One way of overcoming the Miller problem is to use a cathode coupled stage instead, such as the one in **Fig 168**. The corresponding curves are shown in **Fig 169**.

The cathode follower (the first stage) eliminates the Miller effect and the following grounded grid stage provides the amplification. The cathode follower also returns a wide bandwidth along with a low output impedance to the cathode of the grounded grid stage. The output impedance is high, but not nearly as high as that of the cascode.

When calculating the grid bias resistor (R2), one must bear in mind that cathode currents from both halves of the valve flow through the same cathode resistor.

The curves reveal that the frequency response between the 3dB-points lies between about 0.3Hz and 195kHz. The response is, in other words, wider than that of the cascode stage, due to the isolating effect of the grounded control grid between anode and cathode of the second stage. The amplification is lower,

Fig 168: Cathode coupled input stage

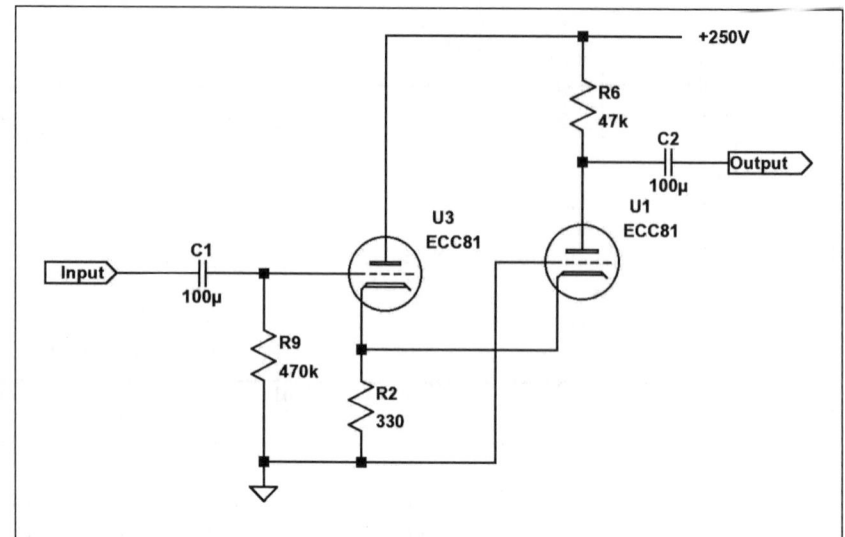

(below) Fig 169: Cathode coupled curves

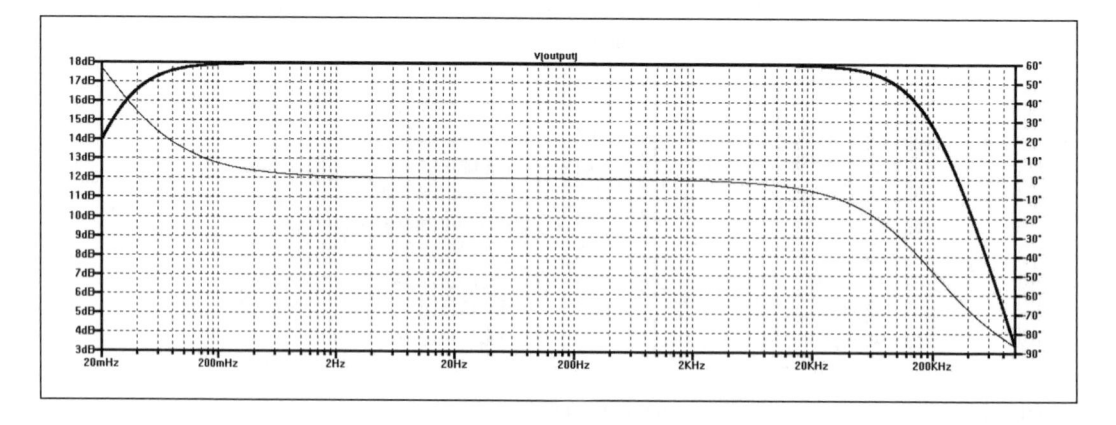

though, about 18 dB as compared to about 39 dB for the cascode. The phase curve shows no phase shift between input and output.

The amplification of the stage is in reality going to be slightly less, due to the damping effect of the input circuit of the next stage.

However, achieving a wide bandwidth alone does not yield the best result. It does sound like a paradox. The fact is, though, that the bandwidth of a stage in a hi-fi amplifier should be wider than that of the previous stage. The bottleneck is the power amplifier, which therefore must be the determining factor for the bandwidth of the preceding stages.

Power Amplifiers

Much effort is being invested in designing the 'perfect' power amplifier. There is, of course, no perfect amplifier! One way of overcoming many of the problems involved in terms of bandwidth and distortion is to adopt the push-pull principle. A non-push-pull amplifier is frequently called 'single-ended' or 'SE'.

The power amplifier's purpose is to provide the loudspeaker with enough power to yield a useful sound level.

CHAPTER 11: HI-FI AMPLIFIERS

Fig 170: Simple single ended ECL86 amplifier

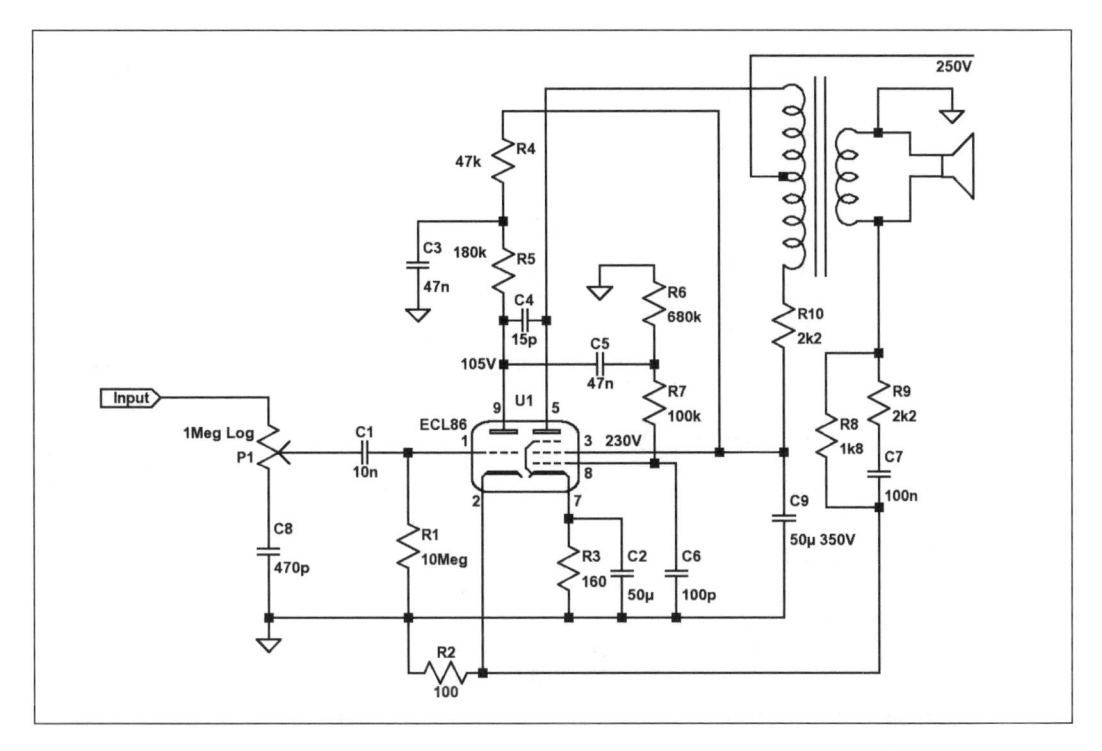

Single ended amplifiers

If you are a beginner at building amplifiers, perhaps you should start with a simple one. It deserves to be pointed out that even a simple valve amplifier yields an excellent sound. The simplest amplifier you can build is one with only one valve, illustrated in **Fig 170**. The ECL86 is a combination of a triode and a power pentode. The triode is used as a pre-amplifier, whereas the pentode is used to drive the loudspeaker. The heater voltage (6.3V AC) is connected to pins 4 and 5.

The left half of the valve (pins 1, 2 and 9) and the associated components constitute the pre-amplifier. R1 is the grid leak and R2 is the cathode resistance.

Ref	Description	Ref	Description
C1	capacitor, 10nF	R3	resistor, 160
C2	capacitor, 50µF	R4	resistor, 47k
C3	capacitor, 47nF	R5	resistor, 180k
C4	capacitor, 15pF	R6	resistor, 680k
C5	capacitor, 47nF	R7	resistor, 100k
C6	capacitor, 100pF	R8	resistor, 1.8k
C7	capacitor, 100nF	R9	resistor, 2.2k
C8	capacitor, 470pF	R10	resistor, 2.2k
C9	capacitor, 50µF	U1	ECL86
P1	resistor, 1M	--	Output transformer
R1	resistor, 10M		with tapped primary
R2	resistor, 100		

Components list for the single ended ECL86 amplifier

Note that there is a frequency dependent feedback chain (R8 / R9 / C7) from the secondary of the output transformer to the cathode of the pre-amplifier. The other end of the secondary winding is connected to ground.

C5 is the coupling capacitor between the two stages. R6 and R7 form a voltage divider. R4 / R5 / C3 form a filter, protecting from positive feedback into the power supply. C4 has a similar function.

C6 protects against RF parasitic oscillations. The pentode's cathode resistor R3 is decoupled by C2 and the two components give the pentode control grid sufficient bias for operation in class A.

The output transformer has a tap at the primary winding. The tap is closer to the bottom end (which is connected to R2). The B+ voltage is connected to the tap.

Single ended ECL82

ECL82 is another combination valve. It contains a triode and a power pentode. The triode can be used as a pre-amplifier, and the pentode drives the loudspeaker, just like the one described above.

This particular amplifier has a frequency dependent feedback chain consisting of R1 and C1 in **Fig 171**. Note that the opposite side of the secondary of the output transformer needs to be grounded for this to work.

A single ended ECL82 would be an excellent amplifier as an audio stage in communications receivers, even though it does have certain qualities suitable for low-end hi-fi applications as well. In a receiver application, R12 would be decoupled and the negative feedback omitted.

Fig 171: Single-ended ECL82 amplifier

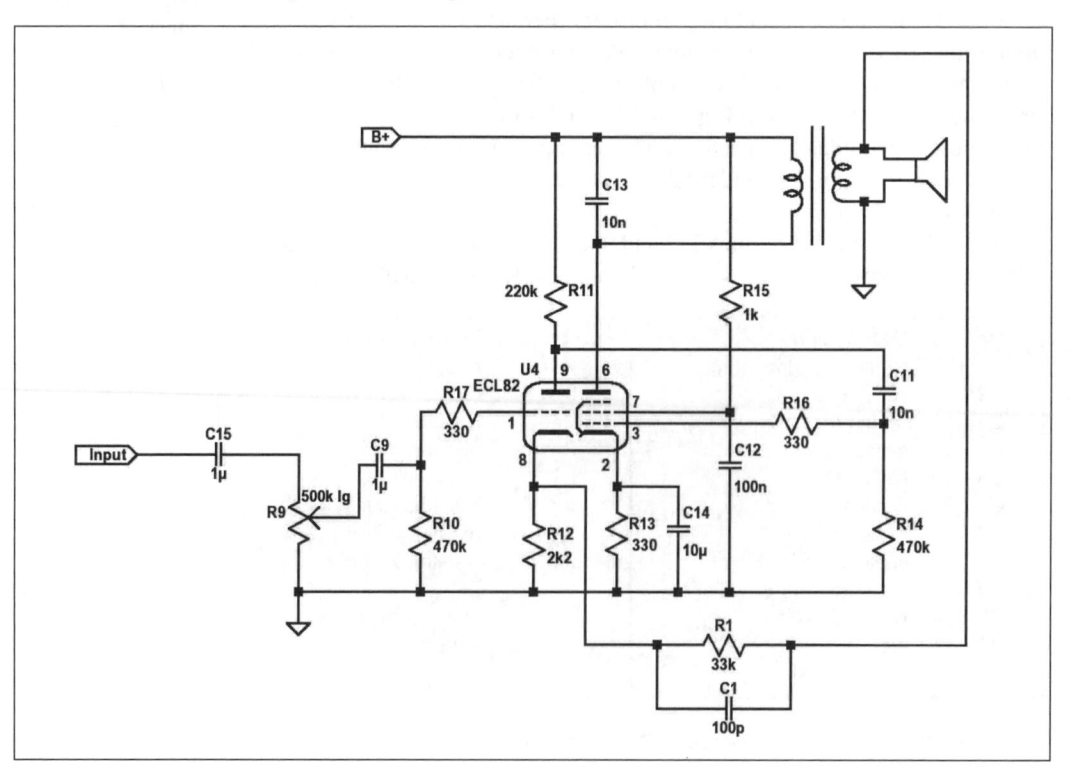

CHAPTER 11: HI-FI AMPLIFIERS

Ref	Description	Ref	Description
C1	capacitor, 100pF	R11	resistor, 220k
C9	capacitor, 1µF	R12	resistor, 2.2k
C11	capacitor, 10nF	R13	resistor, 330
C12	capacitor, 100nF	R14	resistor, 470k
C13	capacitor, 10nF	R15	resistor, 1k
C14	capacitor, 10µF	R16	resistor, 330
C15	capacitor, 1µF	R17	resistor, 330
R1	resistor, 33k	U1	ECL82
R9	resistor, 500k	Output transformer	
R10	resistor, 470k		

Components list for the single-ended ECL82 amplifier

The two capacitors at the input, C15 and C9 protect the volume control potentiometer from DC, thereby preventing it from being noisy when operated. R16 and R17 protect the grids from developing oscillations. R12 is not decoupled, so it becomes a suitable insertion point for negative feedback.

Single ended EL34

A single ended EL34 power amplifier is presented in **Fig 172**. It is a simple amplifier with few components, but should give many hours of pleasant listening. The EL34 is a popular power pentode, present in many amplifiers. According to the data sheet, it is capable of 11W in single ended class A operation. There should be about 288V on the screen grid, and some 17V across the cathode resistor, R5. Input comes from a preamplifier and is connected to C3. The design is merely a matter of providing the correct voltages for the valve,

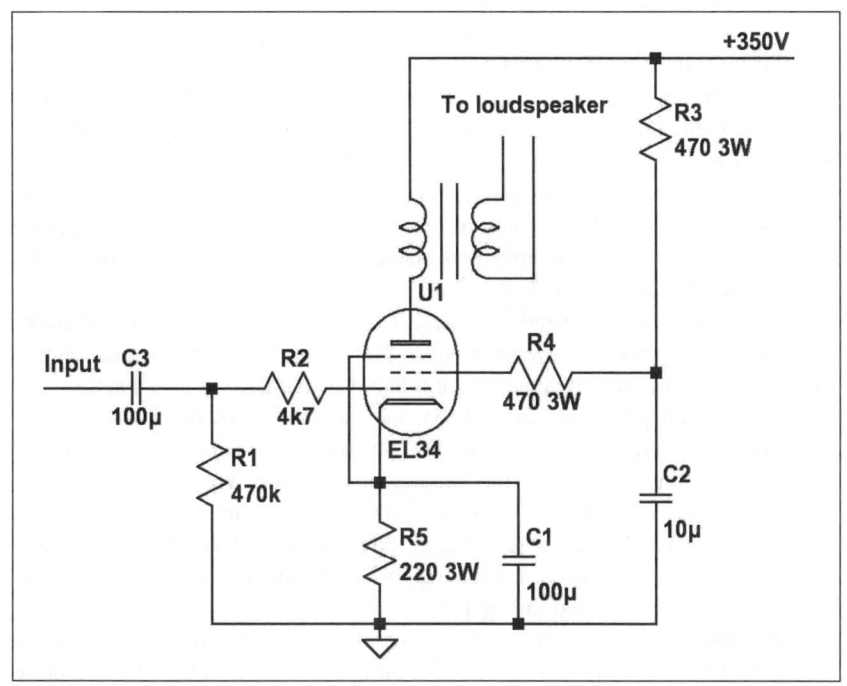

Fig 172: Single-ended EL34 amplifier

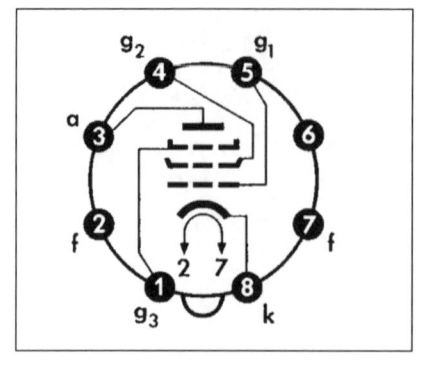

(above) Fig 173: EL34 socket

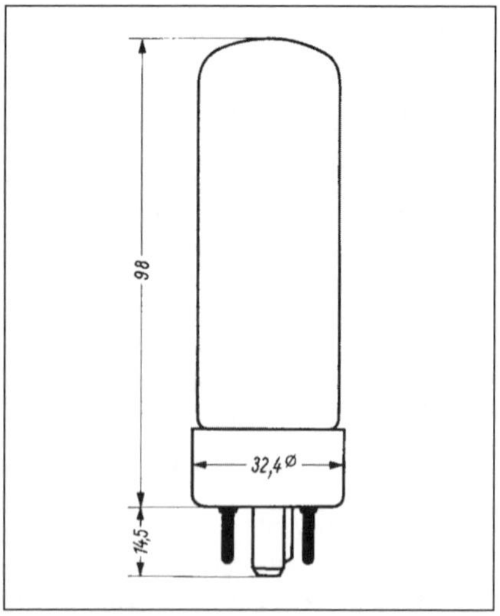

Fig 174: Dimensions of the EL34

Ref	Description	Ref	Description
C1	capacitor, 100µF	R3	resistor, 470
C2	capacitor, 10µF	R4	resistor, 470
C3	capacitor, 100µF	R5	resistor, 220
R1	resistor, 470k	U1	EL34 power pentode
R2	resistor, 4.7k	--	Pentode output transformer

Components list for single ended EL34 amplifier

and just running it. The socket is shown in **Fig 173**. The valve is fairly big; its dimensions are shown in **Fig 174**.

Push-pull Amplifiers

A push-pull power amplifier involves at least two valves. They are connected such that they amplify half of the signal each, in principle. Thus, they are both biased to work in class B, ie near or at the cut-off point. However, there is frequently some overlap, which places them in class AB. The overlap neutralises the non-linearities at the low end of the curves. However, differences between the valves may still cause unwanted distortion at the output. Therefore, some valve suppliers deliver matched pairs.

It is a matter of taste whether the phase inverter should be included in a chapter about power amplifiers. However, no push-pull stage would work without one, so, from that point of view, it is only natural that it be included in this chapter.

The push-pull principle requires that the valves receive the input signal in anti-phase, eg one valve amplifies the positive portion of the input signal, whereas the other valve amplifies the negative portion. This requires a phase-inverter in the signal path. This is a device with one input and two outputs, where one output delivers a zero degree phase shift, and the other 180 degree phase shift. A host of phase-inverters has been developed over the years. One of the simplest ones is shown in **Fig 175**.

This phase inverter follows a simple pre-amplifier. The voltage at the anode of the first triode is low enough to give an appropriate grid bias for the phase

CHAPTER 11: HI-FI AMPLIFIERS

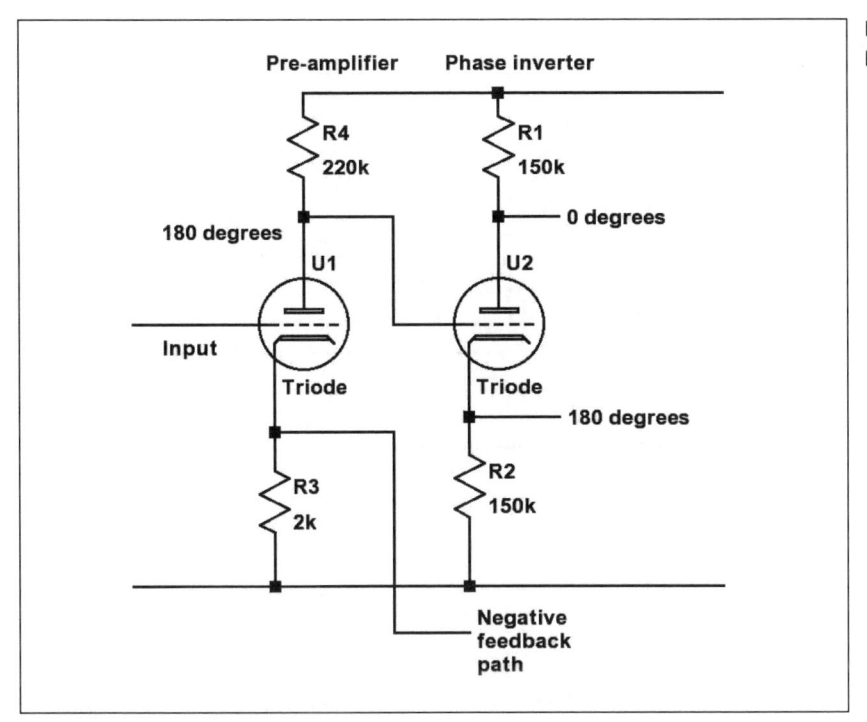

Fig 175: A simple phase inverter

inverter proper. Therefore, it can be directly coupled to the grid. The cathode is not decoupled, so the amplification of the stage is not very high; however, high enough. Consequently, amplification begins at zero hertz.

The cathode is also a good insertion point for a negative feedback path from the loudspeaker to (to some degree) decrease distortion which may occur at the final stage.

The phase inverter proper is valve U2. The two resistors R1 and R2 are of equal value. The outputs should be of equal amplitude, but opposite phase.

The power amplifier anodes are connected to one end each of an output transformer. The transformer has a tap at the mid-point to which the high voltage is connected. The secondary is connected to the loudspeaker. Some data sheets for output pentodes quote information on push-pull operation.

The next example is a slightly different solution. There is still a preamplifier, followed by a phase inverter. R5 at the bottom of the diagram is used to set the amplitude relationship between the two output signals by selection of the resistance.

Looking at the schematic in **Fig 176**, you will find that the anode of U1 does not appear to receive any high voltage. However, as in the previous example, the voltage drop across R4 / R5 is large enough to provide sufficient voltage for the valve to work.

A different approach to phase splitting is a two-stage grounded cathode amplifier, where the amplification of the second stage is restricted through a voltage divider. Since both stages have a phase shift of 180 degrees, the anode of the first stage provides 180 degrees, whereas the second stage anode provides the 0 degree signal.

159

VALVES REVISITED

Fig 177: A different phase inverter

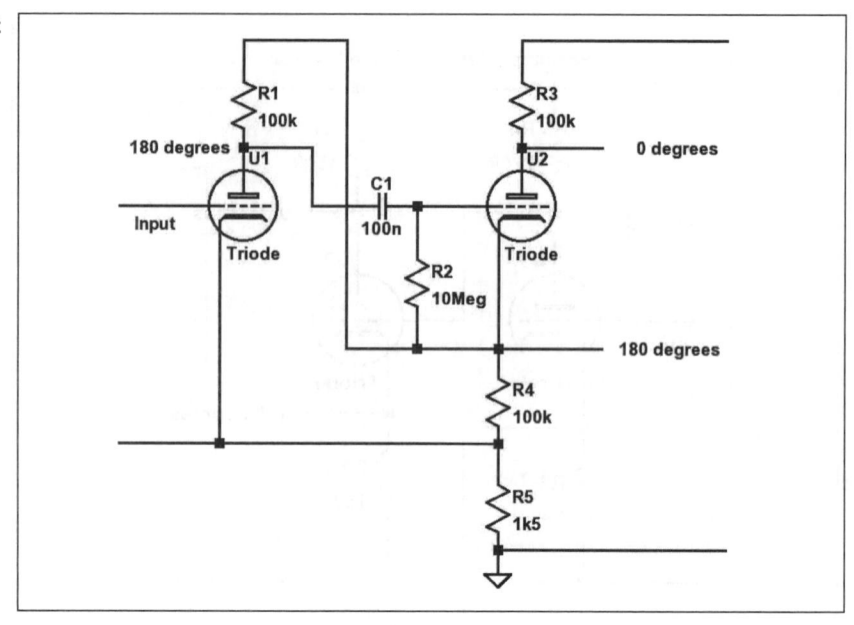

A simple push-pull power amplifier

This amplifier in **Fig 178** (also shown in the general chapter on amplifiers) consists of two EL84s. These are fairly popular valves for audio use.

The two stopper resistors are R1 and R2 (which could be made bigger), whereas R3 and R4 constitute the grid leaks. The cathodes are connected together and fed through the common cathode resistor R5 which is decoupled by capacitor C1. The screen grids are at the same or slightly higher voltage than the anodes, since the power supply is connected to the centre tap of the output transformer.

Fig 178: EL84 in push-pull

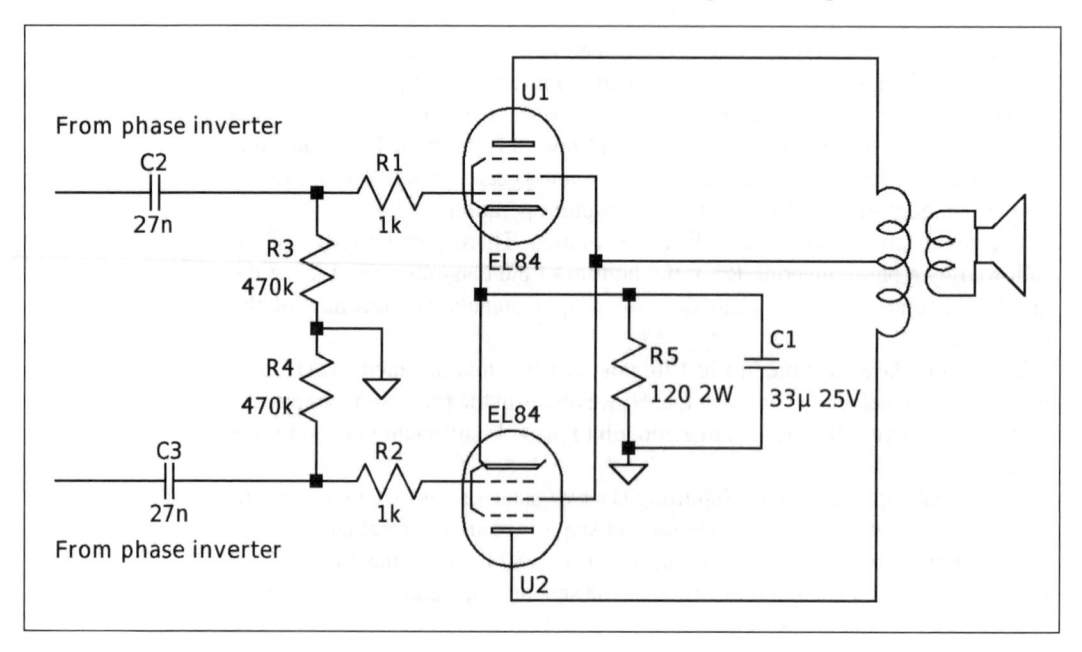

Using individual cathode resistors for each of the pentodes, rather than a common resistor as shown here, allows for fine adjustment of the grid biases, if need be. In some designs, a semiconductor voltage stabiliser, LM317, is inserted at the cathodes to provide grid bias. The amplifier is capable of delivering 10 - 15W.

Triodes in power amplifiers

There is no reason why a pentode should be the only alternative in a power amplifier. A very popular triode, the 300B, is used by professionals and amateurs the world over. Connecting a pentode as a triode is another popular approach to triode power delivery. There are little differences in the practicalities of using triodes in the power stage. On the theoretical level, the anode resistance in a triode is different from that of a pentode. The ra of the 300B is about 1.5kΩ at its lowest, whereas the pentode EL34 has 20kΩ. It is a matter of finding the correct output transformer. Other than that, connect the correct voltages, make sure you have sufficient currents, and you are up and running.

If you don't want much power capability, low power (3W) triodes are available as dual triodes. The benefit is that you are able to build a push-pull stage with only one valve. Such a dual triode is 6189 or 5998 (13W per section). Some people connect triodes in parallel to achieve more power.

A single ended triode amplifier, such as the one shown in **Fig 179**, can be built very simply. The triode is connected as a grounded cathode amplifier with an output transformer in the anode circuit.

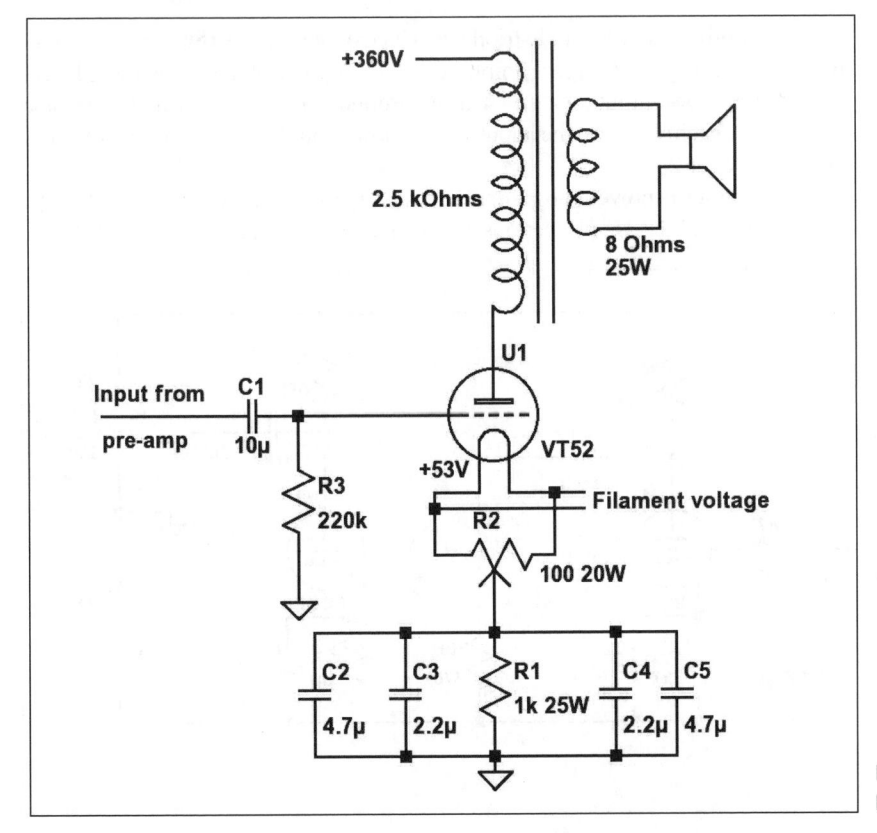

Fig 179: Triode power amplifier

The valve used here is a directly heated one. The filament voltage is 6.3V at 200mA. The potentiometer at the filament is arranged in a 'humdinger' configuration. Its purpose is to neutralise the hum from the power supply that enters the valve through the filament. The hum is neutralised at the electrical centre of the potentiometer.

There are four decoupling capacitors across the cathode resistor. A rule of thumb is: The more capacitors in parallel, the higher the efficiency and the lower the losses.

The input and output circuits are extremely simple. A standard coupling capacitor - grid leak combination connect the output from the pre-amplifier to the grid, and at the anode is an output transformer that drives the 8Ω loudspeaker.

The valve, a VT52, is a powerful one, capable of delivering 25W into the speaker in a single-ended configuration (audio enthusiasts say 'topology').

A Lunar Grid Amplifier

Due to the low amplification of a lunar grid output stage, it needs a huge amount of preamplification. Here is shown a complete OTL (output transformerless) lunar grid amplifier that seemingly contains two valves. However, to achieve the low output impedance required by the loudspeaker, the power stage is in reality ten valves, all connected in parallel. Connecting power valves in parallel yields high power and low impedance.

The pre-amplifier

The preamplifier is one double triode, a 6CG7 shown in **Fig 180**, which is capable of generating 400V signal amplitude at 2V input voltage, a total amplification of 200. The output signal is centred around zero volts, but it still needs a coupling capacitor, since the input of the power amplifier has a negative DC voltage of over 200V.

Fig 180: Lunar Grid pre-amplifier

The amplifier is powered by a dual supply, capable of delivering +/-350V. The resistors R7 and R8 are protections against parasitic oscillations. R1 and R2 are

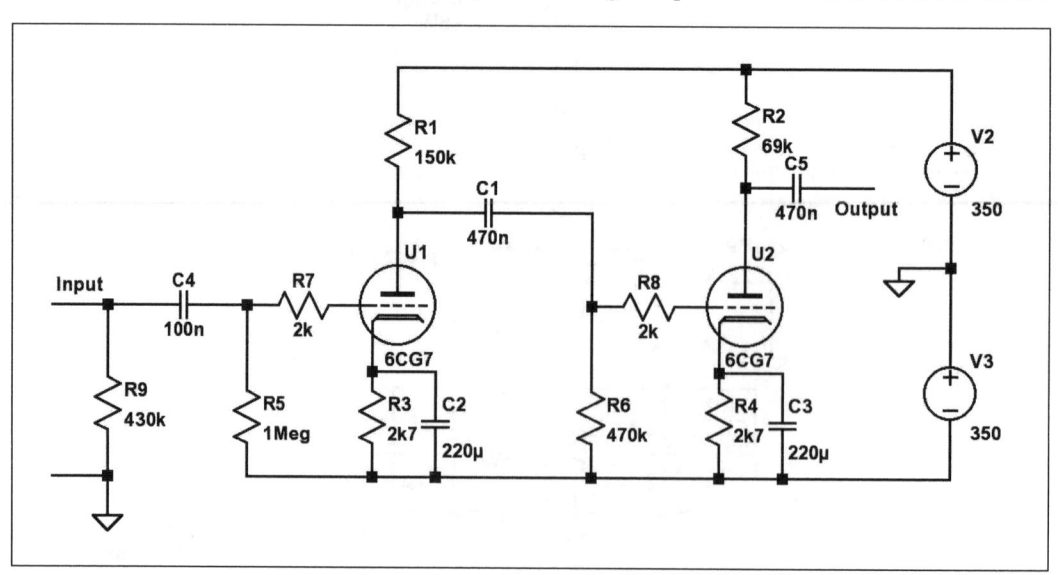

1W resistors, and C1, C4 and C5 are high voltage capacitors, rated at 400V. C2 and C3 are rated at 25V. A reminder: Connecting several smaller capacitors in parallel will reduce capacitor losses in a corresponding degree.

According to simulations, the amplifier is capable of delivering some 48dB gain within 1dB between 4Hz and 80kHz.

The power amplifier

The power amplifier is the lunar grid amplifier proper. No information directly applicable to lunar grid amplifiers can be found in the data sheets. You basically have to make your own data sheet in order to find out the properties of a valve in lunar grid configuration. A few general pointers could be made, though.

The valve should be medium or low μ. The reason for this is that the voltage required to drive the power stage will be far too high to be practical. An upper practical μ limit would be about 15.

Since the anode does not draw any current in this configuration, the power supply can be made very basic. On the other hand, the pre-amplifier needs to be capable of delivering a substantial voltage to enable the anode of the final amplifier to control the electron current to the grid. The grid output impedance is extremely low. Therefore the grid needs to be able to handle some amount of current. This has to be borne in mind when choosing a valve for this purpose.

Due to the low amplification of the power stage, there is virtually no Miller effect to take into account. The consequences are that the stage bandwidth will be wider than would otherwise be possible. Additionally, the need for a bulky, expensive and bandwidth limiting output transformer is eliminated.

The power amplifier consists of several valves connected in parallel. This reduces the available output impedance and increases the available power. The example has ten miniature valves, 5687, which is a medium μ dual triode. Ten valves equals twenty separate triodes. The 5687 has a μ of 17 at 180V anode voltage. In lunar grid configuration, μ becomes about $1/17 = 0.06$.

There is no way the normal parameters given in the data sheet can be used or easily converted for lunar grid operation. The best thing to do to assess the other parameters is to plot a set of curves for the valve by connecting it to a power supply, varying the voltages and measure voltages and currents. It is a tedious work, but it pays off in the long run. You can choose to plot just one half of the valve, or connect both halves in parallel and plot the curves for the combination.

One designer of this particular amplifier, Steve Bench, has done such a plot for 5687, shown in **Fig 181**. Note that the figures in the graph refer to the functionality of the stage, not the actual electrodes. So, the slanted curves refer to the negative voltages connected to the anode, the Y-axis shows the currents at the grids of the valve, and the X-axis shows the voltages at the grids of the valve.

The graph shows the load line for 24 ohms. The horizontal line at 70mA plate current is the quiescent 'anode' current (the current at the output, ie the grid). The load line stays well away from the maximum permissible power so as not to jeopardise the performance or life span of the valves.

The curves are valid for both halves connected in parallel. We find by measuring the curves that μ is about 0.04 and Gm about $60/80 = 0.75$mA/V, so the 'anode' resistance becomes around $0.04 / 0.00075 = 53\Omega$. To get the impedance down to 8Ω and below, we need at least seven valves in parallel (7.5Ω). Ten

VALVES REVISITED

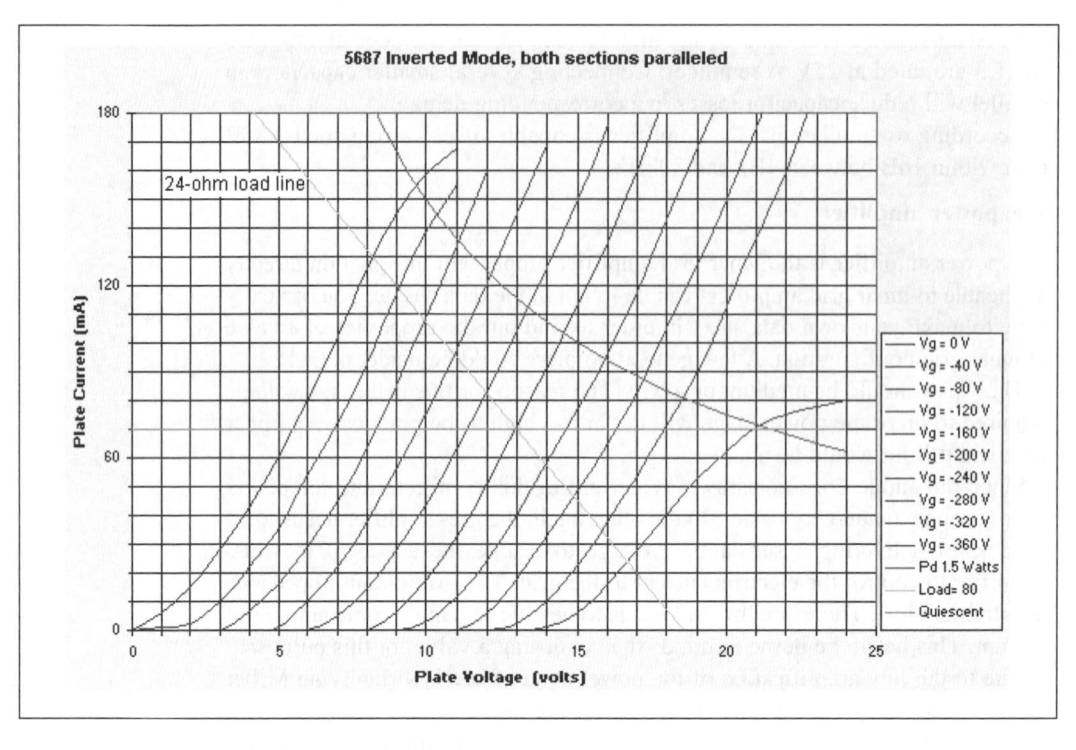

Fig 181: Curves for 5687 in inverted mode

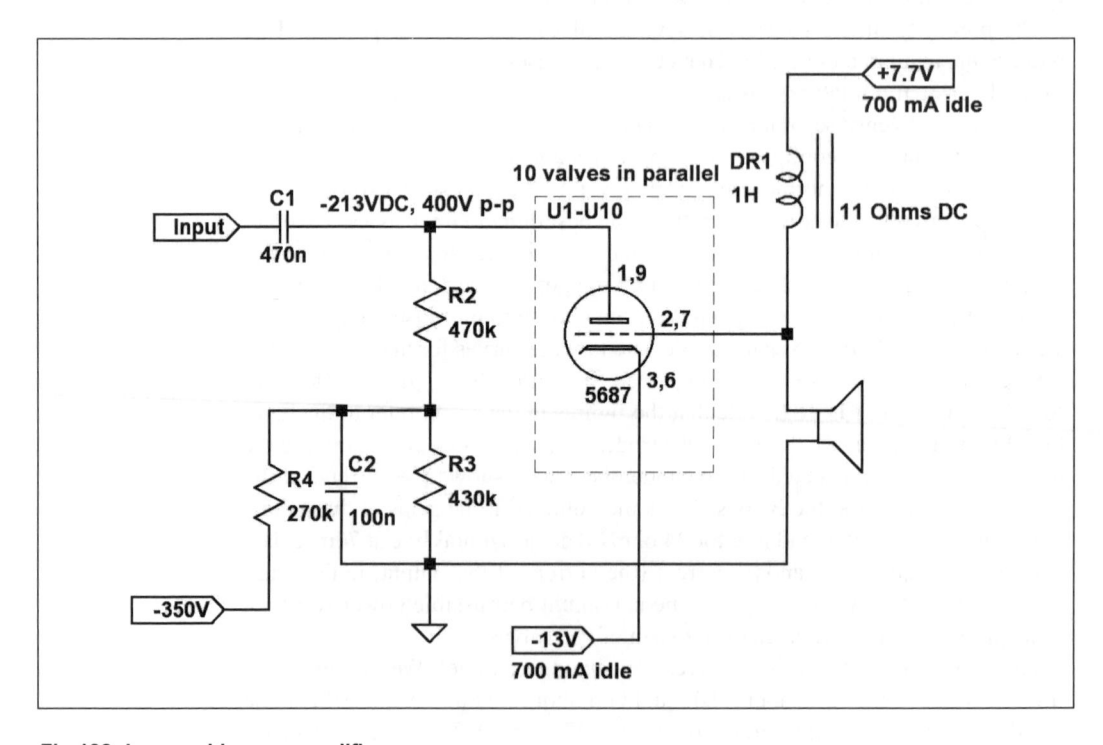

Fig 182: Lunar grid power amplifier

CHAPTER 11: HI-FI AMPLIFIERS

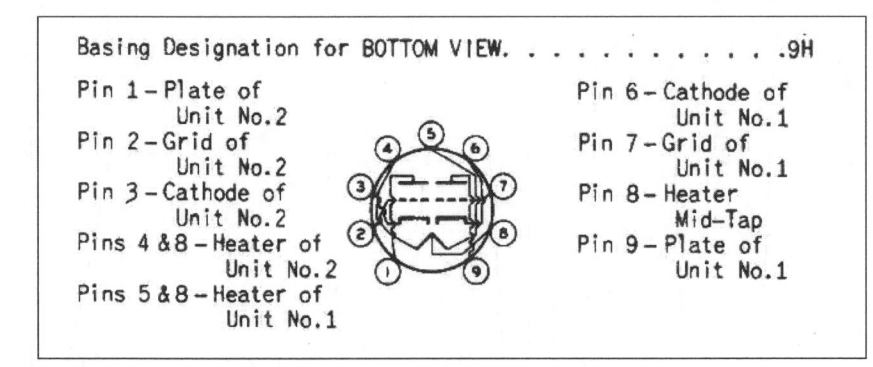

```
Basing Designation for BOTTOM VIEW. . . . . . . . . . . .9H
Pin 1 – Plate of              Pin 6 – Cathode of
        Unit No.2                     Unit No.1
Pin 2 – Grid of               Pin 7 – Grid of
        Unit No.2                     Unit No.1
Pin 3 – Cathode of            Pin 8 – Heater
        Unit No.2                     Mid-Tap
Pins 4 & 8 – Heater of        Pin 9 – Plate of
        Unit No.2                     Unit No.1
Pins 5 & 8 – Heater of
        Unit No.1
```

Fig 183: Socket connections for the 5687. If the heater is to be run at 6.3V, then pins 4 and 5 should be connected together.

valves gives us 5.3Ω, 13 valves would yield 4,07Ω. Actual measurements would, of course, return more accurate figures.

The schematic of the power stage of a lunar grid amplifier is shown in **Fig 182**. All capacitors need to be rated at high voltage, at least 400V. The DC voltage at the output is about 0V, which eliminates output capacitors. Note that the socket for 5687 is different from the usual dual triodes, as you can see in **Fig 183**.

The power supply

Due to the particular requirements of a lunar grid amplifier, the power supply needs to be a bit unusual, as you can see in **Fig 184**. The power supply design offers no surprises, apart from the number of voltages delivered. It is unregulated and generates power to the entire amplifier.

Fig 184: Power supply for lunar grid amplifier

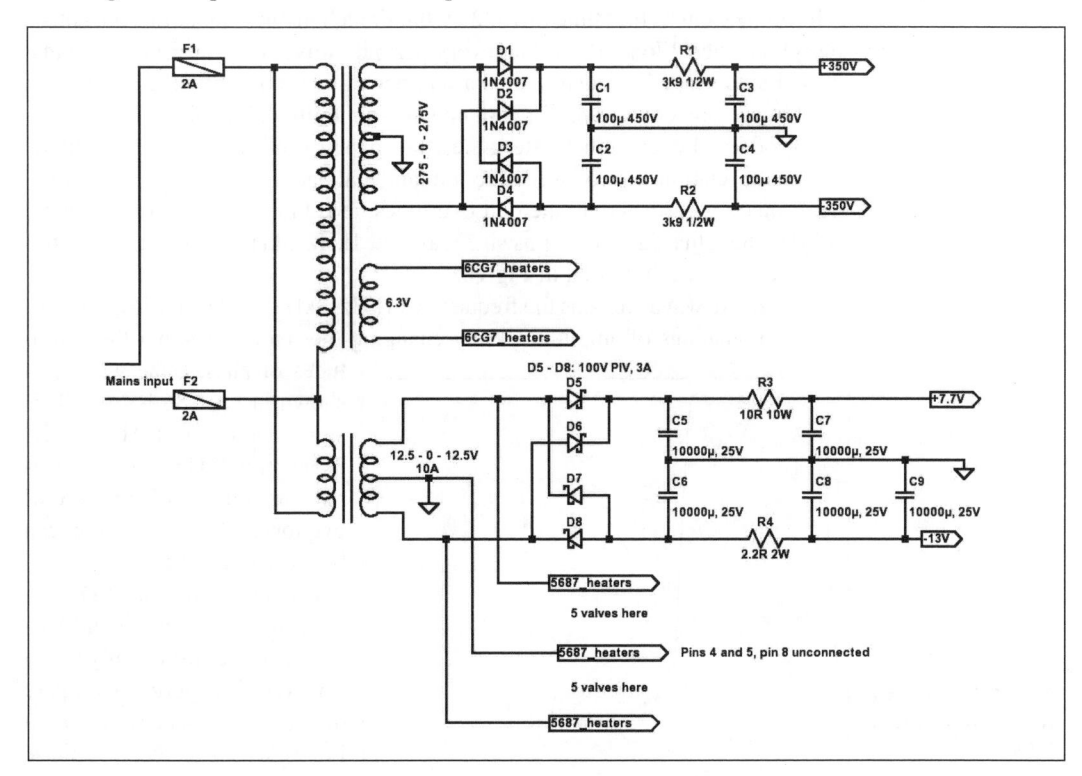

165

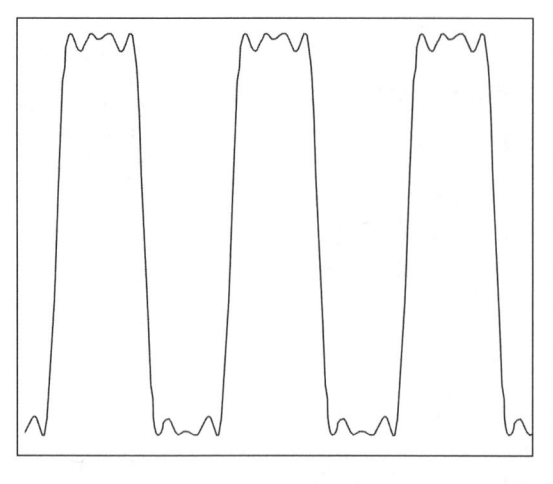

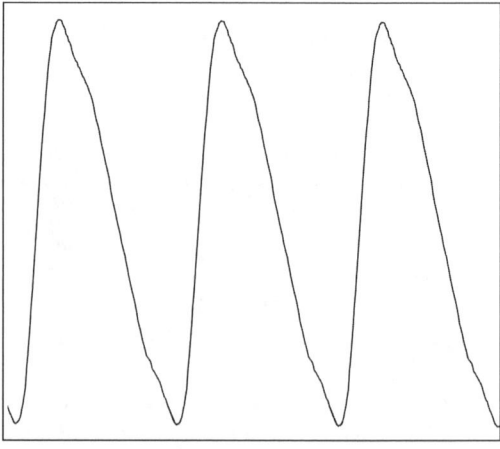

Fig 185: Square wave generated from four frequencies

Fig 186: Triangle wave from same frequencies

Assessing the Qualities of an Audio Amplifier

Plotting the phase and amplitude response of an amplifier is normally a tedious process. Luckily, there is a short-cut, which is used by many manufacturers, by which the response of an amplifier can be assessed. (This is particularly true when data for a modern semiconductor op-amp are listed). Note that the method only gives you a rough idea of how well or badly the amplifier performs, but it is still a valid method.

It can be shown (first time in 1822 by the French mathematician Jean Baptiste Joseph Fourier, 1768 - 1830) that every periodic waveform consists of a fundamental sine shaped frequency plus a number of overtones, all sine shaped. An ideal linear sawtooth wave, for instance, has an unlimited number of sine overtones, odd and even. What determines the shape of the wave is the amplitude and phase relationships between the fundamental and the overtones. A symmetrical square wave has only the odd overtones, ie if the periodicity of the signal is 1kHz, then the signal contains sine waves at 1kHz, 3kHz, 5kHz, etc. Four frequencies have been plotted in **Fig 185**.

This square wave contains the frequencies 1kHz, 3kHz, 5kHz, and 7kHz in suitable combinations of amplitude. Only changing the frequencies, not the amplitudes or phases, the waveform shown in **Fig 186** is built. This signal contains 1kHz, 2kHz, 3kHz, and 4kHz. A pulsed signal has yet a different set of overtones. Any waveform can be created this way.

The curve in **Fig 187** was generated with the second harmonic phase shifted only a tiny fraction of the period. Compare this with the square wave in Fig 185 with the correct phase

Fig 187: Square wave with phase error

CHAPTER 11: HI-FI AMPLIFIERS

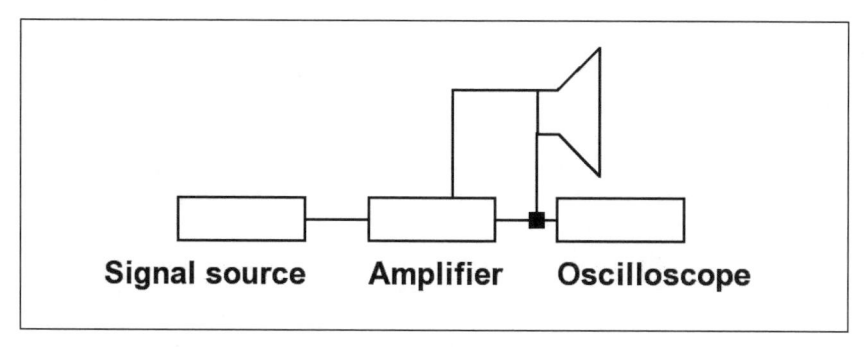

Fig 188: Set-up for testing amplifier

relationships. This shows another aspect of the importance of a correct phase response of an amplifier.

The most common signal to use for assessing an amplifier is a symmetrical square wave. Measurement is performed such that the square wave is inserted at the beginning of the amplifier chain, as shown in **Fig 188**, and the resulting response is picked up at the end of the chain.

Before you begin to measure, though, make sure your oscilloscope itself has a good pulse response.

If your amplifier has tone controls, make sure they are set at neutral. Then connect the signal generator and oscilloscope and increase the amplitude of your signal generator to a point where the amplifier is not overloaded. If necessary, adjust the amplifier's volume control. Now you can study the signal at the output.

The amplifier should be loaded by a loudspeaker. However, the noise might drive you insane, so a resistor in its place will have to do. If you have a dummy load containing the capacitance, inductance and resistance of your loudspeaker, so much the better.

Assume that **Fig 189** shows your signal at the output of your signal generator at 1kilohertz. Further, assume that **Fig 190** shows the frequency response of your amplifier. Then your pulse response at the output might look like the one in **Fig 191**.

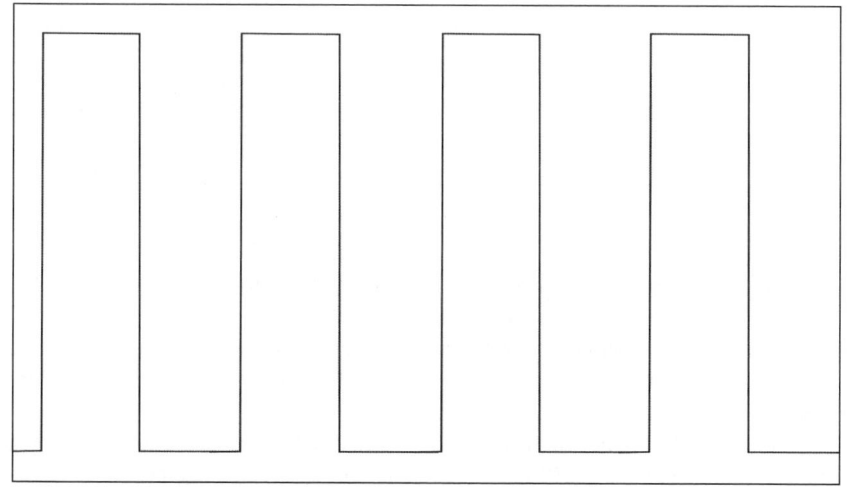

Fig 189: Idealised square wave, input for tests

VALVES REVISITED

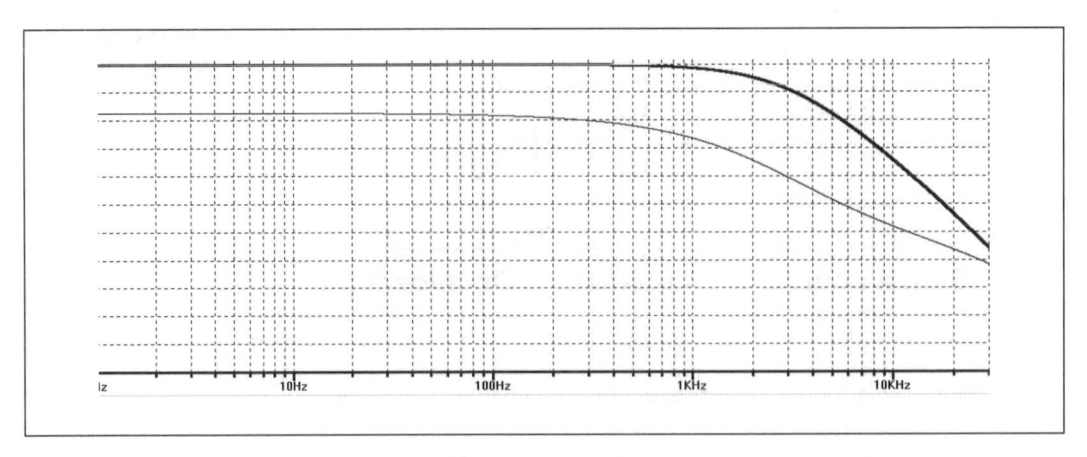

(above) Fig 190: The frequency response for test amplifier

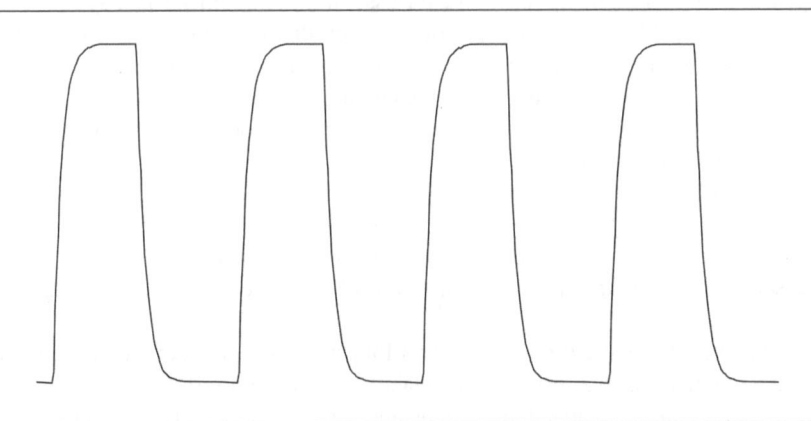

Fig 191: Output of above amplifier

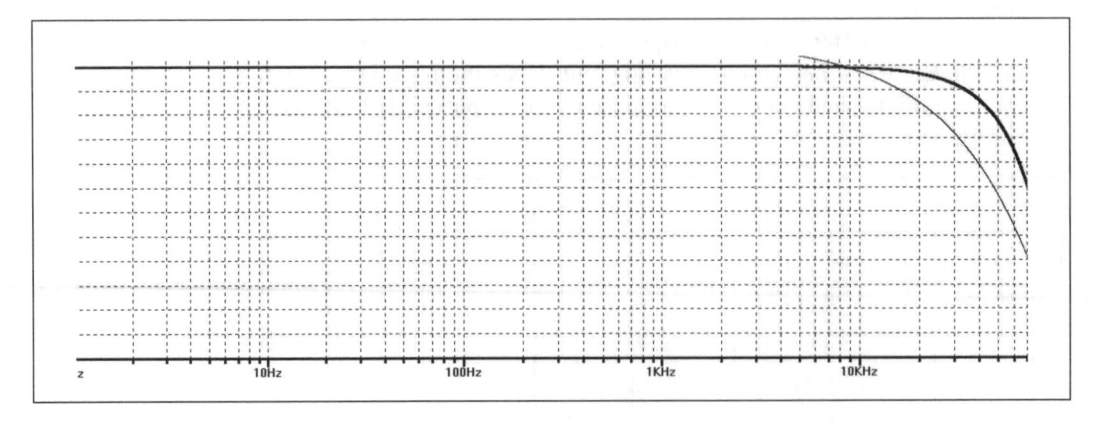

Fig 192: Amplifier with a wider frequency response

As you can see, the sharp edges are no longer sharp, and the short rise and fall times are now longer. The reason for this is that there are overtones missing from the output, in other words, the amplifier's bandwidth is limited in relation to the test frequency. The rise time is measured from the 10% point to the 90% point of the signal's amplitude.

If, on the other hand, the amplifier's frequency response is wider, as in **Fig 192,** the pulse response will be different, as you can see in **Fig 193**.

168

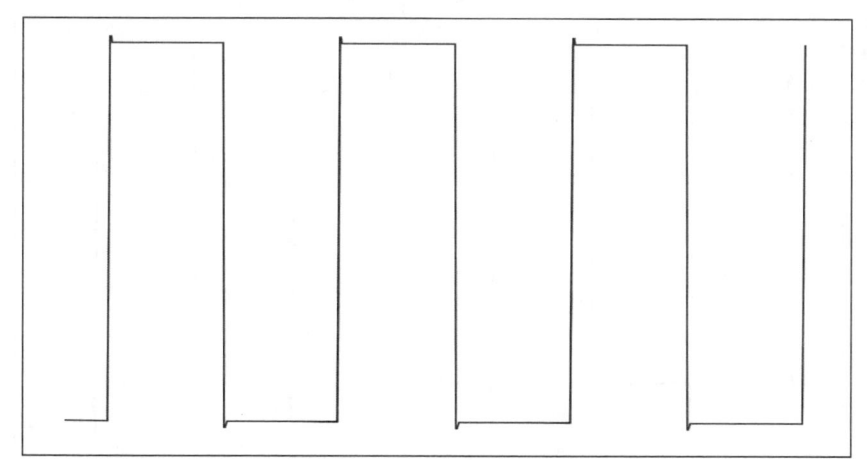

CHAPTER 11: HI-FI AMPLIFIERS

Fig 193: Output of wider response

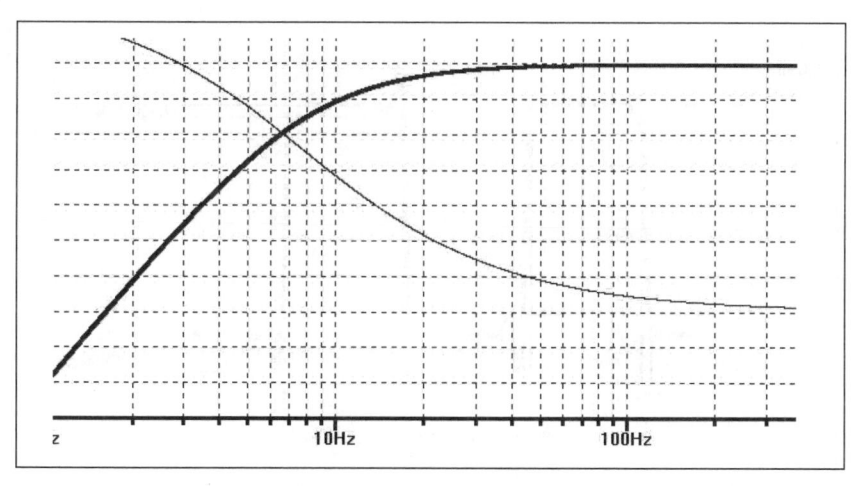

Fig 194: Amplifier with bad low frequency response

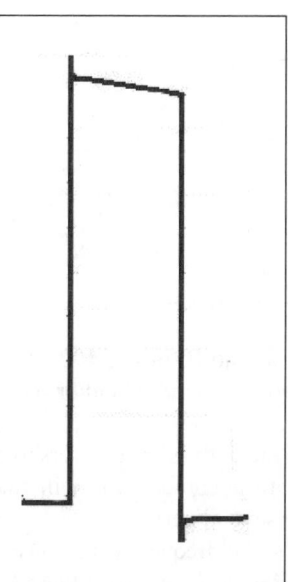

(Left) Fig 195: Output with bad low frequency response

(Below) Fig 196: Exaggerated high frequency response

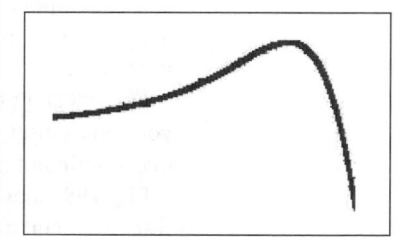

Note how the 'corners' of the response are much sharper and the rise and fall times are much shorter.

There are other phenomena that might arise, such as illustrated in **Fig 194**. The frequency response of the amplifier is bad at the low end of the spectrum. This gives rise to the 'roof' of the pulse falling, as illustrated in **Fig 195** due to the missing low frequencies:

If there is a 'bump' at the high end of the frequency curve, like the one in **Fig 196**, the pulse response will show what is known as 'ringing' or 'over-

169

VALVES REVISITED

Fig 197: Output with exaggerated high frequency response

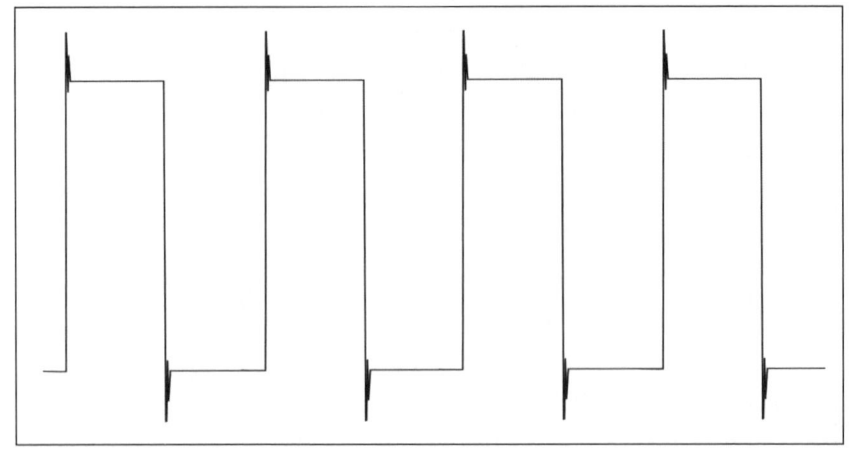

Fig 198: Pulse definitions

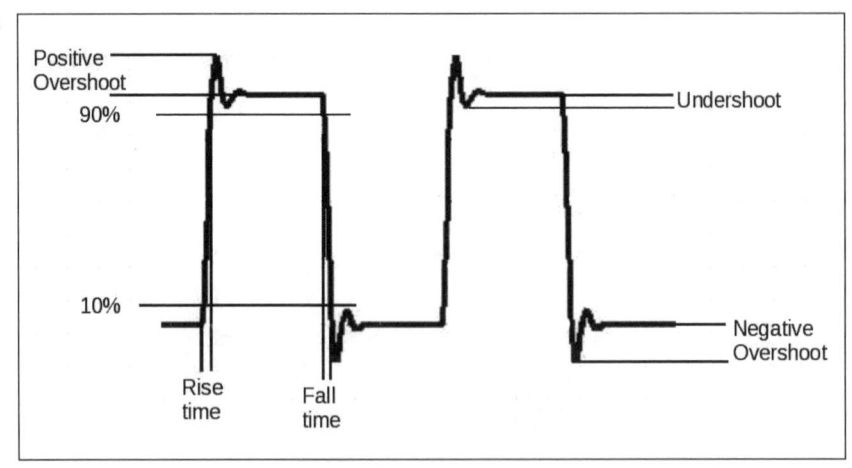

Fig 199: Amplifier with bad mid-range

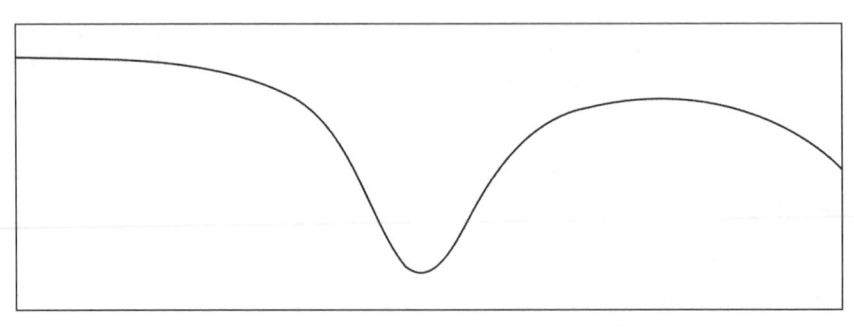

and undershoot', as shown in **Fig 197**. It may also be caused by some impedance matching problem. Under- and/or overshoot is always a sign of distortion in your amplifier.

Fig 198 shows the definitions of overshoot, undershoot, rise and fall times. If you look closer at the corners, you will find that, in some cases, the 'ringing' may continue for several periods.

Fig 199 shows the frequency response and **Fig 200** the resulting pulse response. Ringing may also occur if there is a dip in the frequency response.

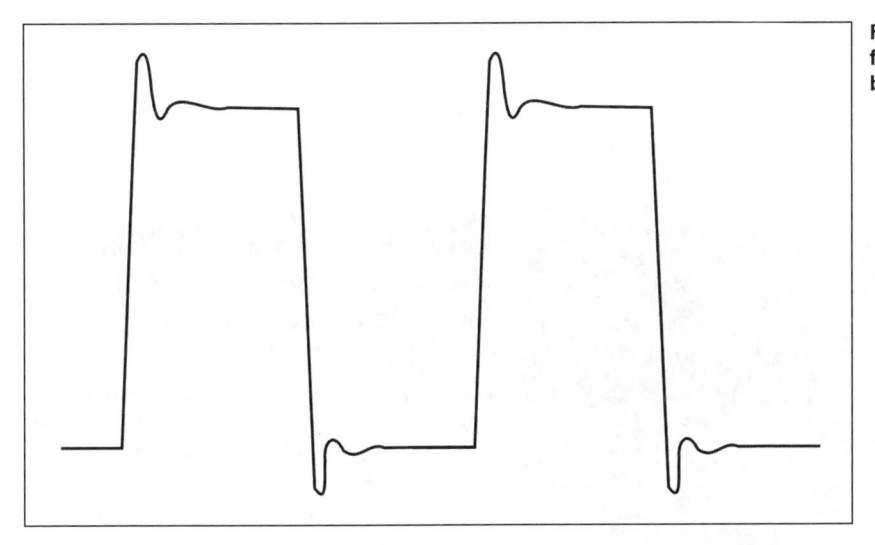

Fig 200: Output from amplifier with bad mid-range

As you can see, the pulse response reveals a good deal of information which is useful when you wish to test your build.

According to Formula 29, there is an approximate relationship between the bandwidth and the rise/fall times.

$$fH \approx \frac{339}{Tr}$$

Formula 29

where:
fH = upper frequency limit in kHz at the -3dB point
Tr = rise time in microseconds

Since the lower limit of the bandwidth only constitutes a tiny fraction of the upper limit, it can be ignored.

Circuit Overload
The bumper book of circuits for radio amateurs
By John Fielding, ZS5JF

This is the book that all keen home constructors have been waiting for! *Circuit Overload* includes 128 circuit diagrams, complemented by an additional 89 other diagrams, graphs and photographs and is a unique source of ideas for almost any circuit the radio amateur might want.

Circuits are provided that cover wide range of topics. Chapters have been devoted to audio, metering & display, power supplies, test equipment and antennas. Chapters have also been included that cover the design of low-pass, high-pass and band-pass filters. Nor are valves forgotten with, two chapters of *Circuit Overload* devoted to this fascinating topic.

Throughout the book, circuits are presented in an easy to understand fashion and many can be inter-connected to make a more complex item if so desired by the reader. If you are interested in home construction this book provides simple circuits and advice for the beginner with more complex circuits for the more experienced.

Size 240x174mm, 208 pages
ISBN 9781-9050-8620-7

ONLY £14.99

RSGB shop

Radio Society of Great Britain
3 Abbey Court, Fraser Road, Priory Business Park, Bedford, MK44 3WH
Tel: 01234 832 700 Fax: 01234 831 496

www.rsgbshop.org

E&OE All prices shown plus p&p

12

Construction with Valves

Building RF circuits is an art, not to mention building oscillators. If you are fairly skilled, though, the satisfaction of a well functioning piece of equipment will be so much higher.

Before you start to build anything, you should plan your equipment. Otherwise you are at risk of placing the stages in the wrong order and there is a risk that your equipment begins to oscillate. Begin your planning with a block diagram. You have to be well composed when you draw the diagram if you want it to be legible and easy to understand. It is in principle possible to build the stages in the same order as you drew your block diagram.

The valves and bigger components, transformers and similar things are mounted on the top side of the chassis, whereas the smaller components are mounted under it. Valves are usually mounted in holders, and there is a host of different holders for various types of valves. The most frequent type encountered is 7-pin (B7G) or 9-pin (B9A) holders, but others do still exist. The best material for valve holders for RF purposes is ceramic. They may be a little more expensive than the cheaper holders made of pertinax or similar materials, but they have better RF properties.

Sometimes practical considerations make the layout decisions for you. For instance, a bunch of coils connected to a switch need to be placed around the switch, which would be accessible from the front panel. You can not place the associated valves until you have placed the switch and the coils. If you use capacitive tuning in a receiver, you must be aware that the capacitor too needs to be accessible from the front panel, and it needs to sit near the coils, etc.

In the olden days, the band switch was of a 'Yaxley' type, where segments of the switch divided the shaft and chassis space around it into segments. Each segment switched a number of coils. This made it possible to place the coils in a natural way along the switch axis. So, RF coils would be in one segment, oscillator coils in another, mixer coils in a third, and so on. Yaxley switches as well as screening boxes for coils are still available for purchase. A different approach was made in some receivers. The coils were changeable. They were consequently mounted in sockets. However, this could be a nuisance in a receiver with many coils. Coils got lost or were placed in the wrong sockets, not to mention the time and effort it took to switch to another band. Additionally, cassettes were made that contained only the coils or tuning circuits. That way you swapped cassettes when you needed to change bands. In some cases, some electronics was also built-in into such cassettes. Each group of coils could thereby be optimised for their respective band.

VALVES REVISITED

Fig 201: Overview of innards of portable radio

There are, of course, many ways of building a receiver. As long as the RF considerations (see the chapter on amplifiers) are observed, only your knowledge, imagination and ingenuity set the limits. The RF and IF stages need, of course, more attention to those rules than the AF and power supply stages.

Below are a couple of examples of the design of the innards of a commercial portable receiver. Portability imposes other restrictions, too. The portable receiver illustrated here is both battery and mains powered. **Fig 201** shows the general overview. It is a good idea to adopt similar principles when planning a receiver.

To the left is the RF portion. Most visible is the tuning capacitor, a two-gang component that tunes both the oscillator and mixer input. Next is the IF amplifier with two IF cans, then the detector and AF pre-amplifier, and the AF power amplifier. The power supply can be seen to the extreme right, with the transformer mounted on top of the chassis. The dark, circular structure on the right is a mains voltage selector. Note how the RF section, IF section and power supply are separated from each other so as not to interact.

Fig 202: Portable radio power supply

174

CHAPTER 12: CONSTRUCTION WITH VALVES

Fig 203: Coils section of portable radio

Fig 202 shows part of the mains power supply. The device with the cooling fins, is the rectifier. The AF output transformer is located next to it, mounted onto the loudspeaker. The big capacitor to the left is the smoothing capacitor.

The RF and mixer stages (shown in **Fig 203**) are located at the other end of the box to avoid influences. The black cylindrical components at the right are the oscillator and mixer coils. Had they been located near the power supply, hum and mains noise would have modulated the signal.

Fig 204 illustrates the RF and IF components. The tuning capacitor (left) and IF transformers (right) are seen from the top of the chassis. The left hand valve is the oscillator/mixer and the middle one is the IF amplifier. The right hand

Fig 204: RF and IF portion of portable radio

175

VALVES REVISITED

Fig 205: Trimmer capacitors of portable radio

valve is the detector/AF pre-amplifier.

Fig 205 shows the trimmer capacitors. For fine adjustment of the oscillator and mixer coils, four trimmer capacitors are provided next to the tuning capacitor, one for the oscillator and one for the RF section per band.

The receiver operates on two bands, MW and LW.

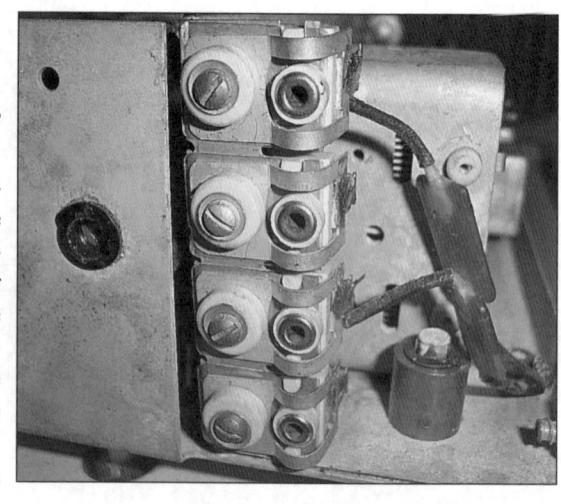

Fig 206 shows that the inside of a portable (or, for that matter, a table top radio) can look quite messy, and it is not always easy to locate the various components.

Neatness has not been an issue here. Note, however, that most components have been mounted onto the valve holders and at an angle to each other, even though these are the AF stages.

Fig 206: Showing functionality before neatness

The Chassis

The chassis is generally made of aluminium and should be as stable as possible. Copper is acceptable too, since copper is a slightly better conductor than aluminium. Iron and steel are not so good for this purpose.

Anyone with some mechanical skills can make a chassis. What you need is the following:

CHAPTER 12: CONSTRUCTION WITH VALVES

Fig 207: Chassis layout before cutting

- Powerful vice
- Two L-profiles, iron, steel or preferably aluminium or brass, which is softer
- A sheet of 5mm plywood
- Sheet aluminium for the chassis
- Work bench
- Hacksaw
- Power drill
- Punch for valve holders and other big things
- Pop rivets or screws and nuts

Most of these can be purchased from your local hardware / diy store.

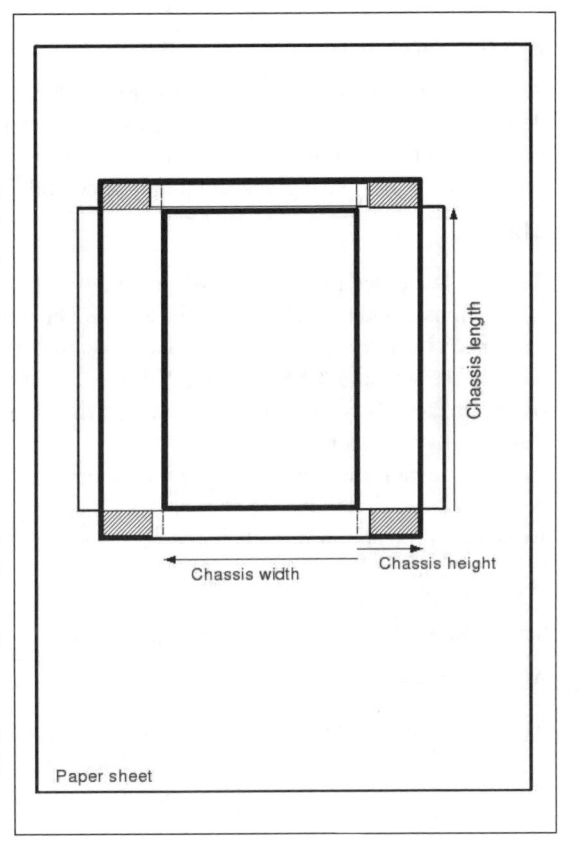

Template

You can start by making a template of the chassis. One example template is shown in **Fig 207**.

- Take a sheet of paper, A3 or A4, squared. Draw the top side of the chassis, scale 1:1.
- Draw according to the sketch and cut out the corners.
- Cut the little 'ears' at the short sides.
- Fold the paper sheet to form a box. This will be your template.
- Do the same thing with your sheet aluminium. The procedure is described below.

Cutting sheet metal

Cutting sheet metal could be done with a hacksaw. However, since the hacksaw leaves rough edges, the following procedure is recommended.

Draw a line along the edge that you want to cut. Use a sharp knife and draw it several times, so it will be sufficiently deep. Then put the sheet between the L-profiles in the vice. Place the plywood along the line and hold it steady. Make sure there is enough space between the line and the plywood sheet. Then bend the sheet about 45 degrees towards the sheet. Move the plywood sheet to the other side of the metal sheet and bend about 20 degrees the other way. Continue bending back and forth until the metal sheet snaps. File the cut smooth.

VALVES REVISITED

Bending sheet metal

Again, draw a straight line along the fold-to-be, using your knife. This time, don't make it too deep. Put the plywood sheet along the line, as described above. Fold the metal sheet in one operation. You will find that the angle now is about 85 degrees. Raise the metal sheet a mm or so and finish the bend.

The walls

Saw the shadowed parts in Fig 207, and then a slit at the corners to get the four 'ears'. Consider the thickness of the sheet. The 'ears' go on the inside of the chassis. Saw two pieces of plywood, one that fits along the sides marked 'Chassis height' and 'Chassis length' above, and the other to fit the top surface of the chassis. Now, fold the edges along the short sides as described above. Use the plywood for support.

Now you can fold the 'ears' that go on the inside. Finally, fold the long sides. Now you need to remove the chassis from the vice. Place the box on top of your work bench upside down. Place the plywood sheet that fits the top area of the box inside it and finish the folding.

Finally, drill holes for rivets or screws through the sides and 'ears'. Rivet or screw. Now you have a sturdy and neat chassis.

Making holes

The best thing you can do is to manufacture a template for the holes of the valve holder. You will find the measures for a B9a (9-pole) socket in **Fig 208**. The centre hole (where the valve goes) should be no more than about 3mm. Place the template on top of the place where you want to place the holder. Remember to turn it the way the pins of the valve are going to be orientated. Drill the centre hole and screw the template in place through that hole to hold it firmly in place. Then drill the two remaining holes. Remove the template.

Now you can drill the centre hole to 10mm, fit the punch and make the hole for the centre of the holder.

A 22.5mm punch fits most 9-pin holders. For 7-pin holders, a 19 mm punch should be used. Other punches should be used for holes through the panel to fit connectors and similar devices.

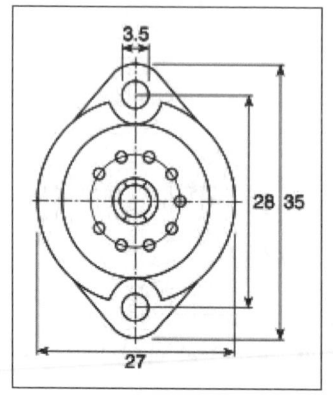

Fig 208: B9A socket measurements (mm)

Sometimes you will need bigger holes. The best way of making those is to drill a series of small holes around the edges of the big one and then file the hole to the desired shape and size.

PCBs

Towards the end of the valve-era, printed circuits began to appear in electronic equipment. However, a certain degree of experience is necessary if you want to avoid the pitfalls involved in RF and hi-fi applications.

13

The Power Supply

THE PURPOSE of the power supply is to provide power to your equipment, using the mains grid as its source. It needs to be dimensioned such that it can handle the voltages and currents needed by your equipment.

A lab supply, however, would need to be a bit more sophisticated than that. Preferably, it would be variable to an extent, insensitive to mains voltage variations, provide 12.6V and / or 6.3V for filaments, and finally, a lower, stabilised voltage for oscillators and the like.

The high voltage at the output of the power supply is denoted 'B+'. This has a historical background from the times when radios were battery powered. The anode battery was the 'B' battery, and the filament battery or accumulator was called the 'A' battery.

For safety reasons, do not make transformer-less power supplies! Do not use auto-transformers for the high voltage(s)!

The Transformer

The most important component in the power supply is the transformer. There are two ways of getting hold of a transformer for valve use: You could buy one from one of the major electronics suppliers, or you could buy a valve radio at a car boot sale for a few pounds.

The transformer needs to provide a high voltage secondary (in the order of 250V) with a centre tap. Also, it needs at least one secondary winding for heater voltages. Preferably, it would have two separate heater windings, one for 12.6V and the other for 6.3V. However, if both voltages are available, the most common configuration is a centre tapped 12.6V winding. If there is a winding for 4V rectifier valves, this is a bonus.

The Rectifier

Again, two alternatives are available - a valve rectifier or a semiconductor bridge. Undoubtedly, a semiconductor bridge is easier to find. Some receivers used metal rectifiers. These are effectively bridges and perform full wave rectification.

Safety Warning
The domestic mains supply has potentially lethal voltages and currents. Also, the voltages produced by a valve power supply can be very high and therefore dangerous. Great care must be taken when working with these supplies.

Full wave rectification is to be preferred, since the resultant hum voltage has double the mains frequency. This simplifies the design of the smoothing filter.

The Filter

A transformer with a centre tap and a valve rectifier yields full-wave rectification. The raw output from the valve needs to be filtered so that the power supply output is smooth. If not, hum is going to be the predominant component in the voltage.

There are several ways of arranging a filter. The most common method is a pi-filter. The filter may contain a choke, but, in cheaper radios, a power resistor was used instead. The choke method gives better smoothing and is to be preferred.

Some sources express concern that valves may be damaged if the high voltage is connected to the valve before it has the chance of heating up properly. Other sources don't. For that purpose, a relay with a delayed on-action is used in many designs. The delay may be built into the relay proper, or generated by electronics.

Voltage Regulation

Different designs may have different requirements regarding B+ voltages. For instance, battery powered portable radios frequently used 45V or 90V. In order to satisfy different requirements, it is beneficial to be able to vary the B+ voltage.

There is series and parallel stabilising. The simplest form of stabilising a voltage where small currents are involved is a glow lamp connected between the output and ground with a resistor (R) in series. The voltage is taken from the intersection between the lamp and the resistor. It works because the voltage drop across the lamp is essentially the same even if the current through it varies (within limits, of course). The same principle applies to the special voltage stabilising valves that are used in connection with anode voltage stabilisation for oscillators, eg the 0B2 (that is 'zero'-B-2) and similar devices. More than one lamp or valve can be used if connected in serial to increase the voltage. These are not variable.

The resistor R is calculated by Formula 30:

$$R = \frac{Vi_{max} - Vs}{Is_{max}}$$

Formula 30

where

Vi_{max} is the maximum voltage from the power supply, $Vs =$ is the stabilised voltage and Is_{max} is the maximum permissible current through the valve/lamp.

Fig 209 is the block diagram of a glow tube stabilised power supply.

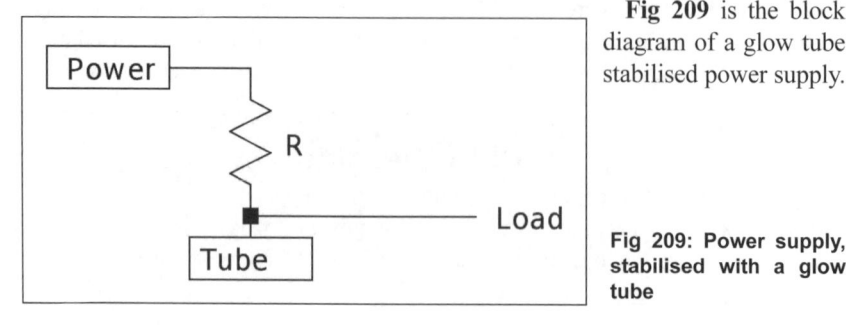

Fig 209: Power supply, stabilised with a glow tube

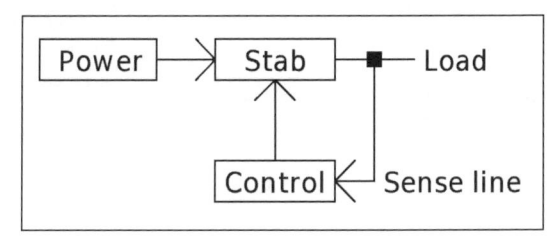

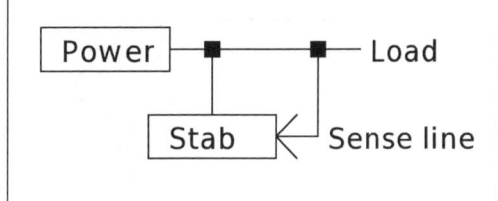

CHAPTER 13: THE POWER SUPPLY

(left) Fig 210: Power supply with voltage stabilisation using a valve

(right) Fig 211: Voltage stabilisation with valve parallel with the output

The load current should not be more than (Ismax - Ismin), preferably more to allow for variations in the unstabilised voltage.

A more active form of stabilisation uses a valve which is connected in series with the output of the power supply (**Fig 210**). It functions like a variable resistor, whose value increases if the voltage tends to increase. Its grid bias is controlled by a control valve which senses variations in the voltage. The series valve needs to be able to handle the load current.

Another way is a valve connected in parallel with the output, loading it more when the voltage tends to increase (**Fig 211**). There is no control valve involved.

Series stabilisation is the one most commonly used, even though it may be a bit more complicated.

Bleeder Resistor

Due to the fact that an electrolytic capacitor can hold a considerable charge for some time after the power supply has been disconnected and turned off, a bleeder resistor is necessary for security reasons. The value of the resistor is fairly high but should empty the capacitors within reasonable time.

Negative Supply

In valve technology it is not very common to use high negative voltages. However, a dual supply, one positive and one negative and both isolated from chassis would allow you to choose whether you run your equipment on one single supply, the sum of them both, or one positive and one negative supply. All you have to do is to build two separate supplies according to the description above, make sure that nothing is connected to chassis and incorporate the appropriate controls and meters. The stabilisation circuits remain the same.

In a dual power supply you need to watch out for the electrolytic capacitors that are mounted in cans. In many cases, an internal connection is done between the negative side and its mounting bracket. Using such capacitors would ground the power supply, even if everything else is floating. Avoid them, if at all possible.

Also, each power supply should be fitted with a fuse for safety and to protect your equipment.

Practical Implementations

For the rectifier part of the power supply, there are four alternatives:

- Dual valve diodes
- Semiconductor diodes
- Semiconductor bridge rectifiers
- Metal rectifiers

VALVES REVISITED

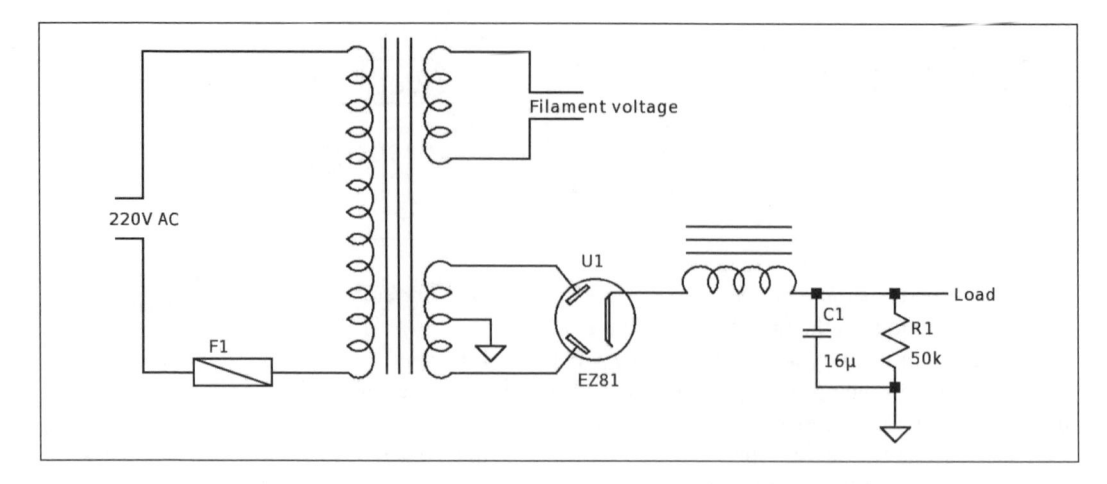

Fig 212: An unregulated power supply

The first alternative delivers far less current than the next two can do. However, the filter choke and transformer also need to be able to deliver more current if the full capacity of the semiconductors is to be used. Metal rectifiers are hard to come by these days.

Valve equipment doesn't normally require much anode current. The suggested approach is therefore that, even if semiconductors are used, the rest of the power supply be dimensioned as if a valve were used.

A choke is preferred even for smaller supplies, as they filter the voltage better than a resistor would do. Additionally, a choke drops the output voltage less.

A standard valve power supply may look as illustrated in **Fig 212**. The transformer can be purchased from larger electronics component suppliers, or taken from an old valve radio. If you cannibalise an old radio, make sure there is an inductor in the filter. Some manufacturers used a resistor instead, which gives less ripple suppression.

Fig 213: The semiconductor equivalent of above supply

Fig 213 shows a semiconductor diode rectifier circuit which is virtually identical to the one above. The only difference is that semiconductor diodes are used

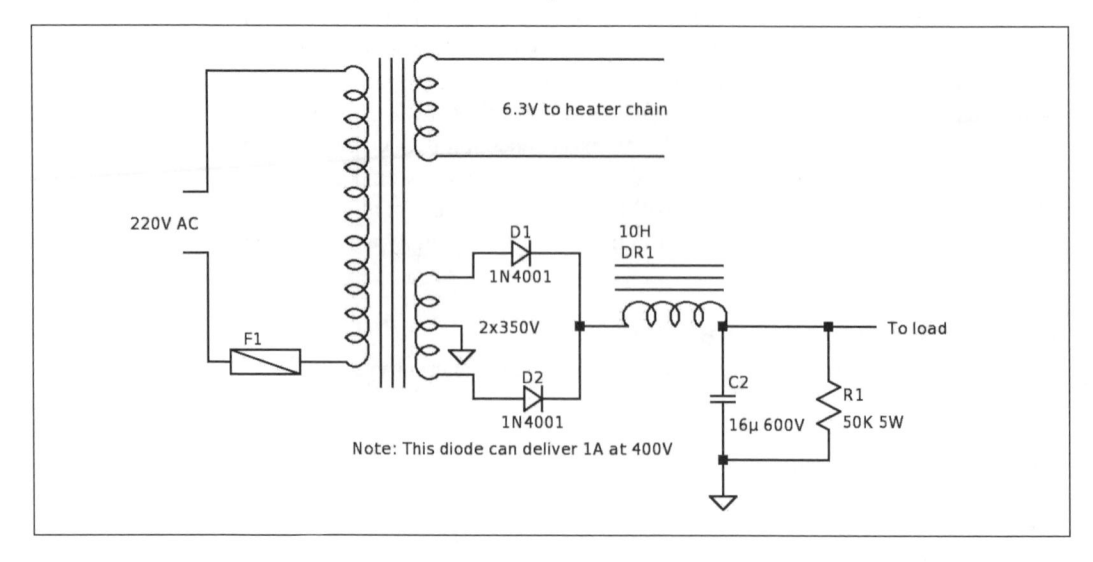

182

CHAPTER 13: THE POWER SUPPLY

Fig 214: Using a semiconductor bridge

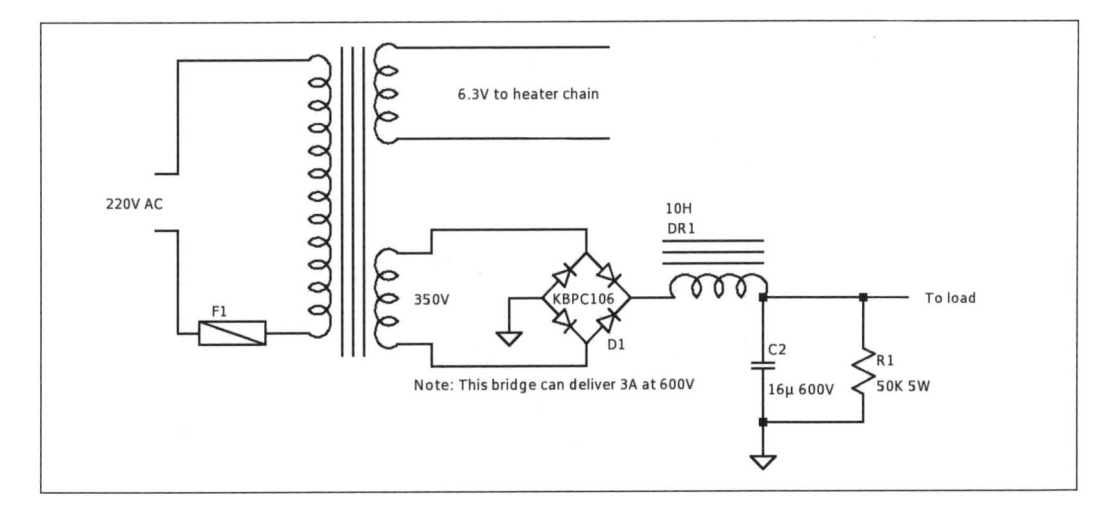

instead of the valve. Also note that the centre tap at the secondary of the transformer is needed for valve operation.

The electrolytic capacitor (C2) is polarised. Make sure the plus pole is connected to the output and the negative pole to the negative side (ground).

Finally, the third alternative, shown in **Fig 214**, the bridge rectifier. This configuration does not require a transformer centre tap.

It is possible to use valves in a bridge configuration if the centre tap is missing, but it is a high price to pay. It would require at least three separate valves and gains nothing compared to the full-wave rectifier circuit at the top.

All three circuits assume the same output voltage from the transformer.

A filter with a capacitor at the input can also be used. However, if a valve is used for the rectifier, there are restrictions on the value of the capacitor. A choke input at the filter reduces the power-on peak current, hence reducing the stress on the rectifier. Besides, a choke input filter is capable of delivering a higher current to a heavier or varying load. **Fig 215** shows such a filter.

Two chokes and capacitors in the filter would give a better filtering.

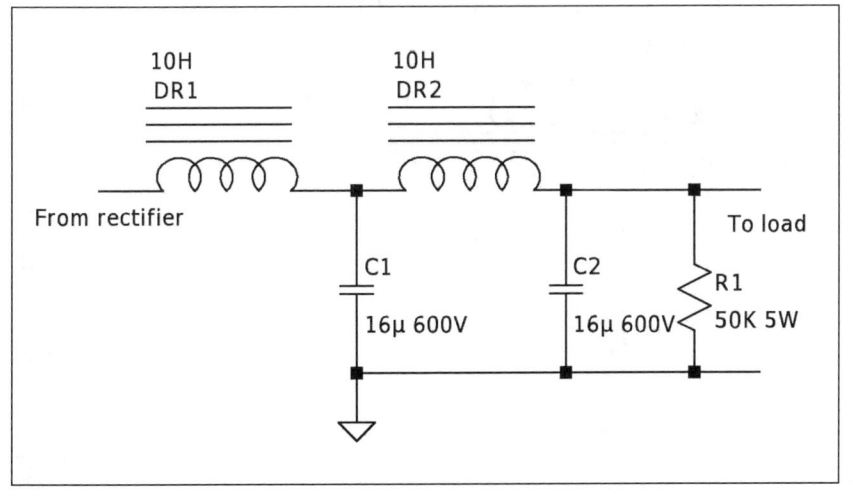

Fig 215: A double pi-filter output

183

VALVES REVISITED

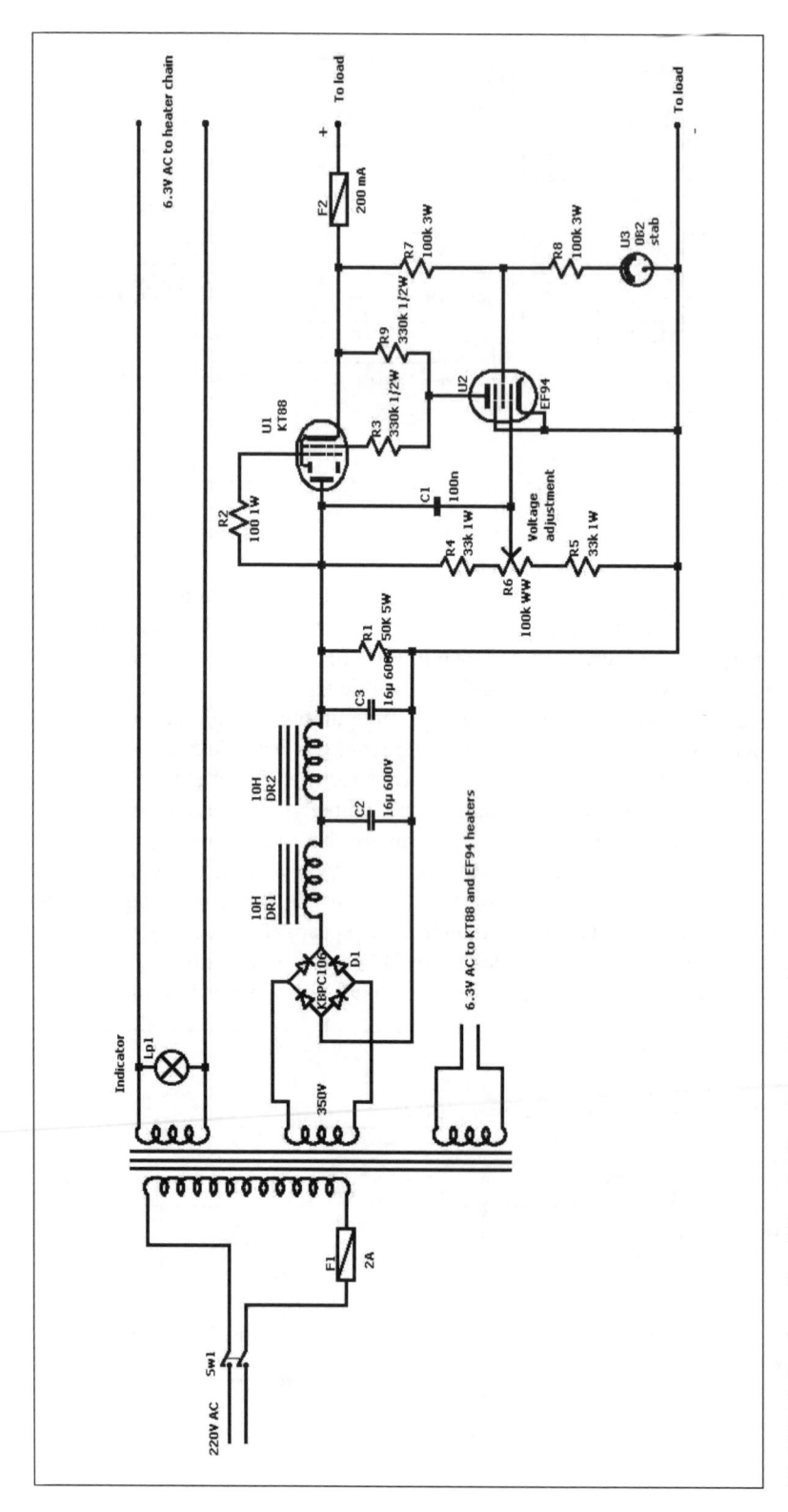

Fig 216: Full schematic of the stabilised power supply

CHAPTER 13: THE POWER SUPPLY

The reason for the better performance is that each pi-filter is in effect a low pass filter. Two filters connected in series yield a steeper filtering curve, dampening the unwanted frequencies more. In this case, the unwanted frequencies are the 100Hz, resulting from the full wave rectification, and its overtones.

Building the power supply

This power supply has a bridge rectifier and a single valve stabiliser. There should be no problems in finding the components even today. The complete schematic is shown in **Fig 216**. A commercial power supply is illustrated in **Fig 217**.

Ref.	Type no	Description
C1	--	capacitor, 100nF
C2	--	capacitor, 16µF 600V
C3	--	capacitor, 16µF 600V
D1	KBPC106	diode bridge
Lp1	--	indicator lamp, 6.3V AV
R1	--	resistor, 50k 5W
R2	--	resistor, 100 1W
R3	--	resistor, 330k 1/2W
R4	--	resistor, 33k 1W
R5	--	resistor, 33k 1W
R6	--	potentiometer wire wound, 100k
R7	--	resistor, 100k 3W
R8	--	resistor, 100k 3W
R9	--	resistor, 330k 1/2W
U1	KT88	power tetrode
U2	EF94	pentode
U3	0B2	stabiliser valve
Dr1	--	choke, 10H 150mA
Dr2	--	choke, 10H 150mA
F1	--	fuse 2A slow
F2	--	fuse 200mA fast
Sw1	--	two-pole mains switch
Tr1	--	transformer primary: 230V AC 50Hz secondaries: 2 x 350V, centre tapped 2 x 6.3V (12.6V)

Components list for the stabilised power supply

Construction

Building the power supply is straightforward. A standard valve technology metal (aluminium) chassis is used. Terminals and controls are on the front panel, along with instruments to show current and voltage, and a mains socket on the rear.

The chassis can be used for grounding components. Components are best mounted onto tag strips, as shown in **Fig 218**.

VALVES REVISITED

Fig 217: Commercial power supply for lab use

Fig 218: Components mounted onto tag strips

Stabiliser circuit

The power supply uses series stabilisation. The voltage can be set from 0 to full output minus the voltage drop across the stabiliser valve, if a negative control voltage is used for the control valve. However, this requires a separate supply for the control voltage.

A power triode is sometimes used for stabilising purposes, but a power pentode or tetrode works just as well. Due to the high amplification factor, a pentode is best used as the control valve.

CHAPTER 13: THE POWER SUPPLY

The main requirements of the stabiliser valve are:

- Low ra. The voltage drop across the valve needs to be kept as low as possible.
- High Gm. The Gm of the valve determines the degree of stabilisation that can be achieved.
- High Pa. Particularly at low voltages, the current could be rather high. The valve needs to be able to dissipate the power drawn from the power supply.

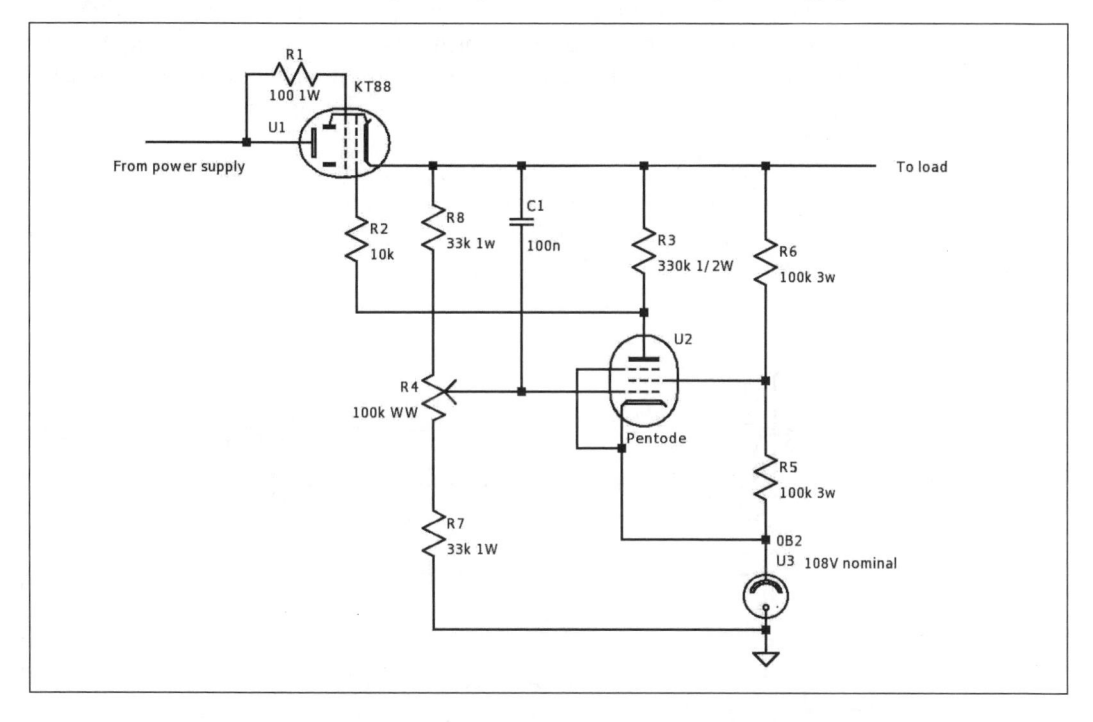

Fig 219: Series stabiliser for power supply

The stabiliser circuit is shown in **Fig 219**. A popular audio output valve, the KT88, performs the stabilisation. Its screen grid voltage is taken from the unregulated voltage, and its control grid bias comes from the anode of the control pentode.

The EL34 is another popular power amplifier valve still in use that could be used to regulate a power supply. It would be connected in a similar way.

The control pentode's cathode voltage is taken from the stabiliser valve U3, which keeps the voltage at a nominal 108V. The control grid receives its bias from the voltage divider R7 / R8 / R4. Virtually any pentode can be used. It is important, though, that the amplification factor of the pentode be as high as possible.

If the output voltage tends to decrease, so does the control grid voltage of the pentode. This change is amplified by the pentode, and its anode voltage increases. This causes the control grid voltage of the stabiliser valve to increase, and its cathode current increases, compensating for the output voltage change.

The change of the screen grid voltage of the pentode has little impact on the anode current.

VALVES REVISITED

The resistors R7 / R8 may have to be adjusted to set the operating point of the pentode control grid. The potentiometer R4 is used to vary the output voltage. Capacitor C1 may not be made too big, or instability in the circuit may occur.

An alternative design

To boost the current capacity of the power supply, two output valves can be connected in parallel. The 12E1 is an old valve that is still available in places. **Fig 231** shows such a stabiliser.

The principle is still the same - a pentode is used to amplify the difference between the desired voltage and the actual voltage. It controls the output valves so that the voltage is brought back to (or close to) the desired value. In addition, this supply has a trip relay for over-current protection. It trips at about 230mA.

This stabiliser requires some calculations before assembling it. The meters can be virtually any type, but you need to know the current for maximum deflection. Also, the trip relay, which in this schematics has a resistance of $1k\Omega$, needs to control the resistor R12 and R13.

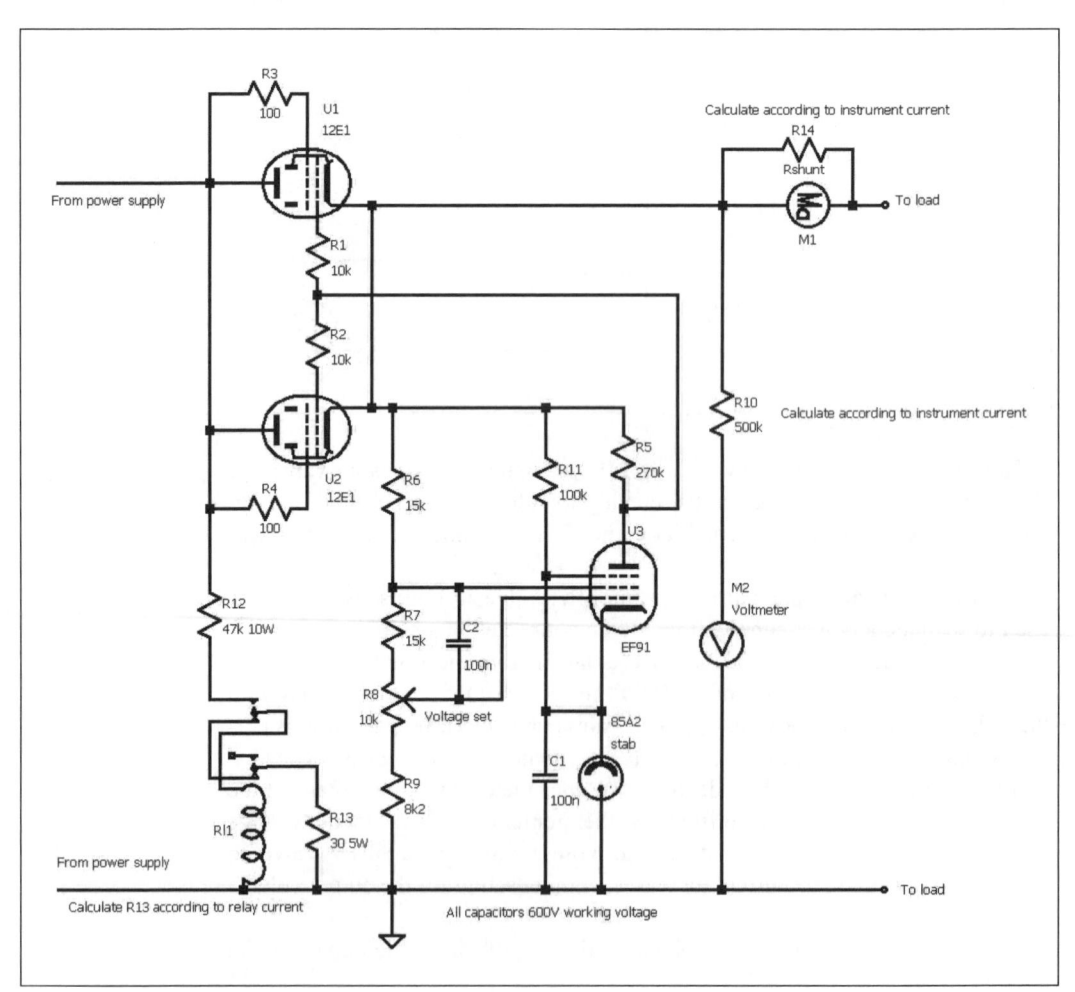

Fig 231: Alternative stabiliser for power supply

188

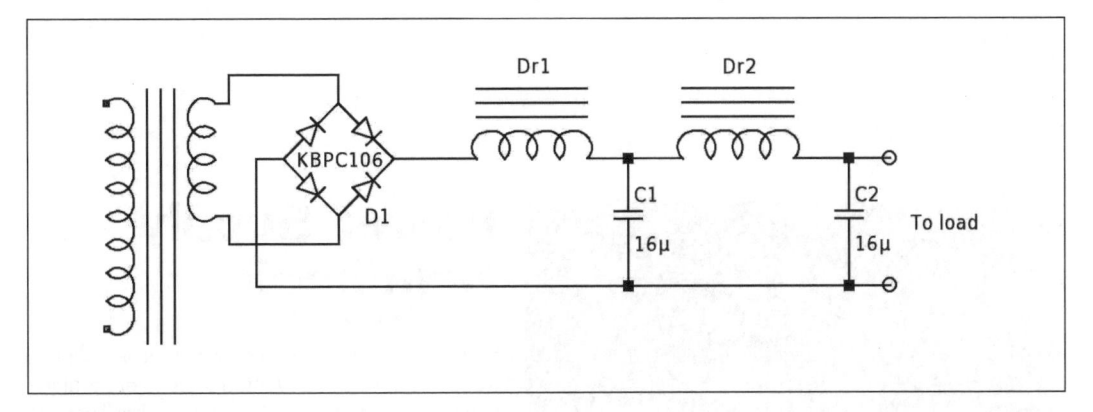

Fig 221: An unstabilised bias power supply using a semiconductor bridge

Bias supply

In some cases it is necessary to generate a separate negative grid bias, in particular in conjunction with class B or C power amplifiers, be they intended for audio or RF transmitters. To this end, it would be convenient to include such a bias supply into the design.

Normally, a grid bias does not draw current. However, it does happen, so provisions have to be made for this.

The most convenient way to create a bias voltage is to use the heater winding of the mains transformer, rectify it, filter it and stabilise it. If your transformer has a 12.6V winding, it is suggested you use that. The supply design is extremely simple, as you can see in **Fig 221**.

Note that the supply does not use the chassis as a ground point. It is a good idea to be able to choose freely whether the voltage taken from the supply is to be positive or negative. The terminals would be mounted on the front panel, isolated from the metal. A stabiliser IC may be fitted after the last capacitor if desired. The rectifier bridge shown would in the past have been a dual valve diode.

Power Supplies for Test Equipment

You have a number of alternatives, and they are about how to set up your research and development area power needs.

You can choose to:
- Let each instrument have its own, built-in power supply
- Make provisions in your lab supply to connect heater and anode voltages to external equipment
- Build all instruments in one big box and let them all share the same built-in power supply

The first alternative is undoubtedly going to be the most expensive one. The second alternative may lack plugs for all instruments you need to run at any one time. It appears as if the third is the most economical and flexible one, provided the box is big enough to house all instruments you need, and the power supply is powerful enough.

Power Supply Handbook
By John Fielding, ZS5JF

Have you ever wondered how your power supplies work? Have you ever wanted to build or modify a power supply but haven't had the confidence? Do you need a supply that is difficult to find or expensive to buy? The *Power Supply Handbook* answers all of these questions, and much more.

This book provides all that is required to understand and make power supplies of various types. *Power Supply Handbook* is written in an easy to approach style by the electronics guru John Fielding, ZS5JF. This book explains how to select each of the components in a power supply, how the various types of circuit work and how to measure the finished supply. *Power Supply Handbook* covers a range of converters and also deals in detail with the ubiquitous switch-mode supplies. There are also chapters on high voltage supplies, batteries and chargers, and test equipment. Packed with this and much more, *Power Supply Handbook* teaches the reader how to be confident with building, maintaining and using power supplies of all types.

From the new home constructor looking for a straightforward guide through to those seeking a practical reference book, all will find this book a useful and a must for their bookshelf.

Size 240x174mm, 288 pages
ISBN 9781-9050-8621-4

ONLY £15.99

RSGB shop

Radio Society of Great Britain
3 Abbey Court, Fraser Road, Priory Business Park, Bedford, MK44 3WH
Tel: 01234 832 700 Fax: 01234 831 496

www.rsgbshop.org

E&OE All prices shown plus p&p

14

Oscillators

NO OTHER UNIT in a receiver or transmitter has been designed in as many variants as the oscillator. It is not surprising, though, because the oscillator is the heart of every transmitter and receiver since the 1930s when the superheterodyne was invented. Besides, different oscillators have somewhat different qualities. There is Armstrong, Butler, Colpitts, Hartley, cathode coupled, ECO, and a host of other circuits. It is impossible to cover them all in a single chapter, but we will take a look at some of them.

Every oscillator (including those built with semiconductors) consists of an amplifying element, a tuned circuit and components to achieve positive feedback. The amplification of the stage only needs to be sufficient to compensate for the losses of the components.

The tuned circuit may be a coil and a capacitor, or a crystal, or some other selective device. The Q of the tuned device needs to be as high as possible to ensure stability. For comparison, a crystal may have a Q of 2000 or more, whereas a highly stable tuned circuit has a Q in the order of a couple of hundred.

Building Oscillators

The rules stated in the chapter on construction apply to oscillators as well. However, due to the high requirements of a good oscillator, certain additional rules need to be followed. These points are almost all related to stability. Each one is as important as any of the others. It only takes one of them to fail for the oscillator to have less stability than you intended.

The oscillator should have low anode voltage, about 150V or lower. Higher voltages may cause impurities in the signal, which causes harmonics to be generated. These can not easily be eliminated elsewhere in the design. The anode voltage should be stabilised. If you wish your oscillator to be really stable, you need to rectify and stabilise the heater voltage too because the cathode emission varies with the temperature of the filament.

Build the oscillator into a die cast box. These are mechanically very stable. Keep the oscillator screened from other stages. You can put a tiny 'computer-processor fan' inside the box. This helps keep the temperature low and evens out temperature gradients inside the box. Make sure the box has air holes at the opposite wall to help the air circulate. Use rigid wires.

Choose a low frequency range for your oscillator. It gives you a good degree of stability, measured in Hz (which is an audible amount of drift). Normally, instability is measured in ppm (parts per million).

It is imperative that the tuned circuit has a high Q. The higher the Q, the better stability. It is also important that the tuned circuit faces an environment of high impedance. Choose a design that is known for good stability. Different designs have different degrees of inherent stability.

Gluing the components around the tuned circuit is a good idea. Electrical stability depends heavily on mechanical stability. By carefully moving a screwdriver close to the various components while the oscillator is running, and at the same time listening to the signal, you can easily find out which components are most sensitive. Also, you may find that one side of a capacitor is more sensitive than the other. Turn the least sensitive side towards the chassis when you glue it.

If you need fixed capacitors above 200pF, use metallised ceramic capacitors with zero temperature coefficient, and connect them in parallel until you get the desired value. This also gives you the benefit that the losses in the capacitors will be connected in parallel, thereby decreasing them.

Keep the oscillator away from loudspeakers and transformers in your design. The oscillator signal may otherwise be modulated with hum or sound vibrations. The valve itself can be screened. It would not protect it from magnetic influences, but it will protect against electric fields. You can still find valve holders with screens on the market.

If you use capacitive tuning (which is simpler), use the best tuning capacitor you can find. After all, a tuning capacitor is a mechanical device, and its stability influences the stability of the oscillator. Thick vanes, ball bearings, ceramic fixtures, and a wide distance between the vanes are some of the aspects to consider.

If the oscillator design allows for large capacitors in the tuned circuit, you will gain the benefits of being able to swamp stray capacitances (including valve capacitances), and using a small high-Q coil (fewer turns and hence less resistance) and thereby improved stability.

Try to avoid things like PLLs to improve stability. Such techniques tend to increase frequency modulated noise. Also, try to avoid varicaps for tuning purposes. Most semiconductors such as diodes and bipolar transistors are inherently noisy. Besides, the voltage across the varicap will vary with the RF oscillations, causing the capacitance to vary.

Sometimes it is beneficial to connect the coil to its environment from a tap part way down the windings. This can lower the impedance, and hence disturb the circuit less. Be aware, though, that a tapped coil has somewhat different properties from a coil without taps.

Do not use overtones (harmonics) from the VFO to increase the frequency. As you multiply the frequency, you also multiply the instabilities. It is better to mix the VFO signal with a signal from a crystal oscillator and use the sum frequency.

Avoid trimmer capacitors. They are mechanically unstable and easily affected by temperature variations.

Place the components such that they don't affect each other electrically. Overrate components. If you need a quarter-watt resistor, choose a half-watt or one-watt resistor instead. Higher wattage resistors generate less heat.

Keep heat generating components away from the tuned circuit and components connected to the tuned circuit.

Switches are mechanical components. and therefore difficult to use successfully in VFOs. If you need to switch frequency range, mix the signal with a crystal controlled oscillator.

If you are unlucky, the switch could affect the crystal oscillator too, but to a lesser degree.

Under 'frequency measurements' you will find a description of a GDO (grid dip oscillator). It is a very useful tool when building oscillators, easy to construct, easy to make stable and has few components. It is recommended for beginners.

Spectral Purity

Choose an oscillator design that is known to be spectrally pure. It is important where you take the output signal from, and where you connect it to.

If you can take it from the tuned circuit, you will get the purest signal your design allows for. However, do not load the circuit. A small coupling coil, loosely coupled, to a cathode follower would be a good choice.

A cathode follower can be built with an exceptionally high input impedance but with a low output impedance and is therefore a good driver. The follower should be located in the immediate vicinity of the oscillator. (There is a section of this book dedicated to cathode followers).

The coupling coil has to be built as stable as the primary coil. Avoid, if you can, direct couplings from the tuning circuit to the output.

Finding the spectral purity of the oscillator

There is a very simple way of finding out how pure the signal is. Make a series tuned circuit out of a coil and a variable capacitor. Connect it to the output of the oscillator with a big resistance in series - 1MΩ or so should do. Connect an oscilloscope to the intersection between the coil and the resistor as shown in **Fig 222**.

Tune the circuit to the fundamental frequency of the oscillator. You should see the signal almost disappear when the circuit is correctly tuned.

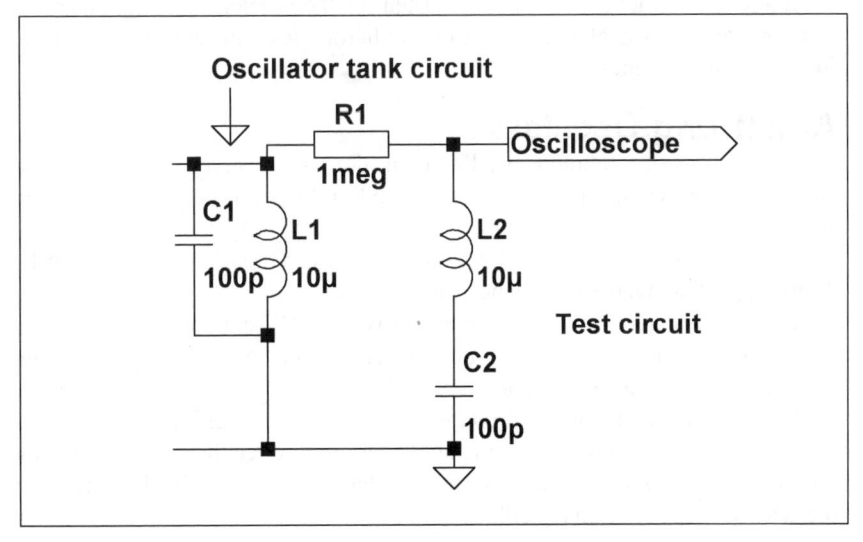

Fig 222: Set-up to test spectral purity of an oscillator

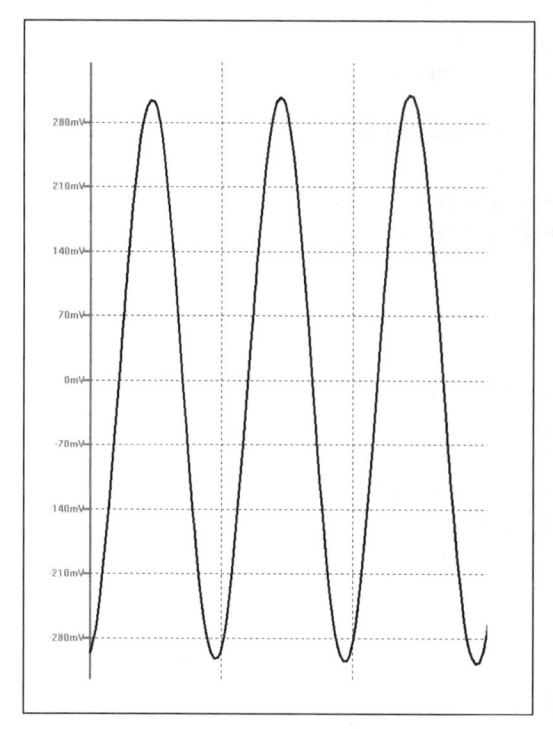

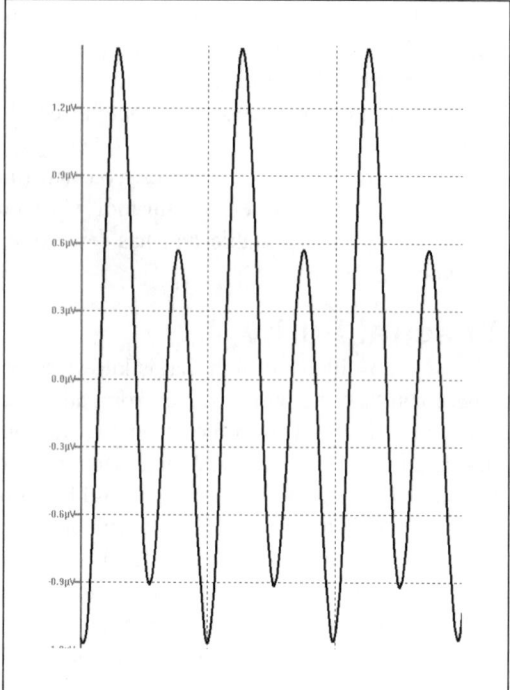

Fig 223: The oscillator output containing the fundamental as well as some harmonics

Fig 224: Having filtered out the fundamental, these lower-level harmonics are revealed

What remains on the 'scope are the overtones (harmonics). The level of these should be as small as possible.

The original signal is illustrated in **Fig 223**. The harmonics with the original filtered out are shown in **Fig 224**. Note from the scale on the left that the level of these is much lower than the fundamental.

You can find out the purest output point of the oscillator the same way.

An alternative method would be to listen for the harmonics using a receiver and compare the signal strengths from the harmonics with the strength of the fundamental frequency.

Armstrong Oscillator

By far the most common oscillator in domestic valve radios was the Armstrong, shown in **Fig 225**. It is simple to build but was never known for its stability.

The oscillator was invented by Edwin Howard Armstrong (1890 - 1954). Armstrong also invented frequency modulation.

Domestic receivers were used to tune in LW and MW stations, and sometimes short wave, but they were all intended for reception of AM transmitters, so the requirements for stability was not very high.

The oscillator works on the principle of an amplifier with inductive positive feedback. The triode constitutes the amplifier and the feedback is arranged by the two coils L1 and L2. It is important that the coils be connected as shown in the schematics, or it won't oscillate.

CHAPTER 14: OSCILLATORS

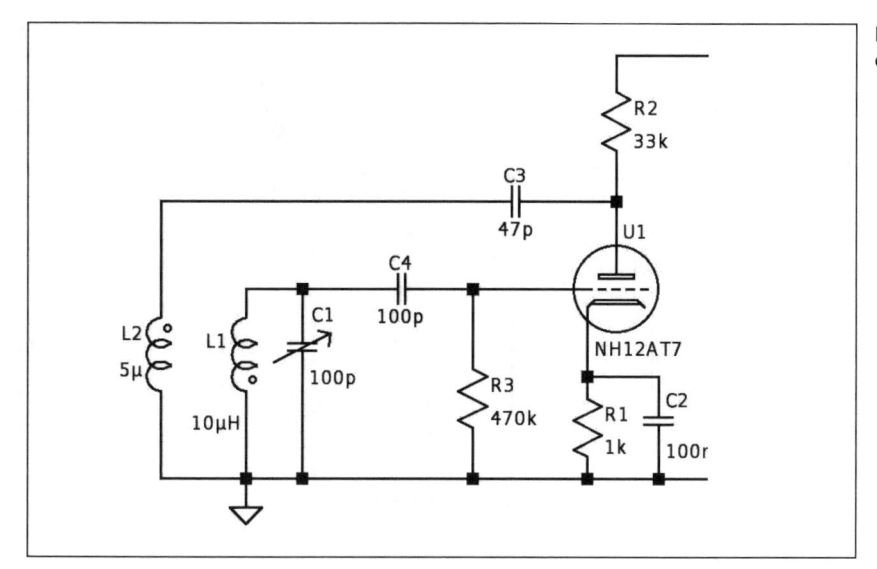

Fig 225: Armstrong oscillator

The design was generally incorporated in a mixer valve, such as an ECH81 or similar. Therefore, the signal was internally taken from the grid of the triode, where the spectrum is reasonably pure.

Colpitts Oscillator

Edwin Henry Colpitts (1872 - 1949) invented the Colpitts oscillator. It consists of two capacitors in series across the tuned circuit. The intersection between them is connected to the cathode, which is grounded via a choke. Note that the grid bias is created by R1 and C4 in series with the grid. This means in turn that the grid draws current at the positive peaks of the signal. **Fig 226** shows a Colpitts oscillator.

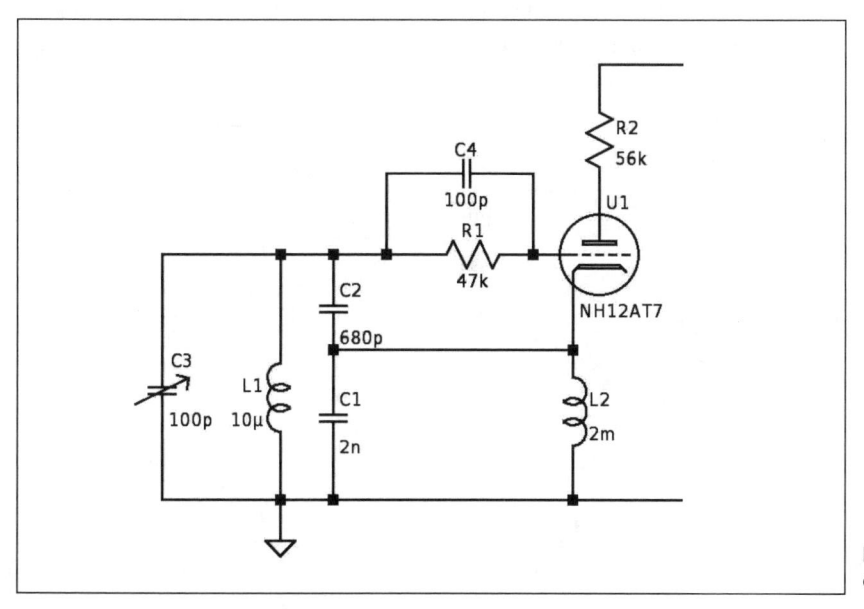

Fig 226: Colpitts oscillator

195

Fig 227: Seiler oscillator

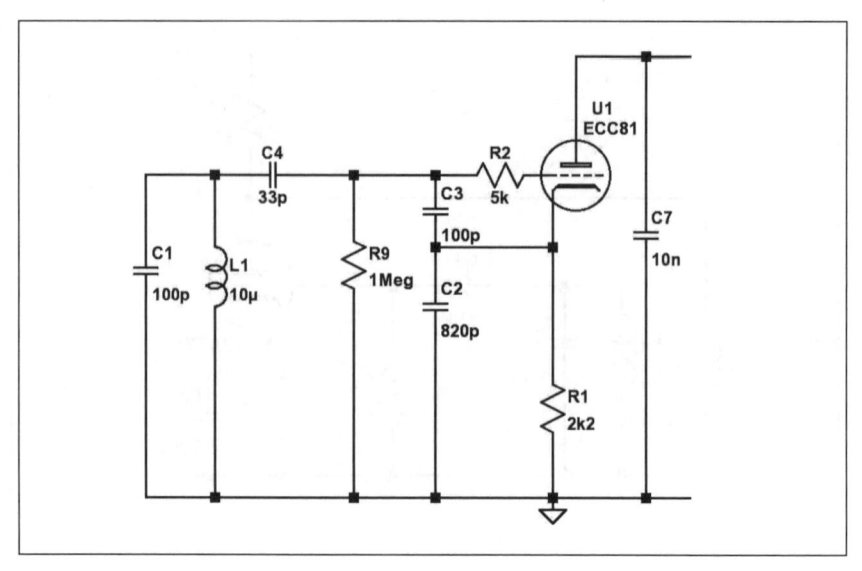

Seiler Oscillator

The Seiler oscillator is basically a Colpitts with an additional coupling capacitor between the grid circuit and the tuned circuit, as you can see in **Fig 227**. This isolates the tuned circuit from some of the stray and valve capacitances. The valve is connected in a cathode follower configuration, which makes the input impedance high and the capacitances low. A two capacitor voltage divider provides the positive feedback. First described by E O Seiler in the 1941 issue of *QST* magazine.

Hartley Oscillator

Ralph V L Hartley (1888 - 1970) invented the Hartley oscillator (**Fig 228**), and patented it by 1915. It resembles slightly the Colpitts above in that the cathode is not connected in the usual way to ground, but connected to a tap on the inductor.

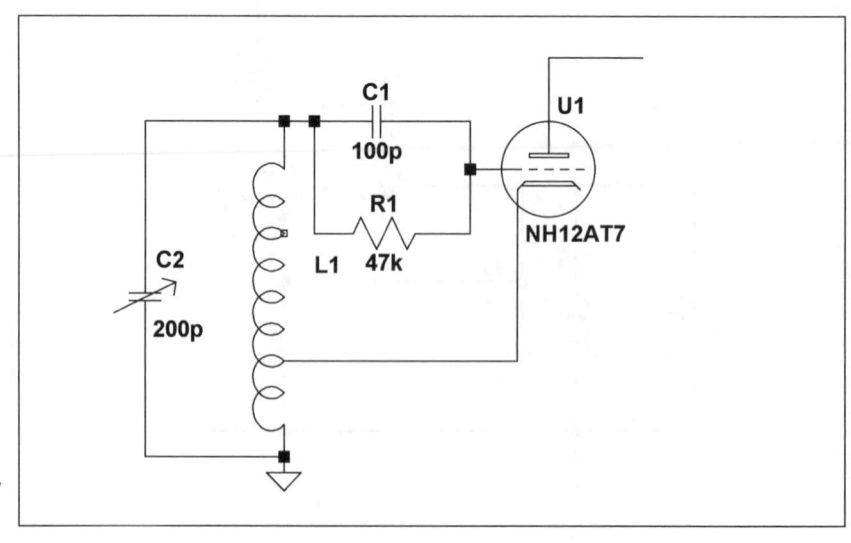

Fig 228: Hartley oscillator

The Hartley output is not particularly pure in its spectrum. However, its output level is constant over a fairly wide tuning range.

Clapp Oscillator

This oscillator is similar to the Hartley oscillator. The difference is that the the Clapp uses a series tuned circuit. It was invented in 1948 by James Kilton Clapp.

The Clapp oscillator, seen in **Fig 229**, can be made very stable within a limited tuning range, and has a very good spectral purity.

The capacitors in the voltage divider should be in the order of 1nF and equal. They swamp the stray capacitances as well as the internal capacitances of the valve.

Note that the grid bias arrangement is different from that of the Hartley and Colpitts oscillators.

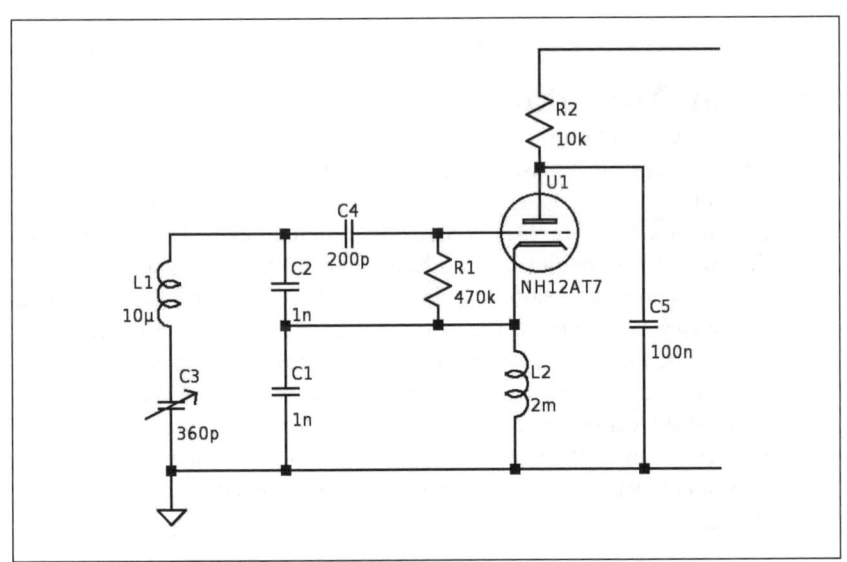

Fig 229: Clapp oscillator

Vackar Oscillator

The Vacker oscillator is named after the Czech engineer Jiri Vackar, who presented it in 1949. (It is also known as the Tesla oscillator, after Nikola Tesla, 1856 - 1943).

The oscillator is more complicated than those previously shown. It is a one-band oscillator, since four capacitors (C1, C2, C3 and the tuning capacitor) and the coil need to be changed in order to change bands.

The tuned circuit is connected in series, as in the Clapp. Vackar enthusiasts praise it for its stability and spectral purity.

The Vackar oscillator exists in a number of configurations. **Fig 230** shows one of them. The frequency determining components, albeit in various degrees, are L1, C1, C4, and C3. Any one of those components can be used to tune the oscillator.

The higher the amplification factor of the valve, the better is its frequency stability.

Fig 230: Vackar oscillator

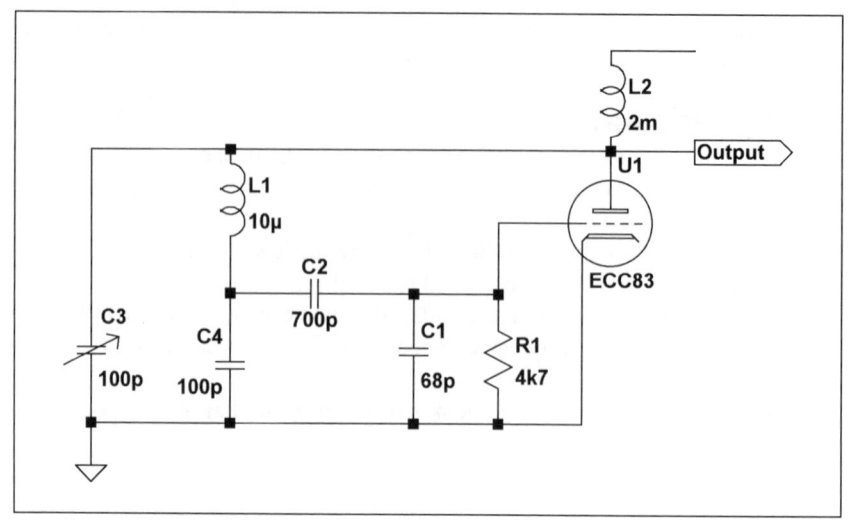

Franklin Oscillator

Charles Samuel Franklin (1879 - 1964) invented the Franklin oscillator. It is interesting, because it is based on a totally different principle - the bistable multivibrator.

A multivibrator (see **Fig 231**) is basically a digital device consisting of two amplifiers where both outputs are fed back to the other device's input:

Its output has no resemblance to a sine wave whatsoever, because both amplifiers are driven into saturation. The capacitors and resistors form time constants, which determine the frequency. What Franklin did was to connect a tuned circuit at the feedback path of the multivibrator, as shown in **Fig 232**.

Note the small value capacitors connected to the tuned circuit. Two wires twisted together will suffice. However, stability depends on the values of those capacitors, and twisted wires can not be expected to be very stable. Increasing

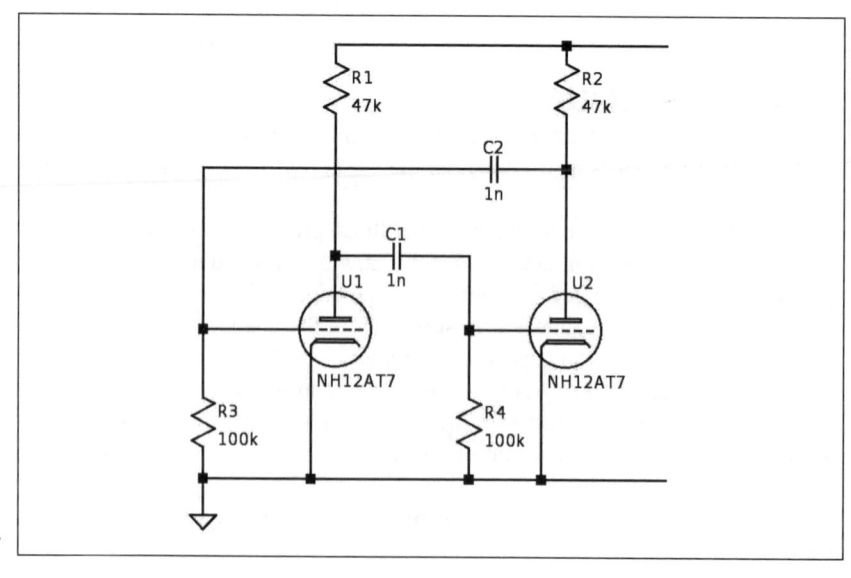

Fig 231: Multivibrator

CHAPTER 14: OSCILLATORS

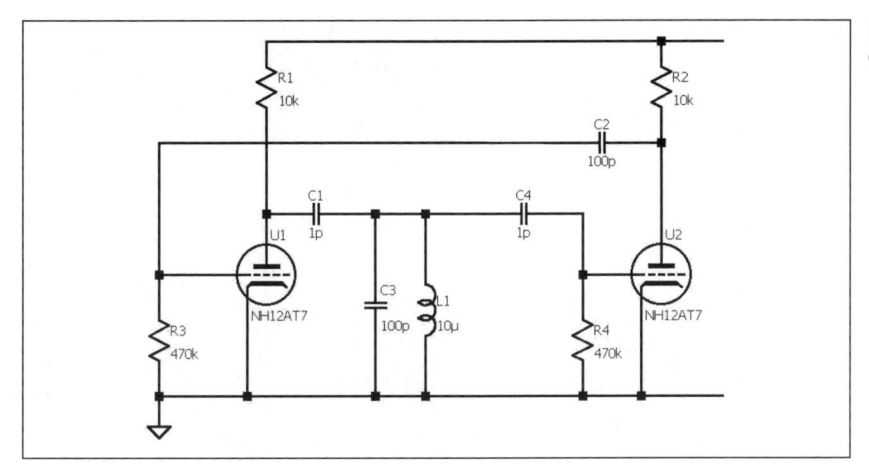

Fig 232: Franklin oscillator

their values will decrease the impedance seen by the tuned circuit, hence decreasing stability of the oscillations.

The signal needs to be tapped from the tuned circuit.

Cathode coupled oscillator

Towards the end of the 1990s an investigation was launched in the UK. Its objective was to find out how stable a valve VFO can be built using modern standard over-the-counter components.

A variable frequency oscillator (VFO) needs to be able to read and write a tuned circuit. As we know, a tuned circuit is a device which is very sensitive to loads. Therefore, both the input and the output of the amplifier need to be high impedance. One amplifier with those properties is the cathode coupled amplifier, and this was the design chosen

The oscillator, illustrated in **Fig 233,** consists of a dual triode with the two halves joined together at the cathodes and with a common cathode resistor. The tuned circuit consisted of a number of fixed capacitors, a variable capacitor and

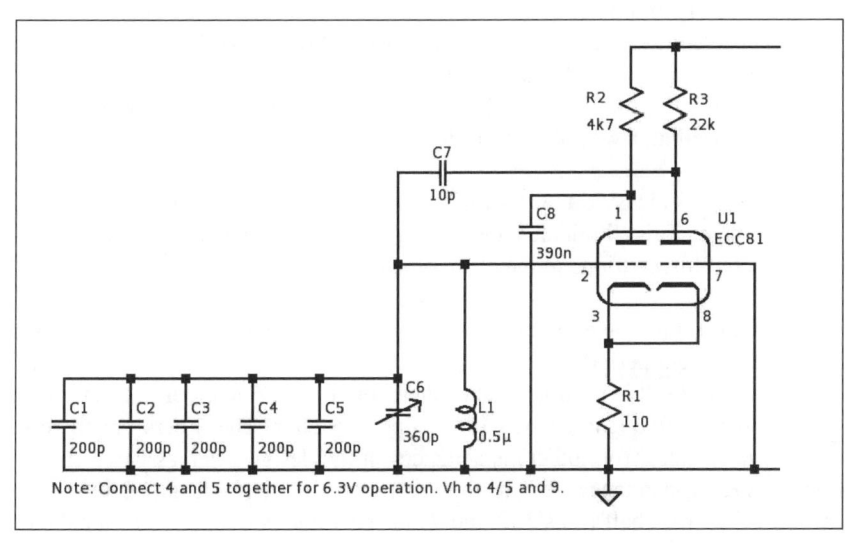

Fig 233: Cathode Coupled oscillator

199

VALVES REVISITED

Ref.	Part No.	Description
C1	--	capacitor, 200pF
C2	--	capacitor, 200pF
C3	--	capacitor, 200pF
C4	--	capacitor, 200pF
C5	--	capacitor, 200pF
C6	--	variable capacitor, 360pF
C7	--	capacitor, 10pF, high voltage!
C8	--	capacitor, 390nF
L1	--	inductor, 500nH
R1	--	resistor, 110 1W
R2	--	resistor, 4.7k 1W
R3	--	resistor, 22k 1W
U1	ECC81	dual triode

Components list for the cathode coupled oscillator

a coil. Component values were varied during preliminary tests to ensure greatest possible stability and spectral purity. Once all values had been established, serious testing began.

The supply high voltage was stabilised at 110 volts. The oscillator was allowed a warm-up period of at least 30 minutes before measurements began. The frequency was measured using a frequency counter with a resolution of 0.1Hz. A reading was taken every ten seconds. Each test period lasted for 30 minutes.

Some additional measures were taken in order to additionally increase stability after each test period. Some of the measures above and beyond normal RF considerations were:

- The coil windings were cemented to the coil former using epoxy cement.
- Those components (capacitors) that were directly connected to the tuned circuit were similarly cemented together and fixed to the chassis.
- Very thick (1 mm diameter) copper wire was used for connections.
- The oscillator was placed inside an enclosure to protect it from drafts in the room.
- The final measure that was taken was to rectify and stabilise the heater voltage. This eliminated those frequency variations that appeared connected to variations in the grid voltage.

The final result: The oscillator instability was reduced to less than 10ppm over the measurement period!

Let us analyse this oscillator. The capacitors C1 - C6 and coil L1 determine the oscillating frequency. C6 is the tuning capacitor. The others are there to swamp influences from other capacitances in the design, such as valve capacitances, stray capacitances, etc, as they participate in the frequency determining process. The first half (pins 1, 2 and 3) of the valve is connected as a cathode follower. This reduces the input capacitance of the valve and yields a high input impedance.

Adding all tuning capacitors together gives a total capacitance of 1.36 nF. This is very high! On the other hand, the larger the tuning capacitance, the more stray and valve capacitance will be swamped. There is another benefit in high capacitances in the tuned circuit: The coil! It can be made very small, thereby reducing the number of turns in the winding and the length of the wire required. This in turn means that its resistive as well as inductive losses are small, and the resulting Q of the coil can be kept high, adding to stability.

Capacitors C1 - C5 are disc ceramics. The reason there are so many of them is that disc ceramics are made with a zero temperature coefficient (TC) in values up to 200pF.

The 360pF value of the variable capacitor compared to the total of 1000pF in the fixed capacitors is very small. Therefore, the tuning range is limited to about 1MHz (from about 6MHz to about 7MHz) with the component values shown. Additionally, this configuration makes the tuning appear nearly linear. If the tuning capacitor is connected in series with a fixed capacitor, the tuning range will, of course, decrease even more.

Connecting several capacitors in parallel is beneficial, since you also connect the losses of the capacitors in parallel, thereby decreasing them significantly. This applies to all applications, be it valve or semiconductor based, RF, IF or AF.

R1 is the common cathode resistor, through which the cathode currents from both halves flows. Therefore it has to be half the value of a single cathode resistor to give the correct grid voltage on both grids.

The second stage is a grounded grid amplifier. Its input impedance is low and the output impedance is high. The tuned circuit is connected to the input of the first stage and the output of the second. Therefore the environment impedance of the circuit is high. This is beneficial to its operation - the Q factor of the circuit can be kept high.

Overall, this is a good oscillator with high stability and good spectral purity. Simulations reveal that the nearest overtone is over 60dB down from the fundamental at the tuned circuit!

Tapping the signal is best done from a secondary winding into a cathode follower to isolate the oscillator from load variations, load capacitances and low impedances. It may be tempting to take the signal from the cathodes, since this already is a low impedance point. However, it is not recommended! The spectral purity at this point leaves a lot to be desired.

The tuning range can be expanded by using a ganged tuning capacitor and connecting the sections together, or by omitting a couple of the fixed 200pF capacitors. This enables you to cover a wider spectrum of frequency ranges with far fewer coils.

Due to the high stability achievable, this oscillator can be used as a local oscillator in a communications receiver or a transmitter. It can also be employed as a signal generator, as described in the next chapter.

Technical Topics Scrapbook
- All 50 Years
By Pat Hawker, G3VA

For 50 years Pat Hawker MBE, G3VA has written his hugely popular Technical Topics column in *RadCom*. *Technical Topics Scrapbook - All 50 Years* brings together the whole 50 years of Technical Topics as a data CD along with the fifth and final compilation book of the column covering the years 2005 to 2008.

Technical Topics Scrapbook - All 50 Years includes all the words, pictures and line drawings from column produced presented exactly as they first appeared. The CD contains easy to use PDF copies of the pages arranged in annual collections from 1958 to 2008. Whilst the book reproduces the pages from the years 2005-2008 in the popular scrapbook format. This set draws together Pat Hawker's blend of clippings and contributed material, linked by his own unique commentary, enriched by a lifetime of fascination for the technical aspects of radio.

This CD and book set provides a fascinating insight into the development of radio and amateur radio in particular. Technical Topics Scrapbook -50 Years is also an invaluable collection of experimental antennas, circuit ideas and radio lore for everyone keen on radio and electronics. If you are interested in Pat Hawker and Technical Topics, this book and CD set is a must have and the definitive collection.

Size 210x297mm, 176 pages
ISBN 9781-9050-8639-9

ONLY £14.99

RSGB shop

Radio Society of Great Britain
3 Abbey Court, Fraser Road, Priory Business Park, Bedford, MK44 3WH
Tel: 01234 832 700 Fax: 01234 831 496

www.rsgbshop.org

E&OE All prices shown plus p&p

15

A Signal Generator

ONE VERY IMPORTANT instrument in the lab is a signal generator. At its heart is a very stable oscillator. The circuit in **Fig 234** is based on the cathode coupled oscillator described in the last chapter.

As a signal generator, it needs to be connected to a host of test objects without the frequency being affected. A high impedance cathode follower at the output of a cathode coupled oscillator has been added, and the fixed capacitors removed for a larger frequency range per coil.

A switch, S1, switches between the frequency ranges. A different coil is switched into the circuit for other bands. Only one coil is shown in the diagram. The signal generator is powered by 150V stabilised.

The cathode follower to the left is of a high impedance type. It picks up the signal directly from the oscillator tuned circuit via a small capacitor, C2. The smaller the value of this capacitor, the better. The grid leak is a 1MΩ resistor connected to the intersection of the cathode resistors R5 and R6.

The output can be taken from a 500k potentiometer in the cathode circuit in place of R8 to enable the output amplitude to be varied.

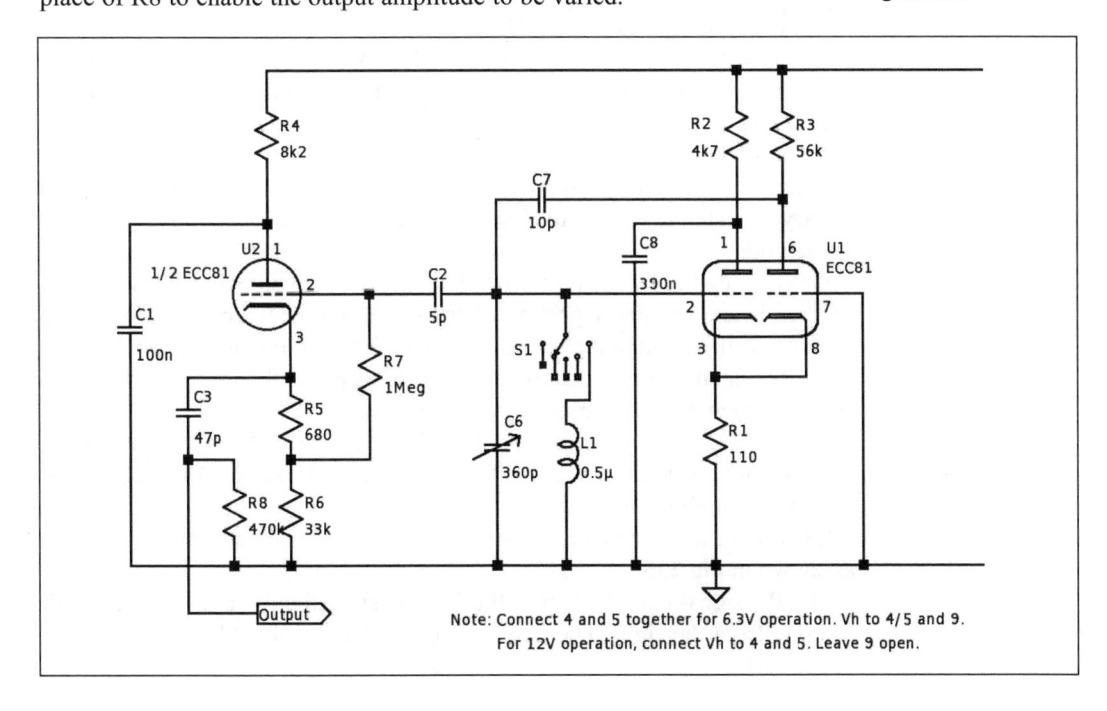

Fig 234: Signal generator

Note: Connect 4 and 5 together for 6.3V operation. Vh to 4/5 and 9.
For 12V operation, connect Vh to 4 and 5. Leave 9 open.

Ref.	Part No.	Description
C1	--	capacitor, 100nF
C2	--	capacitor, 5pF
C3	--	capacitor, 100pF
C6	--	capacitor, 360pF
C7	--	capacitor, 10pF
C8	--	capacitor, 390nF
L1	--	inductor, as needed
R1	--	resistor, 110
R2	--	resistor, 4.7k
R3	--	resistor, 56k
R4	--	resistor, 8.2k
R5	--	resistor, 680
R6	--	resistor, 33k
R7	--	resistor, 1M
R8	--	potentiometer, 500k linear
U1, U2	ECC81	oscillator and cathode follower

Components list for the signal generator

The anode of the first half of the oscillator is decoupled via C1, which needs to handle the high voltage.

The build is straight-forward and should not impose any problems. The oscillator is very easy to put to work.

Modulating the Signal

In some situations it is beneficial to be able to amplitude modulate the output from the signal generator. One such situation is, for instance, during signal tracing (see the chapter on fault finding). To do that, a modulator is necessary, along with an AF signal.

There are several ways of modulating an RF signal. A very simple method is to vary the cathode voltage of a valve with the AF signal as the RF signal is fed to the grid.

If you study the curves of a triode (see chapter 3), you will find that there is an area below the linear portion of the curve, where the curve is bent. This means that if you drive the grid bias voltage into that region, the resulting signal at the anode will be distorted. The bend slants differently from the linear part, and so yields a different, lower, amplification factor.

Choosing a working point near the bend of the curve and varying the grid bias up and down a suitable amount will cause the amplification of the stage to vary - we have amplitude modulation.

It takes one triode and a few components to add a modulator to the generator, as shown in **Fig 235**.

The RF signal comes in from the left, whereas the AF is connected to the cathode from the right. The peak voltage of the AF should be no more than about 1.5V.

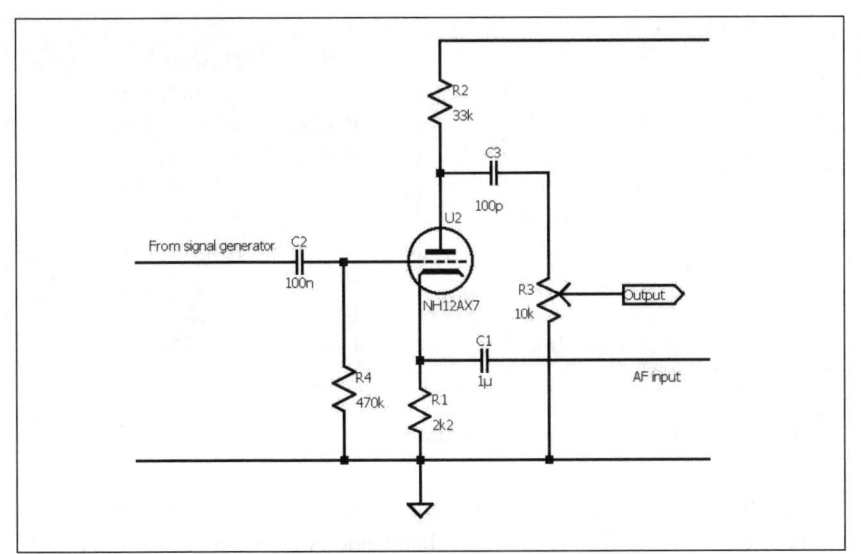

Fig 235: Amplitude modulator for the signal generator

AF Sine Generator

We also need an AF signal to modulate the RF signal. A fairly simple device is the R-C generator. It is an amplifier connected to a network of resistors and capacitors that change the phase by 180 degrees. The resulting phase shift from anode to grid will be 360 degrees, which fills one of the conditions for oscillation.

The output of the network is fed back to the valve grid, and the result will be oscillations with a frequency that is determined by the component values within the network. The values have been chosen to give a frequency of about 400Hz. The oscillator is shown in **Fig 236**.

Fig 236: Sine wave generator for AF

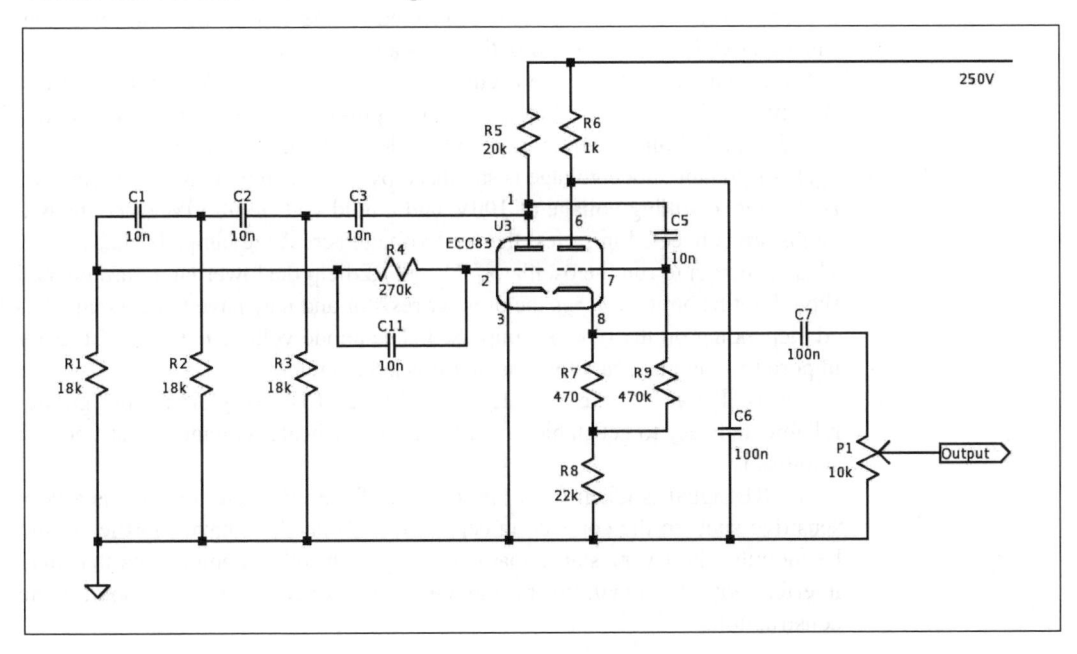

Ref.	Description	Ref.	Description
C1	capacitor, 10nF	R2	resistor, 18k
C2	capacitor, 10nF	R3	resistor, 18k
C3	capacitor, 10nF	R4	resistor, 270k
C5	capacitor, 10nF	R5	resistor, 20k
C6	capacitor, 100nF	R6	resistor, 1k
C7	capacitor, 100nF	R7	resistor, 470
C11	capacitor, 10nF	R8	resistor, 22k
P1	resistor, 10k	R9	resistor, 470k
R1	resistor, 18k	U3	ECC83

Components list for the AF sine generator

One dual triode and a handful of components is all it takes. The network consists of C1 - C3 and R1 - R3. The left hand side of the valve acts as the oscillator, whereas the right hand side is a cathode follower buffer. The output can be fed to the modulator, or connected to a socket on the front panel so the audio oscillator can be used when designing and building AF amplifiers.

To avoid over-modulation, a 10k trim pot should be used to set the modulation depth.

Complete Modulated Signal Generator

A complete modulated signal generator is presented in **Fig 245**. The valves are type ECC81, an RF dual triode. This is one of those valves that can be run on either 6.3V or 12.6V heater voltage. Pin 9 is connected to the centrepoint of the two heaters. If 12.6V is used, it should be connected to pins 4 and 5. If 6.3V is used, pin 9 needs to be used and connected to one end of the heater winding. The other end is connected to both pins 4 and 5.

The schematic contains the oscillator (U1), the oscillator's cathode follower (U2, pins 1, 2 and 3), the AF oscillator (U3, pins 6, 7 and 8), the modulator (U2, pins 6, 7 and 8) and the output cathode follower (U3, pins 1, 2 and 3).

The oscillator anode voltage is stabilised by the two glow lamps Lp1 and Lp2. Both have a striking voltage of 100V, and a hold voltage of 50V. Since the two lamps are connected in series, the total voltage across the lamps is about 100V. This is enough to run the oscillator. R1 and R2 help the lower lamp turn on, and should not be omitted. R7 is the dropper resistor and may have to be re-calculated, depending on the type of lamp used. The anode voltage *per se* is not really important - the important thing is that it is kept stable.

The oscillator is of the cathode coupled type. It is a very simple design, but reliable and easy to get stable, in particular if the heater voltage is rectified and stabilised.

The RF signal is taken from the top end of the tuned circuit. This is a very sensitive spot, so the connecting capacitor (C3) needs to have short leads and be mounted in a very stable manner, away from other components that may interfere with the signal. It is best glued to some fixed non-metallic point in the construction.

CHAPTER 15: A SIGNAL GENERATOR

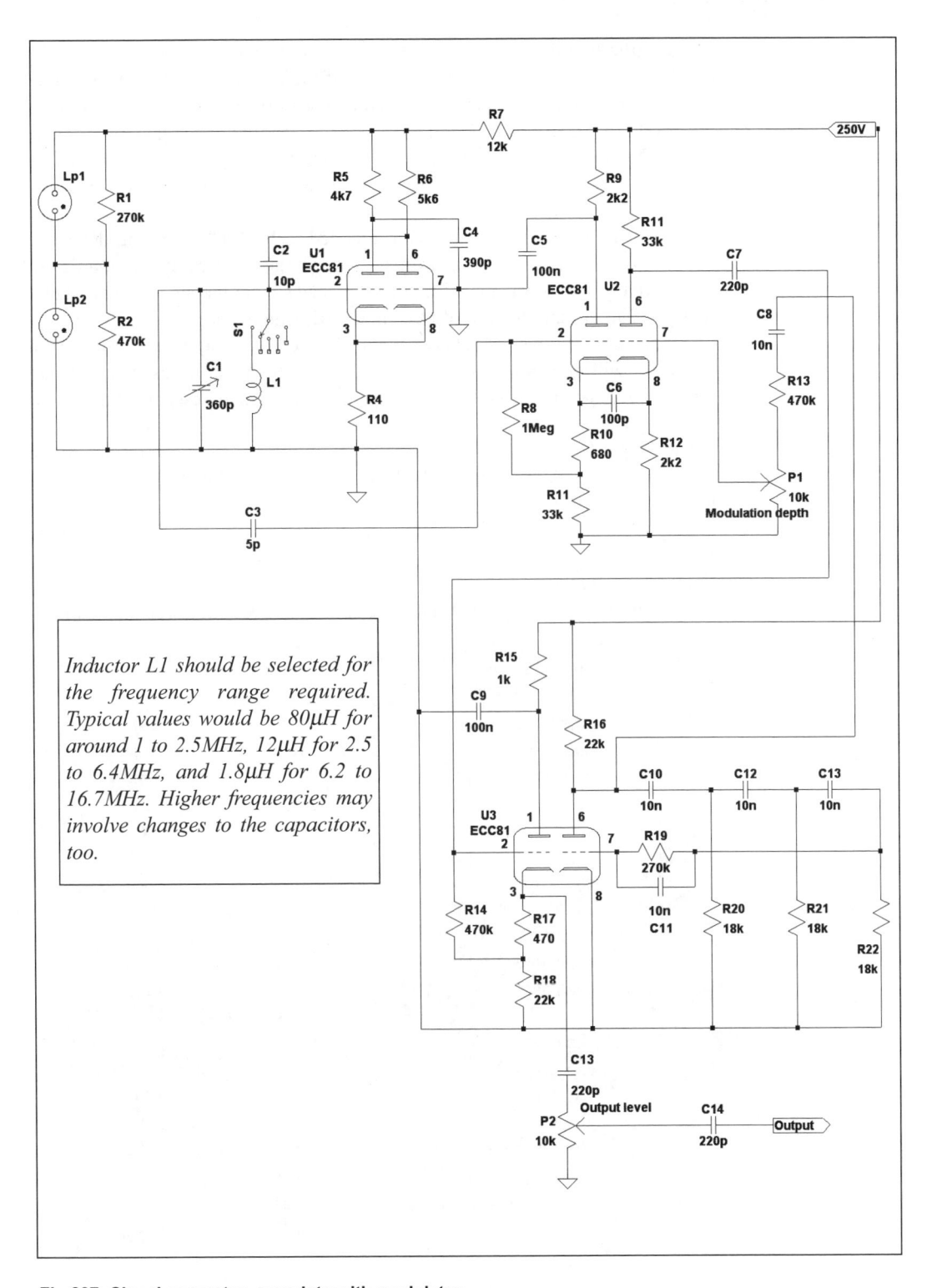

Inductor L1 should be selected for the frequency range required. Typical values would be 80µH for around 1 to 2.5MHz, 12µH for 2.5 to 6.4MHz, and 1.8µH for 6.2 to 16.7MHz. Higher frequencies may involve changes to the capacitors, too.

Fig 237: Signal generator, complete with modulator

207

VALVES REVISITED

The cathode follower has a very high input impedance, due to its design. Its output signal is connected to the cathode of the modulator, which is the other triode in the same envelope. The modulator grid receives the signal from the AF oscillator (U3, pins 6, 7 and 8). This is an oscillator that generates a sine wave. The amplitude is dropped by the combination R13 / P1 and should be about 2V peak to peak. If you get a different result, resistor R13 may be adjusted accordingly.

P1 sets the modulation depth, and intended to be 100%. Turning down the depth to zero enables you to use the signal generator unmodulated. Alternatively, a two-way switch could be placed so that it connects the modulator's pin 7 to ground, instead of passing the AF signal.

Connecting a clipper stage between the AF oscillator output and terminals on the panel will generate a square wave signal, which can be used as a signal injector. Very useful for fault finding radios and also for assessing the quality of hi-fi amplifiers.

The output from the modulator is taken from the anode and connected to the output stage, which is another cathode follower to allow for some load and long leads. The output level is set by P2.

Only the components involved with the oscillator are critical. They should be 1% components, and the resistors should be over-rated. Sensitive capacitors should have zero temperature coefficient.

The unit is powered by 250V DC (see earlier chapter on power supplies).

Ref.	Description	Ref.	Description
C1	capacitor, 360pF	R2	resistor, 470k
C2	capacitor, 10pF	R4	resistor, 110
C3	capacitor, 5pF	R5	resistor, 4.7k
C4	capacitor, 390pF	R6	resistor, 5.6k
C5	capacitor, 100nF	R7	resistor, 12k
C6	capacitor, 100pF	R8	resistor, 1M
C7	capacitor, 220pF	R9	resistor, 2.2k
C8	capacitor, 10nF	R10	resistor, 680
C9	capacitor, 100nF	R11	resistor, 33k
C10	capacitor, 10nF	R11	resistor, 33k
C11	capacitor, 10nF	R12	resistor, 2.2k
C12	capacitor, 10nF	R13	resistor, 470k
C13	capacitor, 10nF	R14	resistor, 470k
C13	capacitor, 220pF	R15	resistor, 1k
C14	capacitor, 220pF	R16	resistor, 22k
L1	see diagram	R17	resistor, 470
Lp1	glow lamp, see text	R18	resistor, 22k
Lp2	glow lamp, see text	R19	resistor, 270k
P1	potentiometer, 10k lin	R20	resistor, 18k
P2	potentiometer, 10k lin	R21	resistor, 18k
R1	resistor, 270k	R22	resistor, 18k

Components list for the complete signal generator

CHAPTER 15: A SIGNAL GENERATOR

Adding Functions

The signal generator's functionality can be expanded. For instance, provisions can be made for:

- Connecting an external AF source to the modulator
- Disconnecting the modulation altogether
- An output from the sine oscillator (via the output cathode follower) to perform AF measurements
- Inserting a clipper stage between the AF oscillator and the modulator and/or output to enable AF bandwidth measurements using a square wave signal

Square wave output

A clipper stage turns the sine wave into an approximation of a square wave. The principle is to feed an amplifier stage with such a large signal that it is over-driven. This causes the amplifier to cut off the peaks of the signal, and what remains is a square wave signal. Such a signal can be used to check qualities of an AF amplifier, such as phase response, input and output impedances, and frequency response. A clipper stage for the signal generator is shown in **Fig 238**.

The output is connected to the modulator or the output cathode follower. The input is taken from the anode of the AF oscillator.

Note the large anode resistance. It reduces the dynamic response of the stage. Resistor R1, between the cathode and B+, raises the grid bias to a level where the output signal is reasonably symmetric. If, instead, pulsed output is desired, R1 may be omitted.

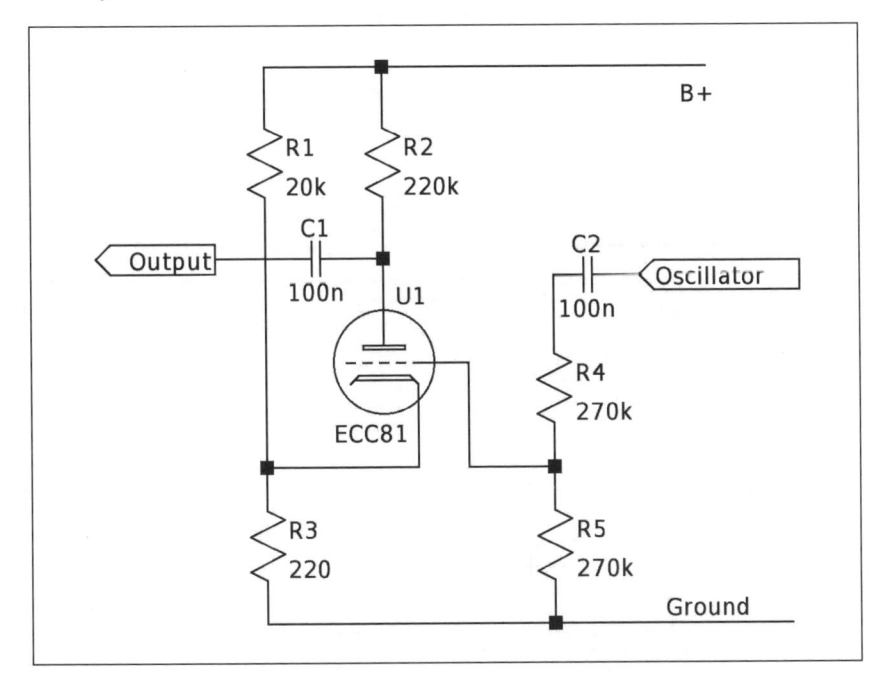

Fig 238: Clipper stage to turn sine into square

VALVES REVISITED

Ref.	Description
C1	capacitor, 100nF
C2	capacitor, 100nF
C3	capacitor, 100nF
R1	resistor, 270k
R2	resistor, 270k
R3	resistor, 20k
R4	resistor, 220k
R5	resistor, 560
R6	resistor, 47k
R7	resistor, 470k
U1	ECC81

Components list for the clipper

Also note that the voltage divider R4 / R5 connected to the oscillator anode now have a different ratio. This is to increase the input amplitude to the clipper stage somewhat. The output from the clipper should look something like the curve in **Fig 239**.

Adding another stage of clipping improves the curve, as you can see in **Fig 240**. The rise and fall times are shorter and the corners sharper. The shorter the

Fig 239: Output from clipper

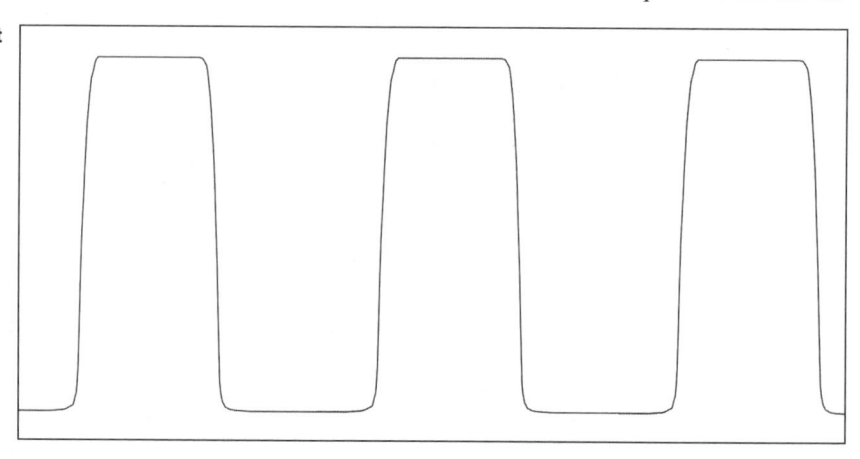

Fig 240: Result of clipping twice

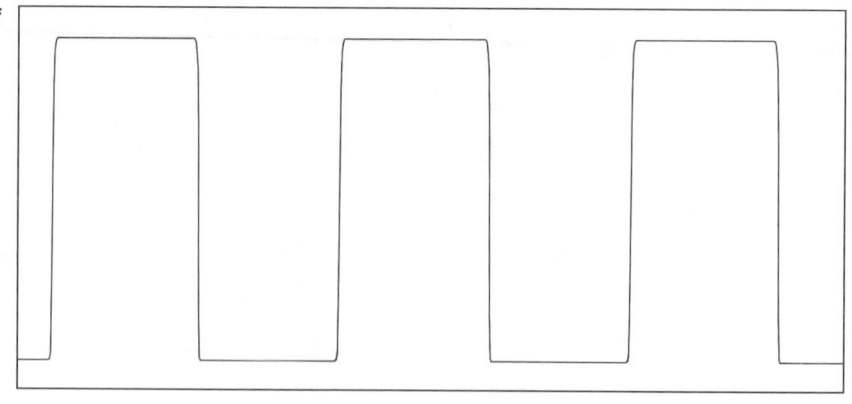

CHAPTER 15: A SIGNAL GENERATOR

Fig 241: Double clipper

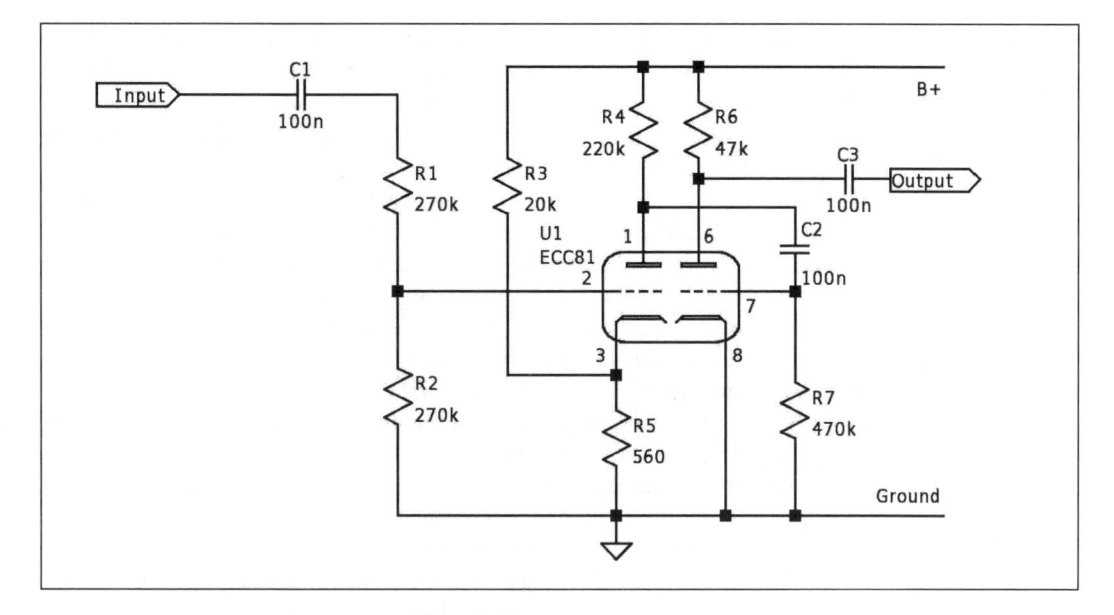

rise and fall times and the sharper the corners, the better signal for measurements. **Fig 241** shows that this is easily done:

Even though the signal has a low fundamental frequency, the sharp corners and short rise and fall times contain the high frequency overtone components. The ECC81 is a high frequency small signal triode, and preserves these high frequencies.

Wobbulator

When attempting to align the IF chain of a receiver, nothing beats a wobbulator! It is basically a frequency modulated oscillator which sweeps a range wide enough to cover the bandwidth of an IF amplifier. The modulating frequency is a sawtooth signal to allow the frequency to sweep in one direction and then suddenly jump back to the start of the sweep. A block diagram is shown in **Fig 242**.

A warning, though, before you start this project: You will also need an oscilloscope so that you can see what you are doing during the measurements.

The signal from the wobbulator is injected into the receiver under test, tuned to the IF frequency and displayed on the oscilloscope. The output from the detector is used for the display. During the process, the AVC chain should be disabled, which is done by simply de-soldering it from the AVC rectifier and grounding it.

Fig 242: Block diagram of wobbulator. The Synch and Sawtooth signals are intended for the oscilloscope

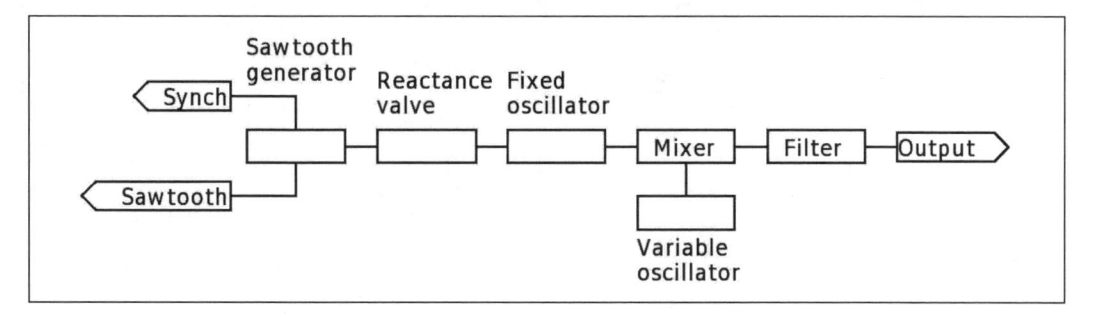

211

VALVES REVISITED

Frequency modulation with valves is done with a reactance valve, which acts like a voltage controlled inductance or capacitance. When it is connected across the tuned circuit of an oscillator, the circuit sees the valve as an additional capacitor, so the resonance frequency shifts. If the added capacitor is varied, the frequency is also varied.

However, the frequency swing is frequency dependent, ie at high frequencies, a certain change of capacitance has a larger impact than it would have on low frequencies. A way of getting around this is to modulate a fixed L/C oscillator and then mix it with a variable oscillator (so that the centre frequency can be set to the desired IF) to achieve a frequency independent swing. For that we need a mixer. A filter separates the signal we want from the others at the output of the mixer. We only need the difference frequency.

We also need a sawtooth generator to modulate the oscillator and with a sync pulse output so we can synchronise the oscilloscope. Alternatively, we can use the sawtooth generator as an external source for the X-axis of the oscilloscope. The Y-axis would be connected to the detector output.

The fixed oscillator will oscillate at about 5MHz and the other between about 5 and 7MHz. Using the difference signal, we get a frequency modulated signal between 0 and 2MHz, because this is where most IF frequencies are located.

Generating a sawtooth

This is an application where using a pentode is essential. The anode current of a pentode is virtually independent of the anode voltage (above a certain threshold). This phenomenon is used in what is called transitron (aka Miller-integrator). It is an oscillator used to generate the sweep in oscilloscopes and the output is very linear, which makes it ideal for the purpose. The principle of a transitron can be seen in **Fig 243**. The pentode needs to be a high gain type, eg EF80 or similar.

As the anode voltage drops, a current flow in C2 causes the control grid to be driven negative, tending to cut off the anode current. Since the anode voltage has

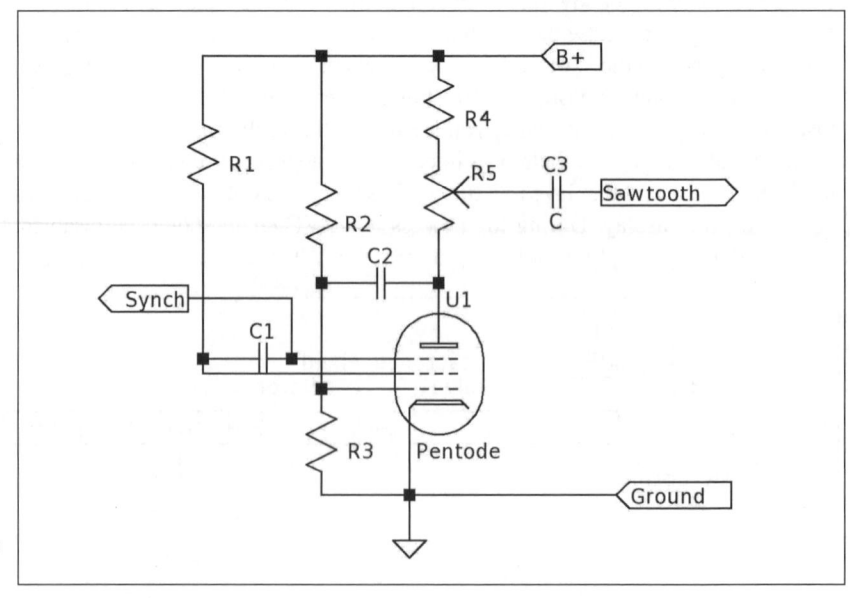

Fig 243: Sawtooth generator

no impact on the anode current, the anode voltage drops at a linear rate determined by the time constant R3 / C2, and the control grid is kept at a constant voltage.

When the anode voltage drops to the point where electrons from the cathode can flow, an sudden increase in screen current occurs which decreases the screen voltage and thereby also the suppressor voltage. This has the effect of suddenly cutting off the anode current. This in turn causes the cathode current to flow to the screen grid. Consequently, a negative pulse appears at the screen grid, and C2 begins to charge until a point is reached where the anode begins to draw current again, the screen voltage is restored, and the cycle is re-started.

The design of the transitron is a bit more elaborate in oscilloscopes - switches and a potentiometer are added to determine the frequency of the oscillator, and provisions are made to insert a synch signal into G3. G3 becomes, in other words, an input and an output simultaneously. Only the output is used in the wobbulator.

Frequency modulating the oscillator

The reactance valve is the frequency determining component in the fixed oscillator. The sawtooth signal from the transitron is connected to its input, and its output is connected to the oscillator. Please refer to **Fig 244**. The valve forms a phase shifter by means of its internal resistance. The voltage across C2 is 90 degrees out of phase, compared to the oscillator tank circuit. This phase difference is fed back to the tank circuit through C3, and the oscillator perceives this as if a capacitor had been added to the circuit. So, the frequency shifts by an amount corresponding to the amount of feedback, which is controlled by the modulation input through C4.

In this case the modulation is a sawtooth, so C4 needs to be big enough so as not to distort the signal. R2 and R3 are bias components. Since the modulation signal is supposed to control the amount of feedback to the oscillator, it is essential that the valve be 'variable-mu'. An ECC82 would do the job neatly, as would any standard variable-mu pentode.

Fig 244: Modulating with reactance valve

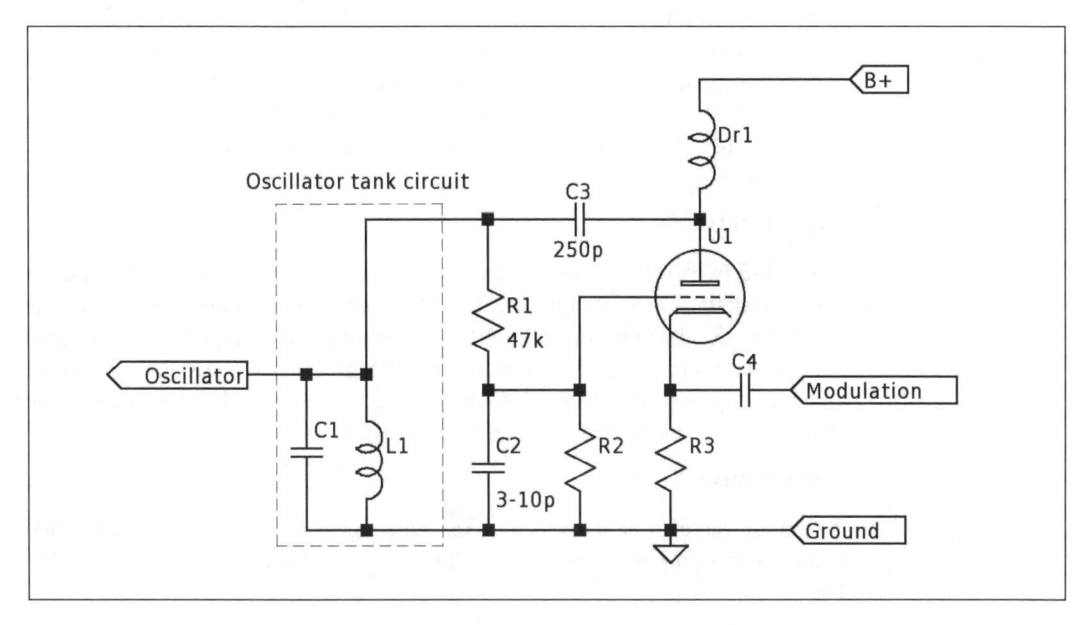

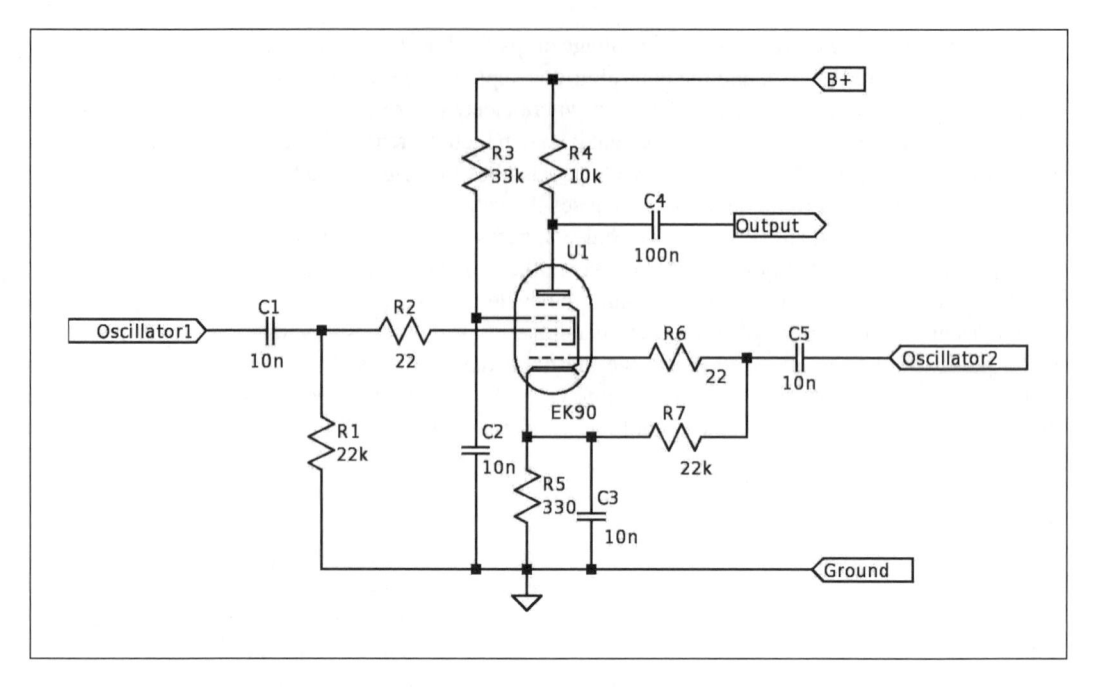

Fig 245: Mixer for wobbulator

Mixer

There are many ways of mixing signals together. The objective here is to mix the two oscillator signals together and make use of the difference frequency. The mixer does not have to be very elaborate, since we do not need excessive sensitivity - we have full control over both signals. Nor is noise a concern, for the same reason.

There is another problem, though, that should be considered. When mixing two signals, the mixed signals, ie the sum and the difference frequencies, tend to be weak in amplitude. The phenomenon is called 'conversion gain'. Valve manufacturers did their best to come up with designs that gave a rather higher conversion gain. So, for instance, the ECF82 can be found in TV sets as an oscillator/mixer stage. The EK90 is a heptode designed for use as both a mixer and oscillator. However, it can be used as a mixer alone in high performance situations.

Heptode mixer

In **Fig 245** two grids are used to inject the signals, G1 from oscillator 1 and G3 from oscillator 2. R3 sets the screen grid voltage based on the screen grid current (from the data sheet), and C2 keeps it filtered. R7 keeps the control grid voltage at zero (connected to the cathode). R5 gives bias to G3. R2 and R6 are there to stop parasitic oscillations in the stage. The signal is taken from the anode via C4. The heptode chosen here is EK90, a seven-pin valve.

Pentode mixer

A pentode can also be used as a mixer. One signal is then normally connected to the control grid, whereas the other to one of the other grids, as shown in **Fig 246**.

CHAPTER 15: A SIGNAL GENERATOR

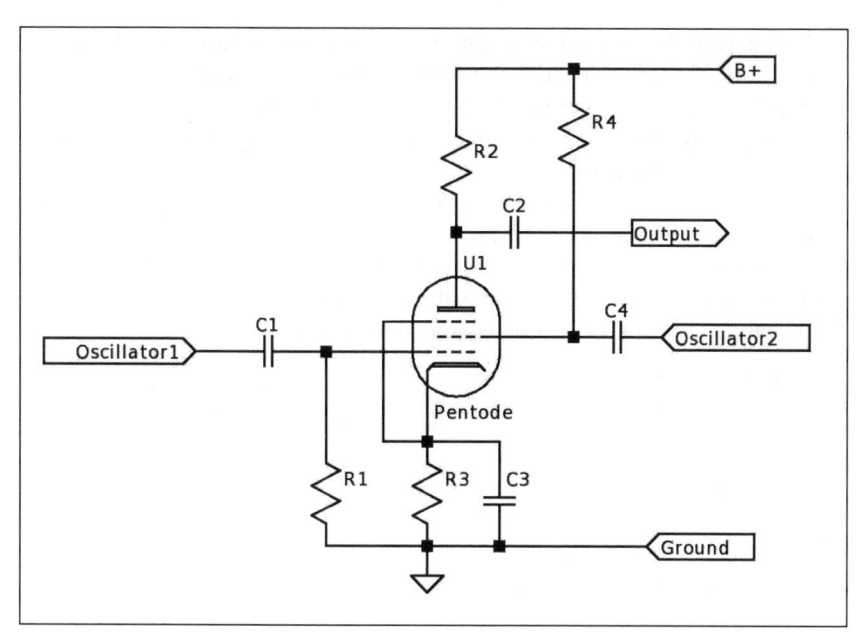

Fig 246: Pentode mixer for wobbulator

A pentode has three grids, oriented in concentric cylinders around the cathode. However, the further away from the cathode the grid is, the less it is able to influence the anode current. Consequently, using a pentode as a mixer requires that the signal that is not connected to the control grid be bigger than the one that is. There are pentodes which were developed as mixers, but they were frequently combined with a triode, which was to act as the oscillator.

Fig 247: Alternative pentode mixer for wobbulator

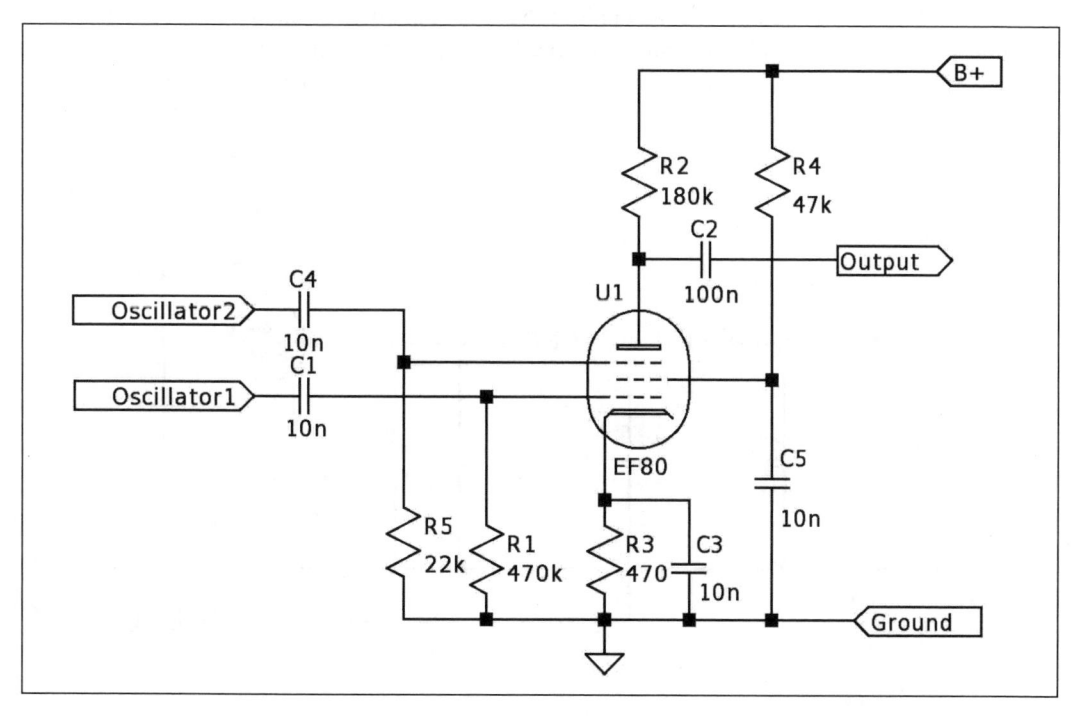

215

VALVES REVISITED

Connecting the oscillator output to the (un-decoupled) cathode resistor instead would work as well. A third alternative is to connect one signal to G3 (through C4) and the other to G1 (through C1 in **Fig 247**).

Alternatively, you can use a triode-pentode, such as the ECF80 or ECF85 as one oscillator and the mixer.

Let us try to keep the valve count down. An ECH81 contains a triode, which can be used as one of the oscillators, and a heptode to mix the signals. A dual triode, such as ECC82, can be used to generate the other oscillator signal as well as the output stage, and the whole wobbulator has been implemented with only three valves. The complete schematic is shown below.

The output filter

At the output from the mixer valve, you will find four signals.

- Oscillator 1
- Oscillator 2
- Oscillator 1 + Oscillator 2
- Oscillator 1 - Oscillator 2

If you don't connect a filter at the output, you could use all four signals. However, the signal we are interested in is the difference signal, which will be between 0 (literally!) and 2MHz.

Many people live under the misconception that the active filter is an invention from the transistor or op-amp time. Nothing could be more wrong! In fact, the transistor / op-amp implementation is frequently a direct adaptation of designs developed during the early days of the valve era. You don't see them very often, because they found applications within the field of measurement and test.

An RC filter could be used for the purpose. The problems are that the flank will not be very steep (some 18dB per octave), and that the resistor values would

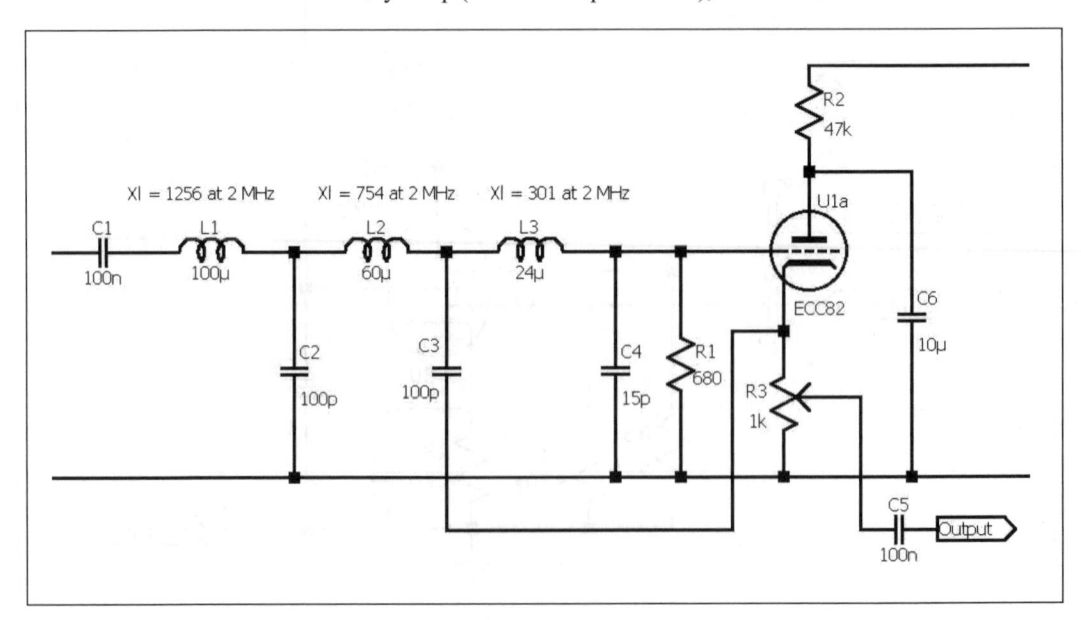

Fig 248: Output filter for wobbulator

be very low. Instead, we shall do the implementation as a pi-network with coils and capacitors, and we shall use a valve as the active element. This active filter has a corner frequency of about 2.5MHz and suits our purpose very well. It is shown in **Fig 248**.

The slope of this filter is about 25dB per octave. The coils should be mounted such that they can not influence each other. R3 influences the frequency curve.

The input at C1 should have a suitable impedance. The design is suitable for measurements because the output has a low impedance.

The input signal is the output from the mixer and connected to C1. The output is taken from the potentiometer R3. This will be the output that will be connected to the test object. 250V is a suitable B+ voltage.

The complete wobbulator

Fig 249 assumes that your oscilloscope has an output where you can take its internal sweep generator's output (the X-axis). Not all oscilloscopes have this facility, though, and in case yours doesn't, you can use the Miller-integrator shown and described below.

The sawtooth is connected to the input named 'Sweep'. Due to its usually very high amplitude, the signal is attenuated by R1 / R2 before it is fed to the modulator's control grid. The setting of R2 controls the amplitude of the signal, and, hence, the amount of modulation of the oscillator. The lack of input capacitor allows for very low sweep frequency with the linearity of the sawtooth preserved.

The modulator (U1) is a pentode here, an EF93. Resistor R4 sets the control grid bias, along with the filtering capacitors C2 and C3. C1 is part of the phase shifting components. R5 sets the G2 bias.

The output signal is fed to the modulated oscillator (U2), which is a Hartley type.

An identical oscillator, which is tunable using C10 generates the other signal. Both oscillators should operate on the same frequency with the sweep control turned to zero and C10 at its highest setting (all vanes in). If not, the iron dust cores of L1 and L6 may have to be adjusted.

U3 is the mixer, which mixes the two signals together. At the anode you should have the four resultant signals. In order to provide a low impedance for the filter, a cathode follower has been inserted. It is U4, which is one half of U6 (double triode ECC82).

The filter proper has a corner frequency of about 2.5MHz and a slope of about 40 dB / octave. At the output you should have a frequency modulated signal (when the sweep control is turned up), which you can tune up and down using the tuning control C10.

When you connect the wobbulator's output to the aerial input of the radio you wish to trim, you should be able to hear the signal when the radio is tuned to the centre frequency of the wobbulator. At the same time, you should be able to see the signal on your oscilloscope with the probe connected to the radio's detector input.

Don't forget to disconnect the AVC chain, though, or you may encounter distortion on your oscilloscope.

VALVES REVISITED

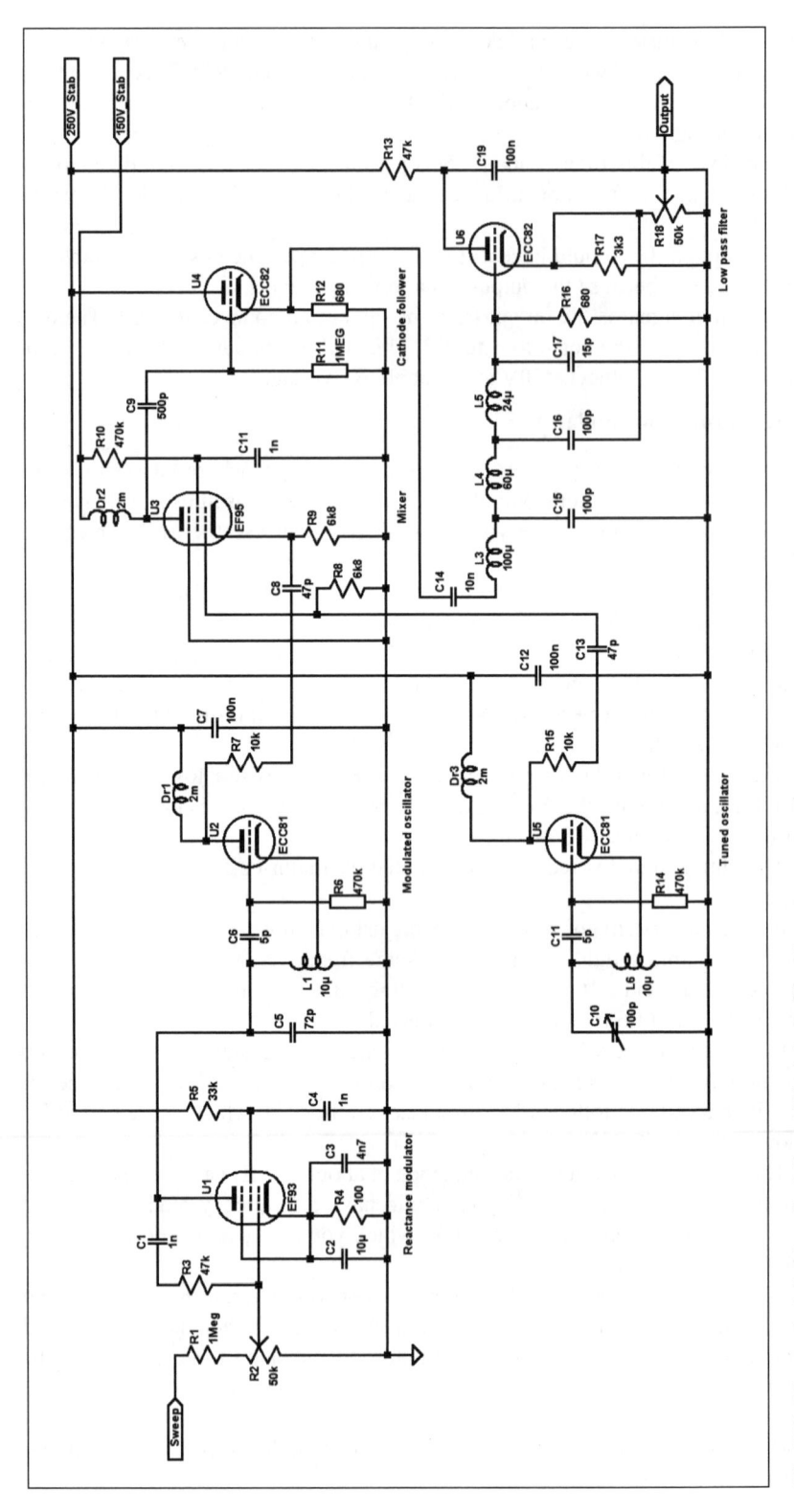

Fig 249: The complete circuit of the wobbulator without sweep generator

CHAPTER 15: A SIGNAL GENERATOR

Ref.	Description	Ref.	Description
C1	capacitor, 1nF	L6	inductor, 10µH
C2	capacitor, 10µF	R1	resistor, 1M
C3	capacitor, 4.7nF	R2	resistor, 50k
C4	capacitor, 1nF	R3	resistor, 47k
C5	capacitor, 72pF	R4	resistor, 100
C6	capacitor, 5pF	R5	resistor, 33k
C7	capacitor, 100nF	R6	resistor, 470k
C8	capacitor, 47pF	R7	resistor, 10k
C9	capacitor, 500pF	R8	resistor, 6.8k
C10	capacitor, 100pF	R9	resistor, 6.8k
C11	capacitor, 1nF	R10	resistor, 470k
C11	capacitor, 5pF	R11	resistor, 1M
C12	capacitor, 100nF	R12	resistor, 680
C13	capacitor, 47pF	R13	resistor, 47k
C14	capacitor, 10nF	R14	resistor, 470k
C15	capacitor, 100pF	R15	resistor, 10k
C16	capacitor, 100pF	R16	resistor, 680
C17	capacitor, 15pF	R17	resistor, 3.3k
C19	capacitor, 100nF	R18	resistor, 50k
Dr1	inductor, 2mH	U1	EF93, pentode
Dr2	inductor, 2mH	U2	ECC81, ½ dual triode
Dr3	inductor, 2mH	U3	EF95, pentode
L1	inductor, 10µH	U4	ECC82, ½ dual triode
L3	inductor, 100µH	U6	ECC82, ½ dual triode
L4	inductor, 60µH	U7	ECC81, ½ dual triode
L5	inductor, 24µH		

Components list for the complete wobbulator without sweep generator

Now, adjustments of the iron dust cores of the IF transformers should show up on the oscilloscope as changes in the curve shape. The aim is as steep sides as possible and a flat top.

Miller integrator

A Miller integrator is a useful device. Its output is very linear. In a valve equipped oscilloscope, it is a Miller integrator that generates the sweep.

A practical implementation of a Miller integrator is shown in **Fig 250**. It consists of a pentode, in this case an EF86 (but other pentodes work just as well).

It generates a sawtooth signal with the aid of C2 / R2 / R3. C2 and the voltage at the potentiometer arm determines the sweep frequency.

C1 and R1 determine the fly-back time, ie the time it takes for the curve to return to its beginning. At G2 is a pulse during the fly-back time, which can be used as a synch pulse for your oscilloscope.

Fig 250: Miller integrator

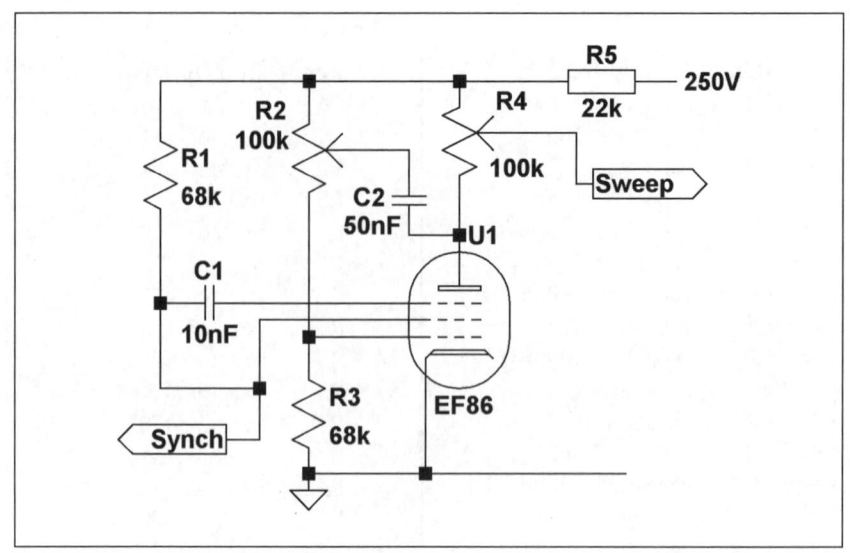

At the same time, it acts as an input, so, if you have the need to synchronise the sawtooth to an external signal, this would be the entry point for that signal. R4 determines the sawtooth amplitude at the output.

16

Measurements

When repairing or renovating old equipment, or constructing your own, it is necessary to make measurements of voltages, component values and performance. Some test gear can be very expensive, but useful measuring equipment can easily be built at low cost.

Measuring Amplifier Impedances

Sometimes it is very important to match impedances to a device, eg an aerial to a transmitter or a receiver, a loudspeaker to an amplifier's output, a microphone to an amplifier's input, etc. Sometimes it is not so important. However, there is a definite need for measuring impedances, and the procedure is not complicated.

Input impedance

Every device has an input and an output impedance, be it a resistor, a transformer, an amplifier or a valve. The input impedance acts like an internal resistor connected across the input terminals of the device, as shown in **Fig 251**. Sometimes the impedance is frequency dependent, sometimes not.

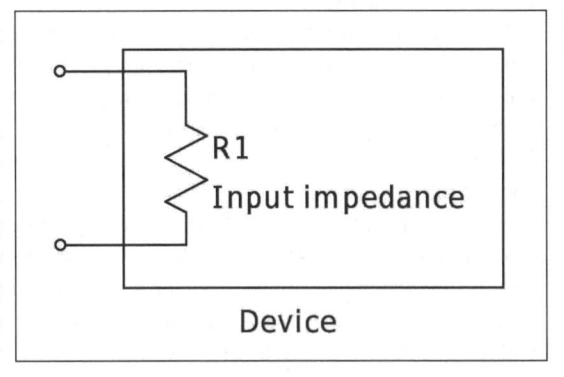

Fig 251: Illustration of input impedance

An example of a non-dependent impedance is a pure resistance. An example of a frequency dependent impedance is a tuned circuit.

The impedance is built-in into the device, and the only way you can change it is generally to change the design.

The principle of measuring the input impedance is by connecting a resistor in series with an input signal and measuring the resulting voltage at the intersection between the external resistor and the input. The external resistor forms a voltage divider with the input impedance. When the external resistor is equal to the impedance, the voltage is half of that of the generator output. So, all you need is a voltmeter, a handful of resistors and a signal source, as shown in **Fig 252**.

In case of frequency dependent devices, you should state the impedance at a specific frequency or, better, measure the impedance at a series of different frequencies and plot the result.

Fig 252: The measurement of input impedance

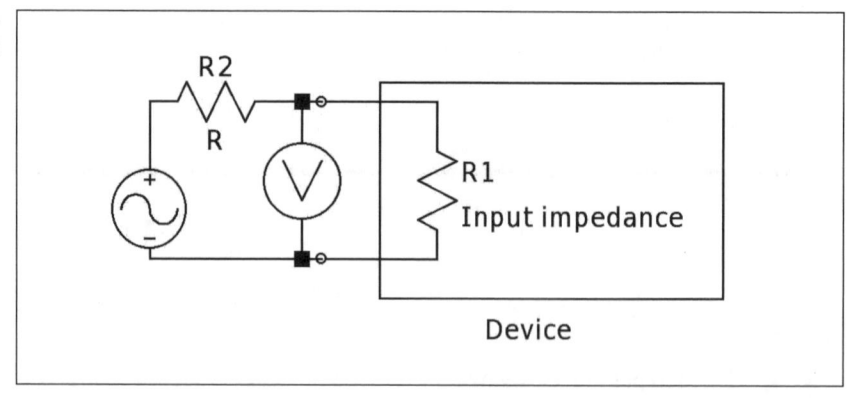

Output impedance

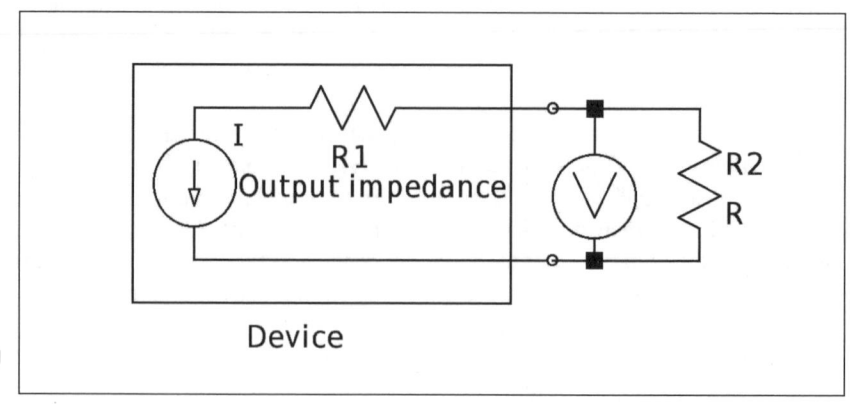

Measuring output impedance is just as easy. The set-up is shown in **Fig 253**. The output impedance acts as a resistor between a current source and the output.

Again, this resistor is internal, and the only way you can change it is to change the design.

A valve or transistor stage has a load at the anode (drain/collector), sometimes a resistor, sometimes another type of load. These components too are regarded as internal to the device.

The procedure for measuring output impedance is similar to that of measuring input impedance. Connect a resistor across the output terminals and measure the voltage across it. When the voltage is half of that without the resistor, the resistor is equal to the impedance.

In most cases you need a DC blocking capacitor between the output terminal and the external resistor.

Needless to say, the higher impedance in your voltmeter, the better the result. Some devices, such as pentode stages or cascode amplifiers have a very high output impedance, so your voltmeter really should not load the signal.

A VTVM (vacuum tube volt meter - see the full description later in this chapter) does an excellent job here with its high input impedance which is in the order of megohms.

Fig 253: Measuring output impedance

CHAPTER 16: MEASUREMENTS

Measuring Capacitance

Capacitance measurements generally use an indirect method. The principle is based on the fact that a capacitance which is subject to AC develops a reactance to the current. The reactance is measured, and, given the frequency, the capacitance value can be calculated. If the frequency is known and stable, an instrument can be calibrated directly in capacitance.

This measurement element is a bridge, consisting of capacitors with known and accurate values. The bridge compares the unknown capacitor (the test object) with the known capacitors plus the capacitance of a variable capacitor and its output goes to zero when the unknown capacitance equals the sum of the known fixed capacitor and the setting of the variable capacitor.

The measurement range of the instrument depends on what frequency is used for the measurements. The higher the frequency, the smaller capacitances can be measured. This instrument uses a frequency of about 1.5MHz, which covers zero to about 1,500pF. This is enough to cover all RF applications.

It is self-evident that stray capacitances will have an impact on the measurements, so good construction practices are necessary.

An oscillator is needed to generate the frequency. In this case a valve oscillator is used. In this case it is a DF91 (a RF pentode), but virtually any valve will do, as long as it oscillates.

Different capacitance bridges use different indicators to indicate bridge null. Some use earphones, some use a panel meter. The interesting thing about the design in **Fig 254** is that it uses a 'magic eye' for indication. If you can not find a 75pF capacitor for C4, connect a fixed capacitor of 100pF in series with a 360pF variable.

The coils L1 and L2 should be wound for a frequency of about 1.5MHz. Start with 50 turns each on a 15mm former.

Fig 254: 'Magic eye' capacitance meter

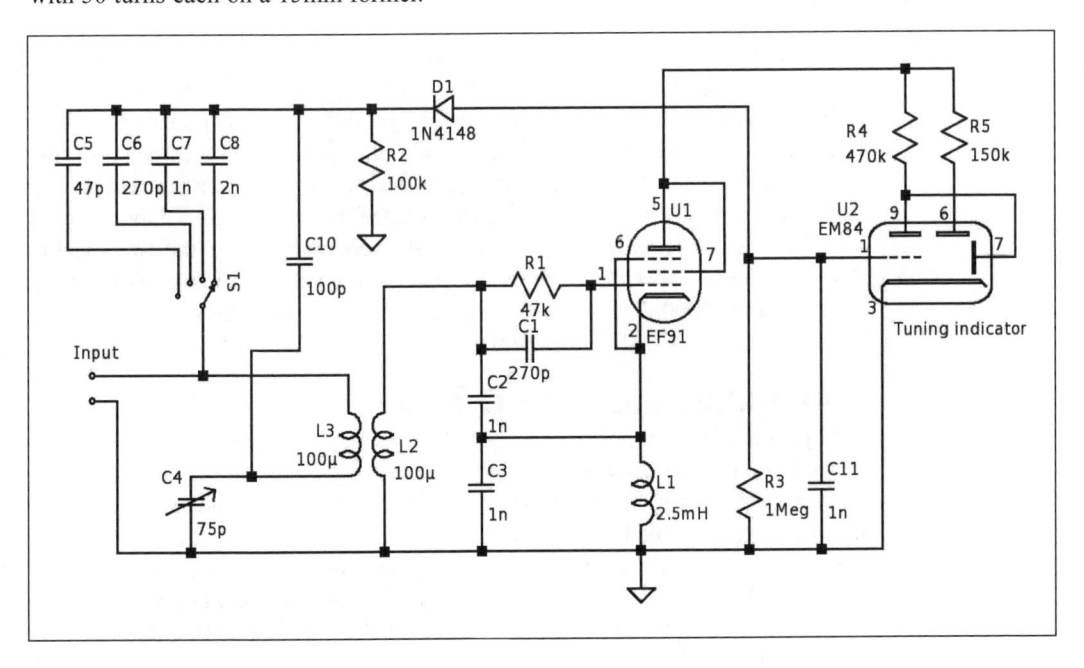

VALVES REVISITED

Ref.	Description	Ref.	Description
C1	capacitor, 270pF	L1	RF choke, 2.5mH
C2	capacitor, 1nF	L2	inductor, 100µH
C3	capacitor, 1nF	L3	inductor, 100µH
C4	capacitor, 75pF variable	R1	resistor, 47k
C5	capacitor, 47pF 1%	R2	resistor, 100k
C6	capacitor, 270pF 1%	R3	resistor, 1M
C7	capacitor, 1nF 1%	R4	resistor, 470k
C8	capacitor, 2nF 1%	R5	resistor, 150k
C10	capacitor, 100pF	U1	EF91, pentode
C11	capacitor, 1nF	U2	EM84, tuning indicator
D1	1N4148, diode	S1	4-way 1-pole switch

Components list for the capacitance bridge

The fixed capacitors C5 - C8 are the reference capacitors. They should be of high quality and, preferably, 1% accurate.

The diode 1N4148 was chosen because it is a readily available small signal diode. A valve diode or a diode connected triode would, of course, do very well (better in fact. See the section on diodes). The meter should be rigidly built, using short and direct leads. Once it is calibrated, everything should remain untouched.

If you can not find a tuning indicator, you can use a panel meter in its place, connected across C11. You need a series resistance to drop the voltage, and limit the current to the meter. The value of the resistance depends on the voltage across C11 and on your meter. Failing an EF91, any other pentode or triode will do.

Calibration

To calibrate the meter, begin by checking the stray capacitances of the instrument. Without anything connected to the terminals and the instrument set to the lowest range, balance the bridge ('magic eye' fully open). If no balance can be achieved, solder a 5pF 1% capacitor across the terminals with shortest possible leads and balance the bridge. Mark the scale on C4 "0". Then connect another 5pF 1% capacitor to the terminals and balance the meter again. Mark this point "5" on the scale. Repeat with a 10pF 1% capacitor. Solder the two together (using shortest possible leads) and repeat the procedure. Go through the entire range until 1500pF has been reached. If there are gaps between the ranges, capacitors C6 - C8 should be replaced.

An Inductance and Q Meter

One of the few components that are usually unmarked are inductors, in particular those that you wind yourself. Winding your own inductances may sometimes be necessary, in particular if they must carry some current. Inductors for RF are commercially available, though.

Here a simple device that not only measures inductance, but also the coil Q. One of the characteristics of a tuned circuit is that it resonates on a specific frequency. Finding the inductance of a coil can thus be done in two ways:

CHAPTER 16: MEASUREMENTS

Fig 255: Circuit diagram of the Q-meter

You can connect it to a known capacitor and find the resonant frequency by sweeping an oscillator while observing the response of the circuit on an oscilloscope or a similar instrument.

Or you can do it the other way around. Your oscillator is on a fixed, known frequency and you sweep the tuning of the circuit by connecting the coil to a variable capacitor.

The latter is better in this case, since we want to measure the Q as well.

This is done by moving the oscillator until the circuit response has gone down from its peak to 0.707 times the peak response.

The difference between the new oscillator frequency and the original frequency is half the bandwidth of the circuit, and from there it is easy to calculate Q. In fact, the capacitor with which the oscillator is tuned can be calibrated directly in Q.

The circuit is shown in **Fig 255**. The left hand half of U1 is the oscillator. The coil under test forms a tuned circuit along with the variable capacitor C6. It is fed from the upper end of the oscillator coil, where the signal is the cleanest, through R2, which provides a high impedance environment for the circuit.

Measurement is done through the cathode coupled amplifier composed of the right hand side of U1 and the left hand half of U2. Since U1 is a cathode follower, the input impedance is high at this point.

The voltmeter is regarded as a low impedance device, so another cathode follower, the right hand half of U2, acts as an impedance converter. The signal is rectified by a diode coupled triode, U3. This can be a valve diode or a triode or pentode with the grids connected directly to the anode.

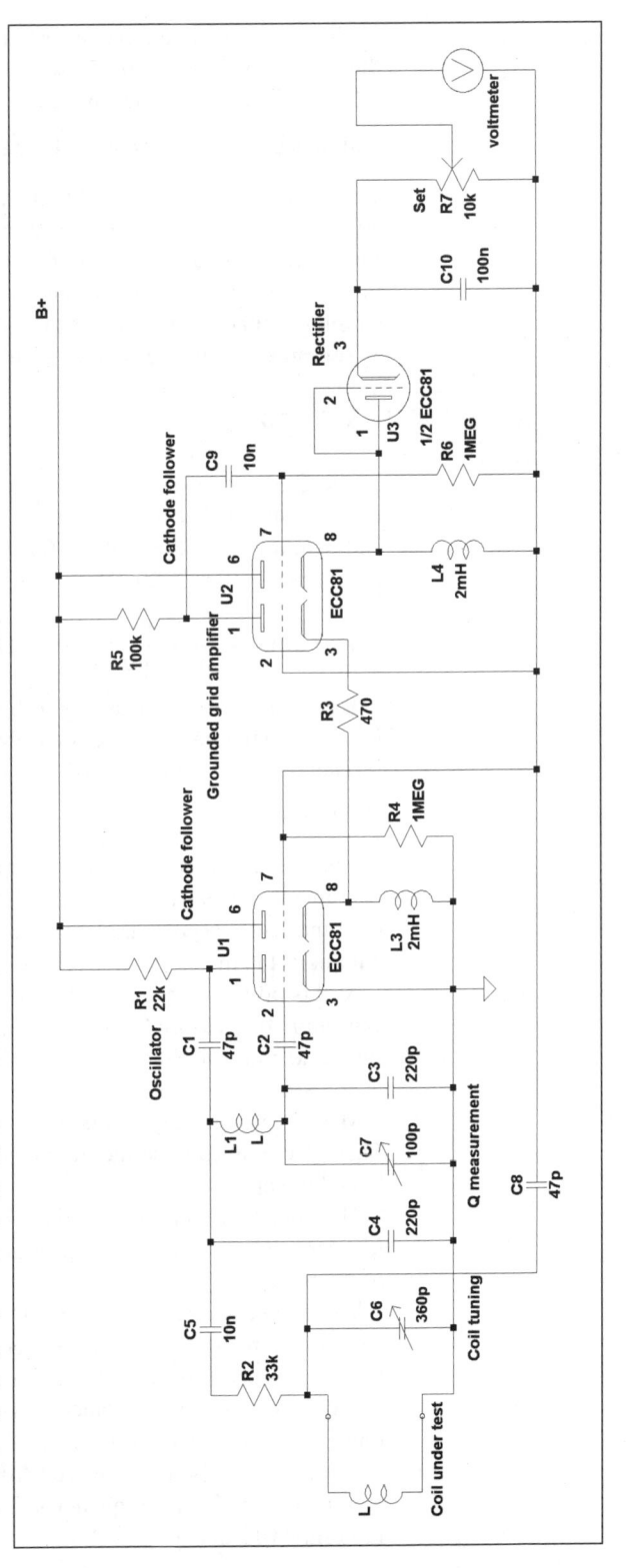

225

The reason a valve diode is used rather than a semiconductor one is to avoid the forward voltage drop of a semiconductor diode.

The leads to the test terminals should be made short and direct.

Calibrating the inductance / Q meter

The capacitors need to be calibrated. The coil tuning capacitor is the most difficult one. One way is to build a cathode coupled oscillator, calculate frequencies for the various capacitances within the capacitor range, and listen for the signal in a calibrated receiver, marking the scale as you go along. This would be the simplest method with good accuracy.

You might remember that Q is defined as

$$Q = \frac{f}{BW} \qquad \text{Formula12}$$

The Q meter capacitor can be calibrated as described above, using the instrument oscillator. Set the capacitor to the smallest capacitance (vanes fully out) and trim the oscillator coil to 3000.000kHz, using its iron core. Then calculate the frequency for Q=200 as:

$$f = 3000 - \frac{3000}{2 \times 200} = 2992.5 kHz$$

Turn the Q capacitor until you hear the oscillator signal on that frequency. Mark this point on the capacitor scale as '200'. Repeat the process until you have reached the lowest Q you wish to measure.

How to use it

Connect the coil to the test terminals. Turn the variable capacitor until the instrument shows an indication. Select the peak reading. Turn the potentiometer until the meter needle shows full scale. No further tuning should be necessary, but this is the most sensitive setting of the instrument, so peak the circuit again.

At this point you can read the inductance of the coil from scale of the tuning capacitor. To measure Q, turn the other capacitor until the meter reads 70.7% of full scale and read the Q of the coil from the scale of this capacitor.

You are done.

Many variable capacitors on the market today are capacitance linear. Therefore, you can use the tuning capacitor as a reference, if the scale is calibrated in capacitance too.

The inductance markings will not be linear. The markings will be tighter at the low capacitance end of the capacitor. This will also be the high inductance end of the scale.

With the component values given and at a frequency of 3MHz, the meter will be able to measure inductances between 8 and 80µH. Shifting the oscillator (which can be done by switching a different coil into the oscillator circuit) to 7MHz gives you an inductance range of 1.5 - 12µH, which means that the two ranges will overlap each other.

Don't be surprised of you get different Q readings for the same coil when measured at different frequencies. The reason for this is the definition of Q as in Formula 12 above.

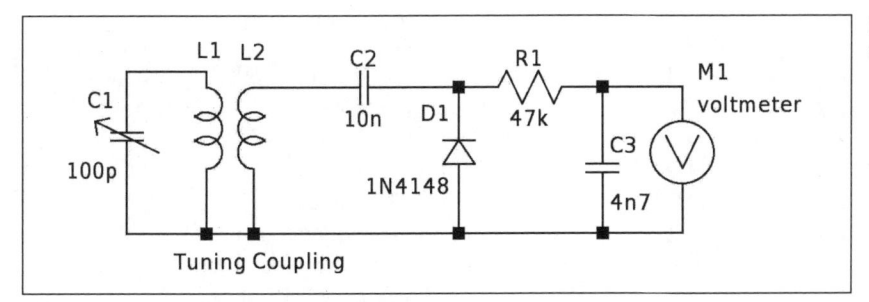

Fig 256: Simple wave-meter

Measuring Frequency

There are several ways of measuring frequency. One way is to use one of the many 'world receivers' that are now on the market, a receiver covering the entire band between (virtually) zero and 30MHz and with a digital display. Another way is to use a frequency counter, capable of handling up to 30MHz. A third way is to build an absorption wavemeter. This is what we are going to do.

The wavemeter in **Fig 256** is simple. It consists of a tuned circuit with changeable coils and a variable capacitor with a scale, calibrated in frequency. A rectifier turns the signal into DC and a panel meter shows when the internal circuit is in resonance with the received signal.

It is so simple, in fact, that it can easily be combined with another instrument, the grid dip oscillator, GDO. Both are used to measure frequency; the wave meter measures oscillator or local transmitter frequencies and the GDO measures resonance frequencies of tuned circuits and aerials.

The grid of an oscillator draws current when it oscillates. If an object approaches, the tuning coil in the oscillator behaves in two ways:
- The frequency shifts
- The grid draws less current, because the oscillator is loaded

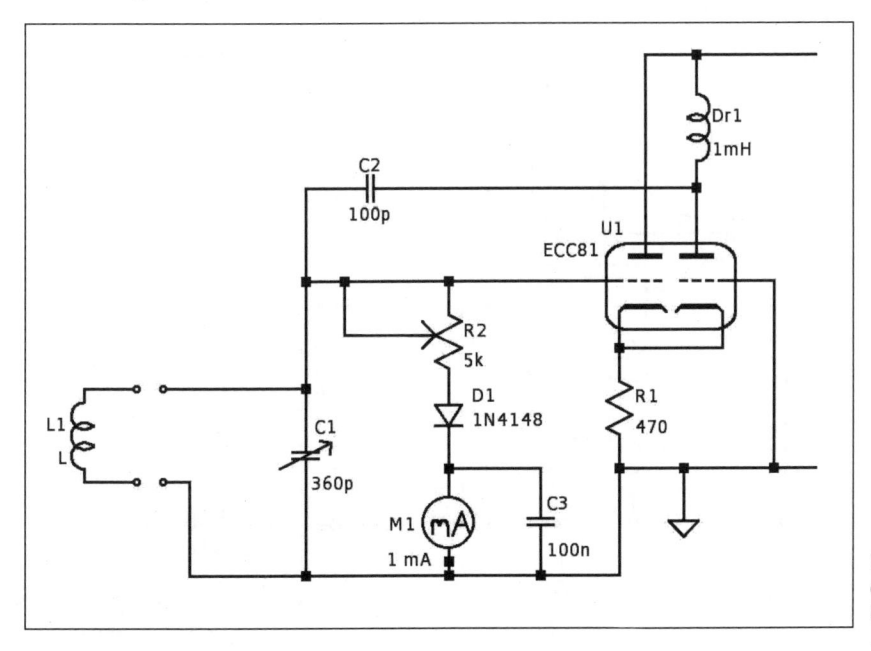

Fig 257: Cathode coupled GDO (Grid Dip Oscillator)

227

VALVES REVISITED

Ref.	Description
C1	variable capacitor, 360pF
C2	capacitor, 100pF
C3	capacitor, 100nF
D1	1N4148, diode
Dr1	RF choke, 1mH
L1	oscillator inductor, as needed
M1	panel meter, 1mA full range
R1	resistor, 470
R2	potentiometer, 5k
DIN males as needed for the coils and DIN female, 1 off	

Components list for the grid-dip oscillator

This second phenomenon is used in the GDO. A small panel instrument measures the grid current and gives a dip when the oscillator is tuned to the resonant frequency of a tuned circuit close to the coil.

It is necessary to make the GDO 'portable' enough to reach the circuits you wish to measure. **Fig 257** shows a cathode coupled oscillator GDO.

L1 is the changeable coil. C1 tunes the oscillator, which is a cathode coupled oscillator. R2, D1, and C3 rectifies and filters the signal to the meter M1. The coils can be fitted onto 3 or 5 pole DIN males. The potentiometer R2 is a sensitivity control. The GDO is best powered by about 150V, stabilised.

In an application such as this, oscillator's stability is of minor importance, but it should still be solidly built and fitted into a sturdy box.

To use it, turn it on and move the coil near the test object. Adjust the frequency of the GDO until you get a dip. If there is no dip, your test object is erroneous or you have chosen the wrong coil.

Fig 258: Alternative circuit for a cathode coupled GDO with RF choke as coupling element

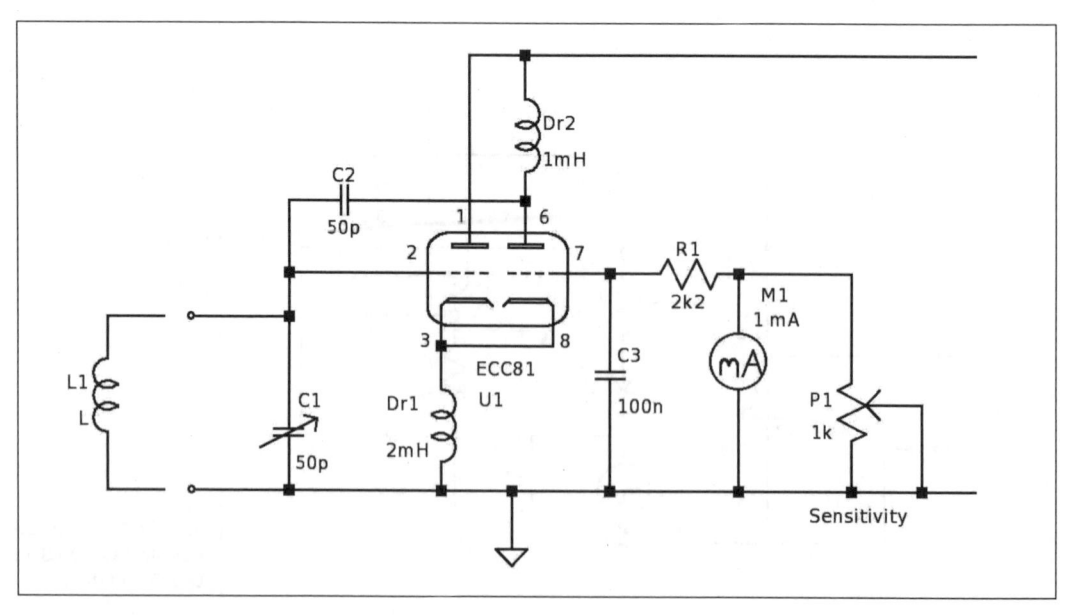

228

When you have a clear dip, gradually move away from the circuit under test, adjusting the frequency, until the dip is barely noticeable. This is the correct resonant frequency. Measuring too closely to the circuit will upset the test circuit due to the high degree of coupling between the coil and the circuit.

When the unit is not powered on, it can measure frequencies from oscillators. Place the coil near the oscillator coil and turn the frequency knob until the instrument shows a current. Slowly move away from the oscillator under test, adjusting the frequency of the GDO and the sensitivity control until the meter barely moves. This is the correct frequency.

An alternative method of connecting the meter is shown in **Fig 258**. The meter is connected in the grid circuit of the second stage so as not to interfere with the oscillator grid circuit. Note that the cathode resistor of Fig 257 is a choke in Fig 258. This gives a more efficient coupling between the stages.

Decoupling capacitor C3 can be made bigger if it turns out that the RF remainder is too high. Virtually any oscillator can be used as a GDO. Also, the GDO could contain a 'magic eye' as indicator.

On this page are a couple of alternative GDO designs. The first (**Fig 259**) is a Hartley oscillator, as can be seen by the coil tap. In this example, the tap is centred. The second (**Fig 260**) is a Colpitts oscillator, recognised by the double

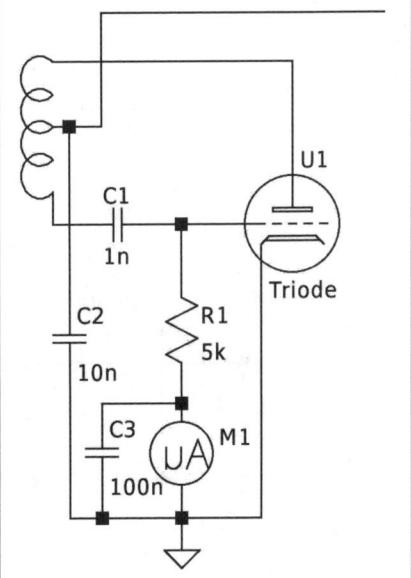

Fig 259: Hartley oscillator as a GDO

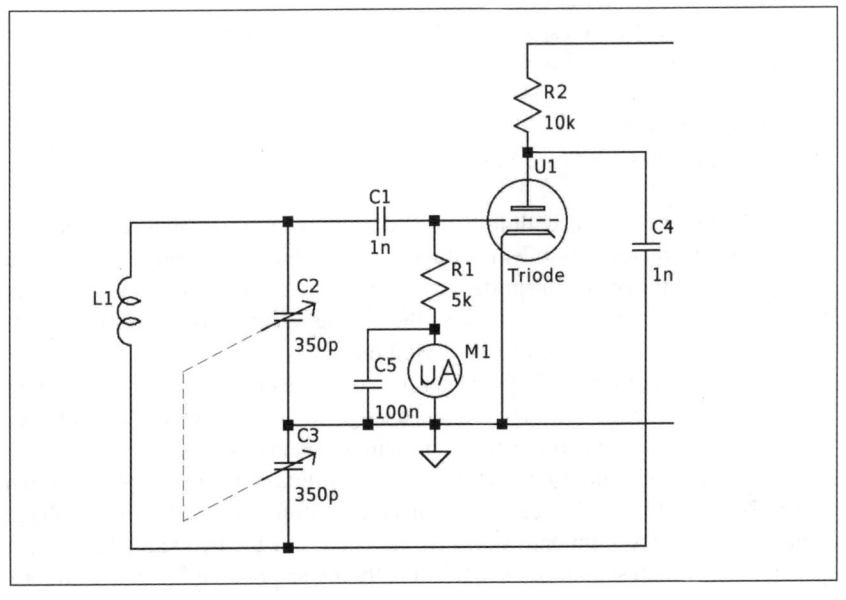

Fig 260: Colpitts GDO

Fig 261: GDO with increased sensitivity

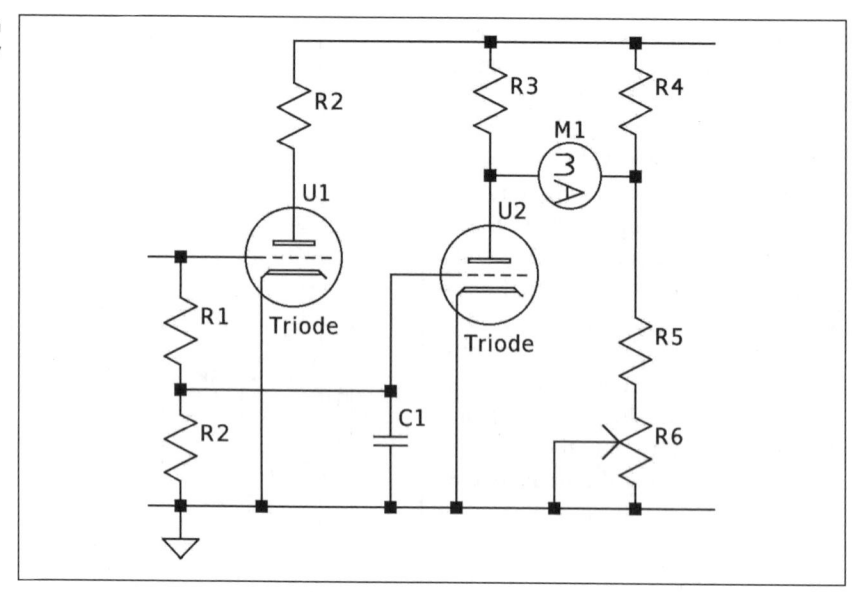

gang tuning capacitor. The Colpitts is capable of oscillating up to and beyond 200MHz, if needed.

Two features are common to GDOs: The instrument that measures the grid current, and the plug-in coil. The sensitivity of a GDO can be increased by inserting a valve bridge circuit, as shown in **Fig 261**. U2 / R3 form one leg of the bridge, whereas R4 / R5-R6 form the other. R6 is used to set the balance of the bridge. Using a dual triode, such as ECC81, still requires only one valve for the GDO.

Whichever alternative you choose, make sure you build it mechanically rigid for best performance.

Measuring Voltages - a VTVM

When working with valves, a modern voltmeter or universal meter is of little or no use. It has two main drawbacks:

- Its display is digital
- Its input resistance is too low

Both points make it difficult to trim filters and other points in a receiver. The second point creates a problem with measuring high impedance points in a receiver or amplifier or transmitter. If a high impedance point is loaded by an instrument, the instrument will show the wrong value. There are many high impedance points in a piece of valve equipment.

When trimming filters, it is far easier to see the result of your actions with an analogue instrument. Watching numbers change completely removes the feel for what you are doing, and you can end up with erroneous settings.

So, a VTVM (vacuum tube voltmeter) is the solution. One example is shown in **Fig 262**. A VTVM is basically a balanced amplifier or a bridge with a voltage divider chain at the front end. Between the anodes of the two valve halves is a panel meter, which should be as mechanically big as possible for easy reading.

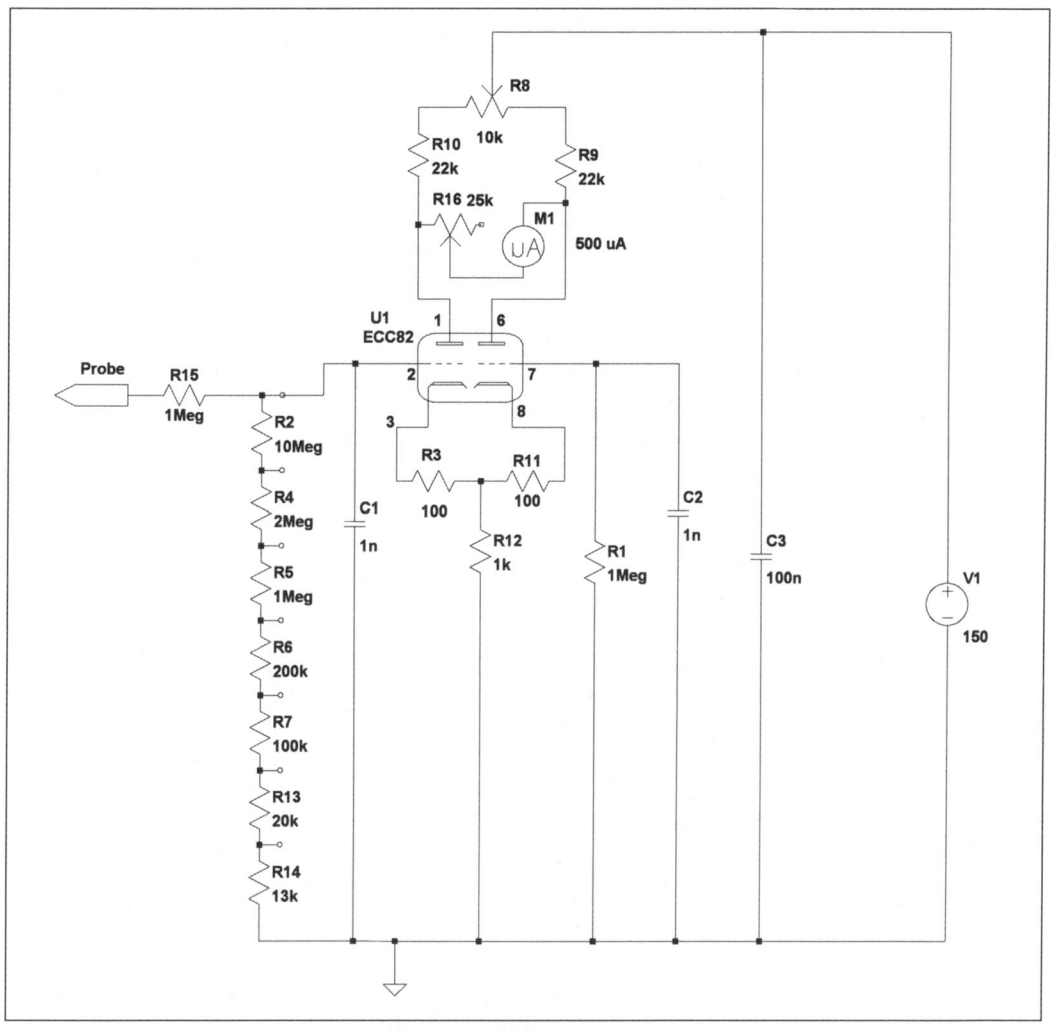

CHAPTER 16: MEASUREMENTS

Fig 262: VTVM (vacuum tube voltage meter)

Ref.	Description
C1	capacitor, 1nF
C2	capacitor, 1nF
C3	capacitor, 100nF
M1	panel meter, 10μA
R1	resistor, 1M, 1%
R2	resistor, 10M, 1%
R3	resistor, 100, 1%
R4	resistor, 2M, 1%
R5	resistor, 1M, 1%
R6	resistor, 200k, 1%
R7	resistor, 100k, 1%
R8	potentiometer, 10k, wire wound (meter null control)
R9	resistor, 22k
R10	resistor, 22k
R11	resistor, 100
R12	resistor, 1k
R13	resistor, 20k
R14	resistor, 13k
R15	resistor, 1M
R16	trim potentiometer, 25k (meter sensitivity control)
U1	ECC82 (or other dual triode)

Components list for the VTVM

231

VALVES REVISITED

Resistance R15 should ideally be built-in into the probe casing. It reduces influences from capacitances in the lead from the probe tip to ground through the leads. R8 should be dimensioned to cope with the anode current of the two valve halves.

VTVM Enhancements

The VTVM is a very useful tool. Due to its high impedance, almost anything can be measured without upsetting the test object. It can be made into an extremely useful tool with a couple of enhancements, such as:

- AC and RF voltage measurements
- Direct reading capacitance measurements
- Direct reading inductance measurements
- Resistance measurements
- Current measurements (DC and AC/RF)

AC and RF voltages

To enable measurement of AC voltages, a RF probe is necessary. It rectifies the signal coming in from the probe tip and can then be measured as a DC voltage. The result is an RMS reading.

For RF measurements, the rectifier should be built-in into a metallic probe casing. The cable between the VTVM and the probe should be coax, and well grounded at both ends.

An RF probe is shown in **Fig 263**. The EA50 is a tiny valve, with very small inner capacitances and intended for measurement purposes. Failing an EA50, other RF diodes may be used in its place. Even a triode or a pentode, connected as a diode may do. Since the valve is to be built-in into a probe casing, its

Fig 263: RF probe for VTVM

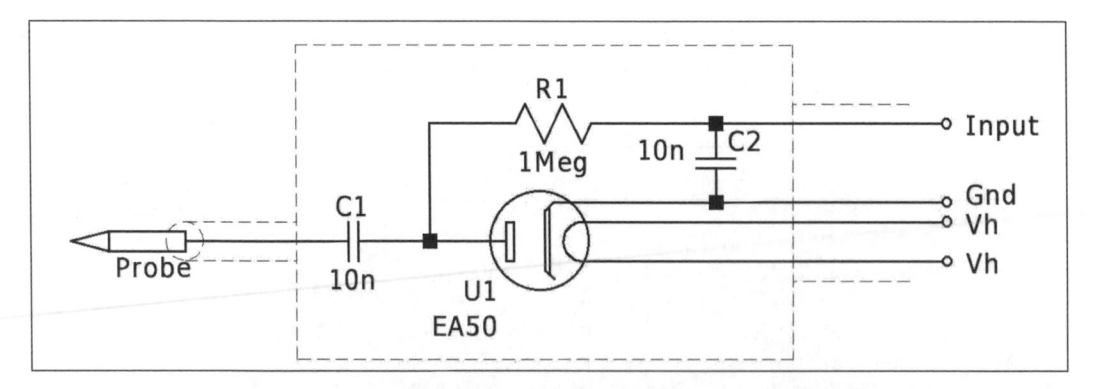

Ref.	Description
C1	capacitor, 10nF
C2	capacitor, 10nF
R1	resistor, 1M
U1	EA50

Components list for the RF probe

physical size is important. A sub-miniature triode still available is the 5703. As a triode, it is capable of working as an oscillator up to some 800MHz, which tells us something about the internal capacitances.

The signal is taken from the anode of the diode. So, during the positive half-periods of the input, electrons begin to flow from the cathode to the anode. Those electrons are captured by the VTVM through R1. The output signal from the probe is therefore negative.

Measuring capacitances

There are other, more complicated, capacitance meters available - among others a bridge instrument previously described in this book.

However, normally, you don't need accuracy down to the third decimal, because there are always going to be stray capacitances present, which affect your capacitor values anyway.

A simple way of measuring capacitors is to connect the unknown capacitor in a voltage divider configuration, impose a signal from an oscillator and measure the voltage drop across the capacitor caused by its reactance, displaying it as a capacitance.

If the oscillator frequency and amplitude are known, the rest is pure mathematics. Let's do the maths:

The reactance of a capacitor is expressed as in Formula 31:

$$Xc = \frac{1}{2 \times \pi \times f \times C}$$ *Formula 31*

where:
$\pi = 3.141592$,
f = the output frequency of the oscillator, and
C is the capacitance.

The things we can control are the oscillator parameters and the series resistor. **Table 17** below shows the full development of formula 31 above.

The output voltage is what we measure with the VTVM. The series resistor has been set to 1kΩ and the input voltage to 5V. All calculations have assumed a frequency of 3MHz. The RF probe is connected to the point marked in **Fig 264**.

Capacitance	Reactance	Output Voltage
10	5305.17	4.21
20	2652.58	3.63
30	1768.39	3.19
40	1326.29	2.85
50	1061.03	2.57
60	884.19	2.35
70	757.88	2.16
80	663.15	1.99
90	589.46	1.85
100	530.52	1.73

Table 17: The relationship between capacitance, reactance and voltage

VALVES REVISITED

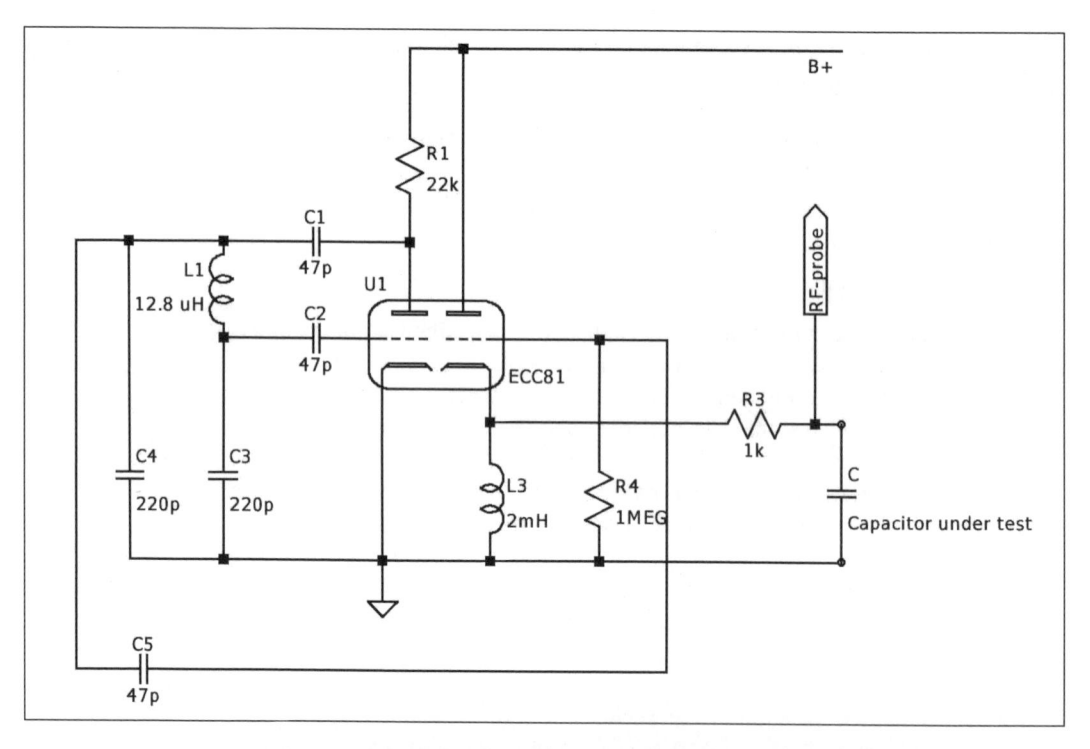

Fig 264: VTVM addition to measure capacitances

Ref.	Description	Ref.	Description
C1	capacitor, 47pF	L1	inductor, 12.8H
C2	capacitor, 47pF	L3	RF choke, 2mH
C3	capacitor, 220pF	R1	resistor, 22k
C4	capacitor, 220pF	R3	resistor, 1k
C5	capacitor, 47pF	R4	resistor, 1M

Components list for VTVM addition to measure capacitances

The diagram is simple. The left hand side of the valve forms the oscillator. The frequency is adjusted to 3.0MHz by means of the iron core of L1. The right hand side of the valve forms a cathode follower to isolate the oscillator from the load and other influences from the capacitor.

With a 1k resistor in series with the cathode follower output, the capacitance range given in the table (10 - 360pF) is achieved. The 5V output voltage is measured across L3 and adjusted with B+ or by varying R1. If those parameters are met, the values in the table can be used and copied straight into the meter scale.

The capacitance range 10pF - 1,800pF will be covered with a frequency of 500kHz, in which case the coil L1 needs to be 460.6µH. Even lower frequencies may be used to measure even bigger capacitances.

The VTVM should be set to a 5V range for best spread of the values. Lower ranges can, of course, be used for a better accuracy of bigger capacitances.

This way of measuring capacitance is by no means new. In fact, many modern instruments have this facility built-in.

Inductance	Reactance	Output Voltage
1	18.85	0.09
2	37.70	0.18
3	56.55	0.27
4	75.40	0.35
5	94.25	0.43
6	113.10	0.51
7	131.95	0.58
8	150.80	0.66
9	169.65	0.73
10	188.50	0.79

Table 18: The relationship between inductance, reactance and voltage

Measuring inductances

Inductances are different from capacitances. The inductive reactance is, as shown earlier, given by:

$$X_L = 2 \times \pi \times f \times L \qquad \text{Formula 13}$$

Again, we have control over f and the output amplitude of the oscillator. The same frequency and series resistor are assumed, which gives an inductance range of 1μH - 1mH, which should be sufficient for most RF applications. **Table 18** shows the relationships.

The same interface as described above can be used. The only difference is that the unknown coil is connected instead of the capacitor under test in the schematic. No changes need to be made to the circuit. Again, a different frequency yields a different inductance range.

Measuring resistance

Measuring resistance is based on the same principle as described above. The difference is generally that, instead of an oscillator signal, a battery is connected as signal source. Using the oscillator principle for measuring resistance does have benefits, though, apart from not having to modify the design.

Different resistors behave differently in DC situations and AC situations. Many wire wound resistors, to use the worst case as an illustration, contain capacitance as well as inductance, apart from resistance. Using the oscillator gives you the whole truth and not only the resistor's behaviour as a pure DC resistance.

Should you prefer to know the DC behaviour instead, connect a battery to the left hand end of R3 instead of the cathode follower, and use a DC probe for measurements.

Measuring currents

A panel instrument is frequently used to measure current and can be very sensitive. To increase its range, it needs shunt resistors. They are calculated as shown in Formula 32.

$$Rs = \frac{R_M}{n-1} \qquad \text{Formula 32}$$

where
Rs is the shunt resistor
R_M is the internal resistance of the instrument
n is the full scale of the instrument.

Suppose you have an instrument with 10Ω internal resistance and 10mA full scale. You want it to measure 100mA instead. The meter needs a shunt resistor of

$$Rs = \frac{10}{100-1} = \frac{10}{99} = 0.101 \Omega$$

If, instead, you want it to measure voltage, you have to connect a series resistance to the meter. The resistor is calculated as in Formula 33:

$$Rf = R_M \times \left(\frac{V}{V_M} - 1\right) \qquad \text{Formula 33}$$

where
Rf is the series resistance
R_M is the internal resistance of the instrument
V is the desired full scale voltage
V_M is the voltage across the meter (which can be found using Ohm's law).

V_M can be ignored to give the simpler Formula 34:

$$Rf = \frac{1000 \times V}{I} \qquad \text{Formula 34}$$

where
I is the current through the meter.

So, a meter with a 5mA range to measure 100V needs a series resistor of:

$$Rf = \frac{1000 \times 100}{5} = 20,000 \Omega$$

A usual method of measuring current is to place a 1 (or 10 or 100) ohm resistor in the circuit and measure the voltage drop across it. Those resistors can be built into the instrument and chosen with a switch.

AC and RF currents would, of course, have to involve the RF probe.

17

Fault Finding

THE FIRST STEP WHEN trying to find a problem is to locate the faulty stage. You will find a block diagram of a superheterodyne receiver in **Fig 265**. The power supply is not included in the diagram.

In this case, an RF amplifier has been included. Some, not all, domestic receivers did have them, and could therefore be a cause of problems.

Of course, the best aid to your measurements is a schematic diagram of the radio you are fault finding. Lacking that, you have to rely on your instinct and logical thinking. Different stages require different approaches, because they are built in different ways. Examples of how the stages may be designed follow below.

The general idea is to find out whether the voltages and currents of the valve are reasonable within the valve specifications. If not, a component at or near the pin is faulty.

By inspecting the innards of the radio, you could spot a suspect component, such a burnt resistor, a loose valve, a valve that is not glowing or warm (it is sometimes difficult to see), or an open connection. Two wires that cross each other, or run close to the chassis, could, even though they are insulated, be worn out or aged such that there is a short circuit between them. Check for those too.

It is advised that you use the VTVM or an instrument with a similarly high input impedance. An oscilloscope is not suitable in some cases.

Safety Warning
The domestic mains supply has potentially lethal voltages and currents. Also, the voltages produced by a valve power supply can be very high and therefore dangerous. Great care must be taken when working on equipment that is switched on, or is switched off but still connected to the mains supply.

Fig 265: Block diagram to aid fault finding

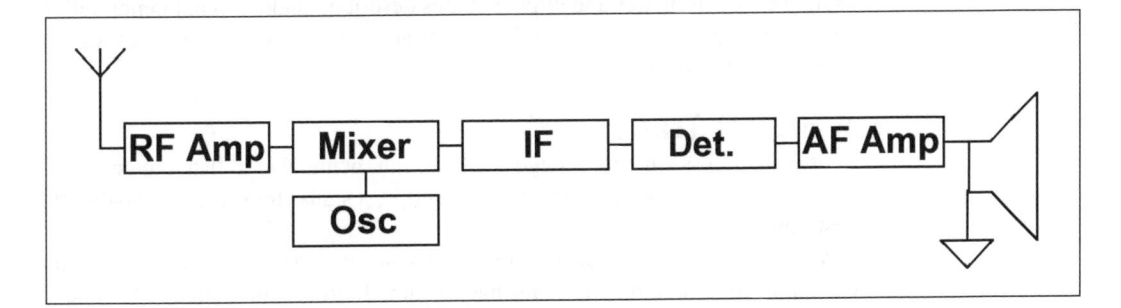

237

These may be causes of a radio having no audio output at all, not even hum.

Once you have established that everything looks as expected, it is time to check the power supply. Power supplies in domestic receivers weren't stabilised, which makes fault finding a little easier. Taking note of the safety notice on the previous page, set the voltmeter to AC measurements and measure the output of the transformer. Remember that most power supplies have a centre tap at the high voltage winding. Measure both halves of the winding.

Also measure the heater voltage. It could be 6.3V AC or 12.6V AC. Look through Appendix 1 at the end of this book, and you will find a list of various heater voltages or currents.

Next find the cathode of the rectifier valve. You should have a DC voltage there in the order of a couple of hundred volts. This voltage is in reality a pulsating DC, which is filtered by a resistor (in some cases a choke) and one or two electrolytic capacitors. Loud hum can be a sign of old and dried out electrolytic capacitors in the power supply. A low HT voltage is also a sign of a malfunctioning power supply filter. The capacitors should therefore be replaced before you continue.

If the power supply appears to be working correctly, it is time to use more sophisticated methods and instruments. We assume in the following that the power supply is fully functional.

Using a Signal Injector

If you own a signal injector or a signal generator, you start from the right in the diagram above (at the loudspeaker). Connect an oscilloscope at the loudspeaker connections. Inject an AF signal into the input of the oscilloscope and check that the signal is there. If not, there might be a short circuit in the output transformer secondary.

Next, inject the signal at the control grid of the output stage. If you had a signal before, but not now, the power amplifier is at fault.

Continue like that, step by step, and work your way toward the aerial input of the radio until the signal disappears or is distorted. Remember to inject a signal on the appropriate frequency, AF for the audio stage, IF for the intermediate frequency stages and RF (on the frequency the radio is tuned to) for the earlier stages. You may need to vary the signal generator's frequency a little to make sure it is correct. Then you know which stage to investigate closer. However, leave the oscillator for the time being.

Remember that when you check the detector stage, you need to inject a modulated IF signal, or a signal with many overtones. A signal injector can be built using the R/C oscillator and clipper stages described under 'Signal generator'. If the output signal is 'sharp' enough, it is capable of generating overtones way up into the short wave band.

A Signal Tracer

In order to assess whether a stage works or not, being able to listen to or see the signal is a great help. Lacking an oscilloscope, a signal follower is probably the best option.

A signal follower is basically a special kind of a radio receiver. It lacks the oscillator/mixer combination and has no tuned circuits to enable it to tune in

CHAPTER 17: FAULT FINDING

Fig 266: Schematic of the signal tracer

anything. A good signal follower can receive signals from RF stages, rectify and display them, as well as showing or making AF signals audible.

The unit presented in **Fig 266** has one wide band RF amplifier, a rectifier, a 'magic eye' as well as an AF amplifier with a loudspeaker. A switch determines what types of signals are being monitored.

The switch is set for RF monitoring. The first valve, U1 (EF91), receives the signal from the probe and amplifies it. At the anode, Dr1 stops the RF. R2 / C2 form yet another RF filter.

The signal is taken from the anode and connected to the cathode of the diode U2, which rectifies it. At the anode of the diode is a pulsating DC voltage, which is smoothed by C5. R5 / C2 filter most of the remaining RF out. What reaches the AF amplifier, ECL82, is the modulation, if any.

The 'magic eye' valve U3 (EM84), receives the rectified RF from the diode and displays a bright bar, whose length is controlled by the amplitude of the RF, modulated or not.

The AF amplifier is pretty standard. It amplifies the modulation and feeds it to the loudspeaker. That way it is easy to hear if the RF signal is modulated or not.

The other position of S1 connects the probe directly to the AF amplifier, bypassing the RF amplifier, the diode and the 'magic eye'. It replaces, in other words, the AF stage of the receiver you are debugging.

Note that the capacitors connected to the probe, C1 and C16, need to be high voltage too.

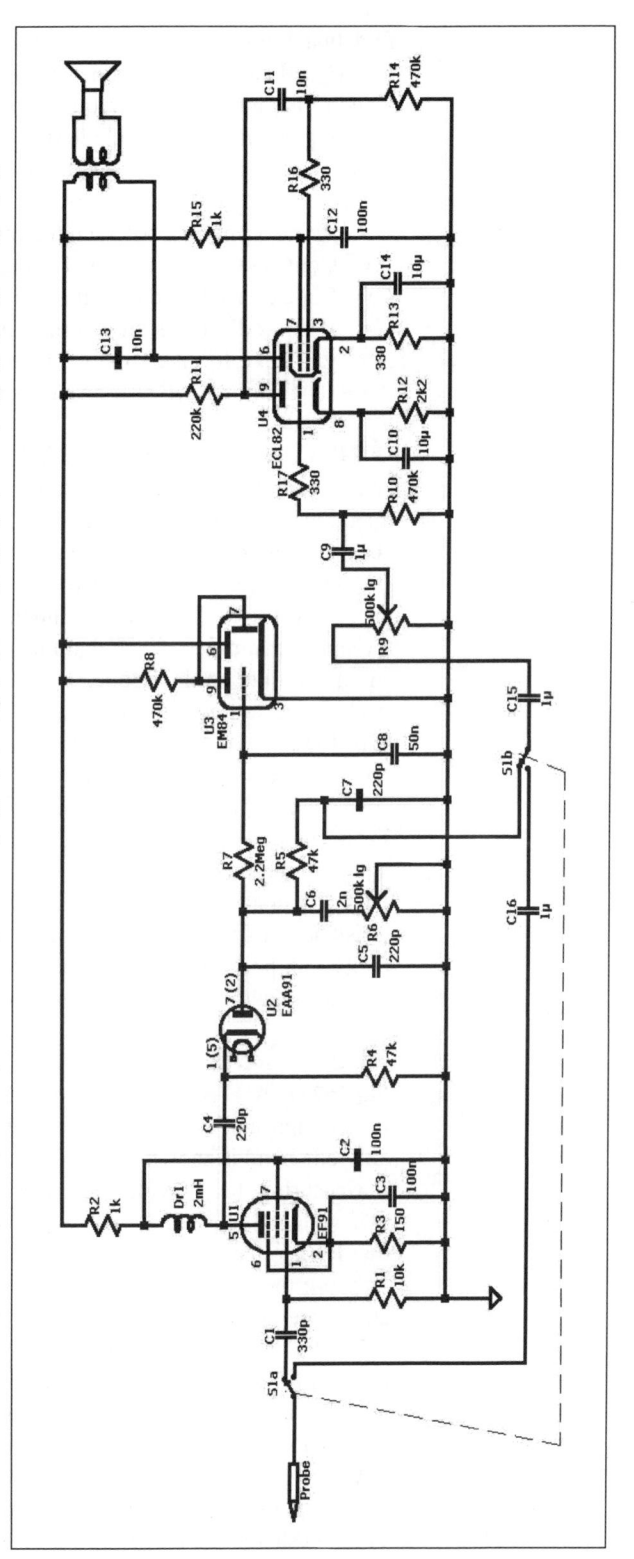

If you compare this circuit diagram with one of a domestic radio, you will find that there are only a couple of small differences. The tracer lacks the mixer/oscillator combination near the aerial, and the IF stage of the radio is tuned to the intermediate frequency. Therefore, re-arranging an old radio could definitely be an economical alternative.

The modifications that need to be done are:

- Remove the link between the mixer output and the IF stage.
- Connect the probe to the IF stage control grid.
- Replace the tuned circuit of the IF stage with a resistor.
- If the radio lacks an external input for a gramophone or a tape recorder, you have to take suitable actions to switch the probe between the input stage and the AF section of your radio.

Using a Signal Tracer

When working with a signal tracer, you do quite the opposite to the signal injection method described above. You begin at the left end of the diagram above and work your way to the output stage.

Connect a signal generator (or a signal injector) to the aerial input and tune the receiver to the generator's signal. Connect the tracer (or an oscilloscope) to the same terminal to check that the signal is OK. It may be very weak, though, and not visible on the 'scope.

Then check the output (anode) of the RF stage, if existing. Leave the oscillator for the time being and continue with the outputs of the other stages, until you have identified the faulty stage.

The oscillator's function can not readily be assessed using a signal tracer. Therefore, after all other stages have been checked, follow the instructions under 'The mixer' below.

Once you have identified the faulty stage, measure the voltages, currents and resistances of the stage according to the descriptions below to identify the faulty component(s). When these have been replaced, check the radio again with the signal tracer to make sure all stages are working as expected.

The RF Stage

Fig 267 shows a typical design of a RF stage. The points to measure are marked with a digit and a bracket. For clarity, the band switch has been omitted, and only one band is shown.

Measuring the DC voltage at point 1) tells you if the screen grid has the correct voltage. The voltage here could be negative or zero. That would imply that R1 is open-circuit or C1 is short circuited. Turn off the power supply and measure the resistance across R1. Also measure the resistance across C1. It should be open-circuit.

Point 7) should have a lower voltage than B+. If not, no current is flowing through R2, which indicates that either the primary coil at the anode is broken or R2 is open-circuit. If the voltage is zero, C5 could be short-circuit. The function of R2 and C5 is to prevent RF from spreading through the radio via the power supply.

Point 3) indicates whether the primary coil in the anode circuit is open-circuit. Point 2) should have a voltage corresponding to the control grid bias, a couple

CHAPTER 17: FAULT FINDING

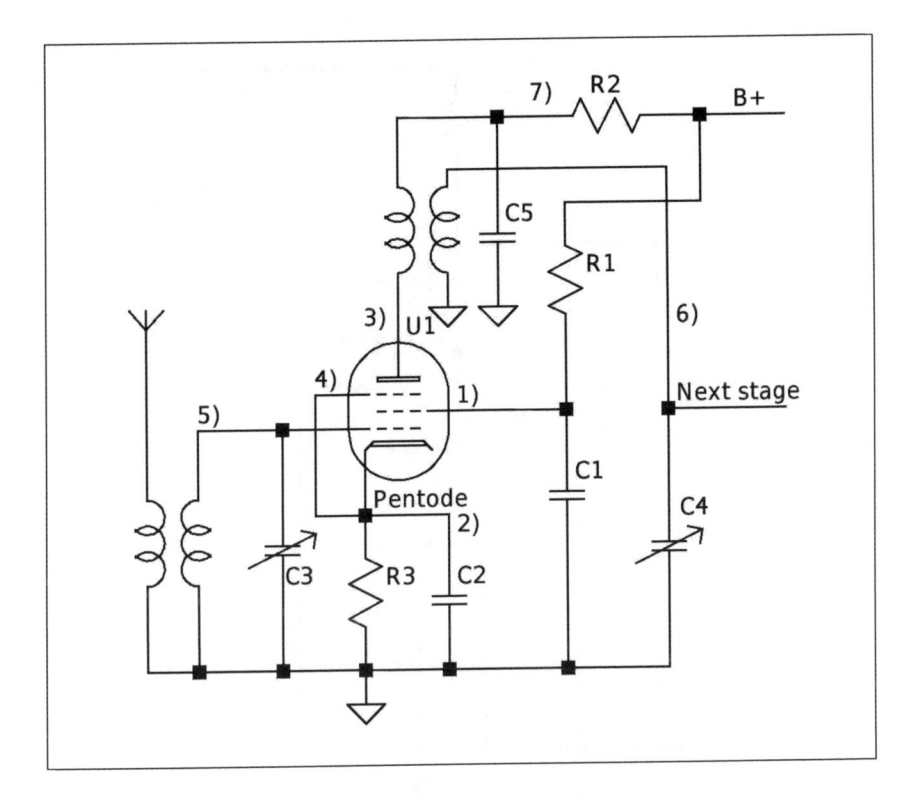

Fig 267: Principal design of RF stage

of volts. If zero, C2 could be short-circuit. If too much, R3 could be open. Point 4) should have the same voltage.

The control grid itself (point 5) should have zero volts. It is grounded via the secondary coil of the aerial circuit. An open control grid circuit tends to collect electrons which have nowhere to go, so they form a negative voltage. Also check for short circuits in C3.

Check the secondary coil of the anode transformer for resistance. Its corresponding capacitor section, point 6) could also be faulty.

The primary coil of the aerial circuit should also be checked for resistance.

The Mixer

A typical mixer circuit frequently contains both the oscillator and the mixer proper. **Fig 268** shows the mixer portion. The oscillator components have been omitted for clarity. The oscillator is described separately below.

Like in the RF stage, there is a resistor / capacitor combination, R3 / C12, which is intended for protection. At point 1) you can see if the resistor and capacitor are functional (see above).

The anode at point 2) should have the same voltage, somewhere in the order of 250V.

Point 3) should have a voltage in the neighbourhood of 100V. Check the valve's data sheet for details. If not, take a closer look at R1. There might be a decoupling capacitor connected to this point, which could be faulty.

241

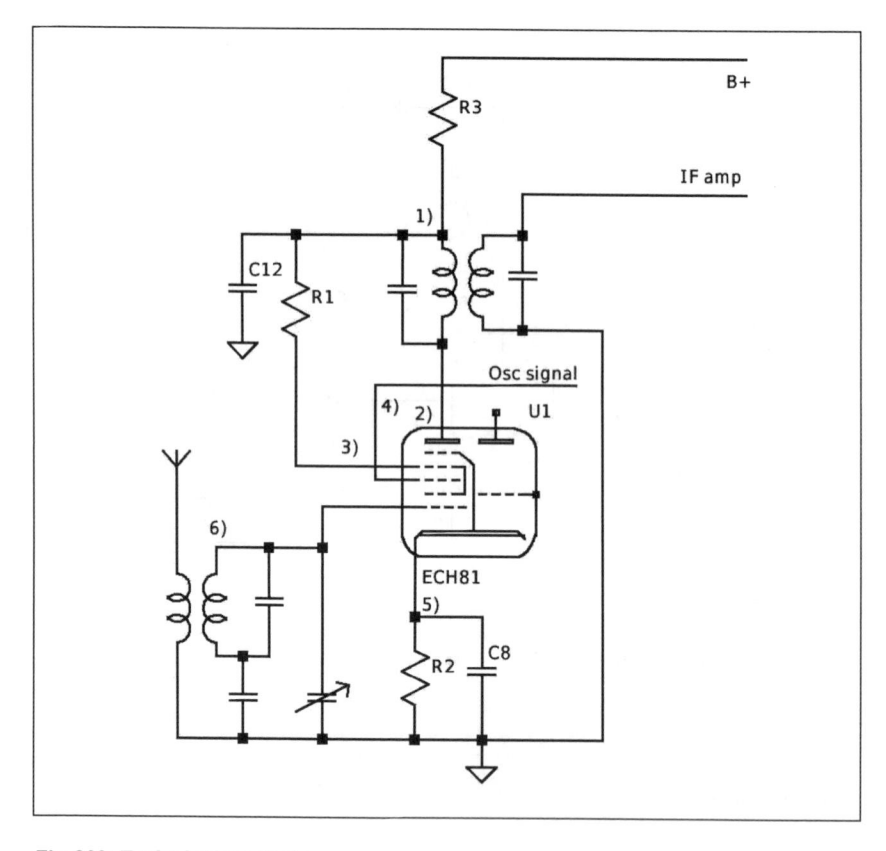

Fig 268: Typical mixer design

Point 4) contains the oscillator voltage and should have a positive voltage of a couple of volts.

Point 5) should have a similar voltage. If not, check R2 and C8. If C8 is open-circuit (that is, it has stopped working as a capacitor), then the oscillator will probably stop working due to the negative feedback across R2.

At point 6) you can check the coils and capacitors at the input. The two capacitors closest to the coil in the diagram could be trimmers.

Watch out for oxidations in wave band switches!

Some mixers are fitted with AVC. Without an input signal, the AVC should not affect the voltages, though. AVC is taken from the detector and frequently connected to the control grid of the mixer via a high-resistance resistor with a decoupling capacitor at the other end.

The Oscillator

The corresponding oscillator section is shown in **Fig 269**. Only the components relevant to the oscillator are shown here.

The cathode has been checked already. What remains here are the oscillator anode (point 1) and the frequency determining components (point 2). R3, C4 and C9 should be checked separately. Point 1) should have a lower anode voltage than the mixer, but still in the order of 100V - 200V.

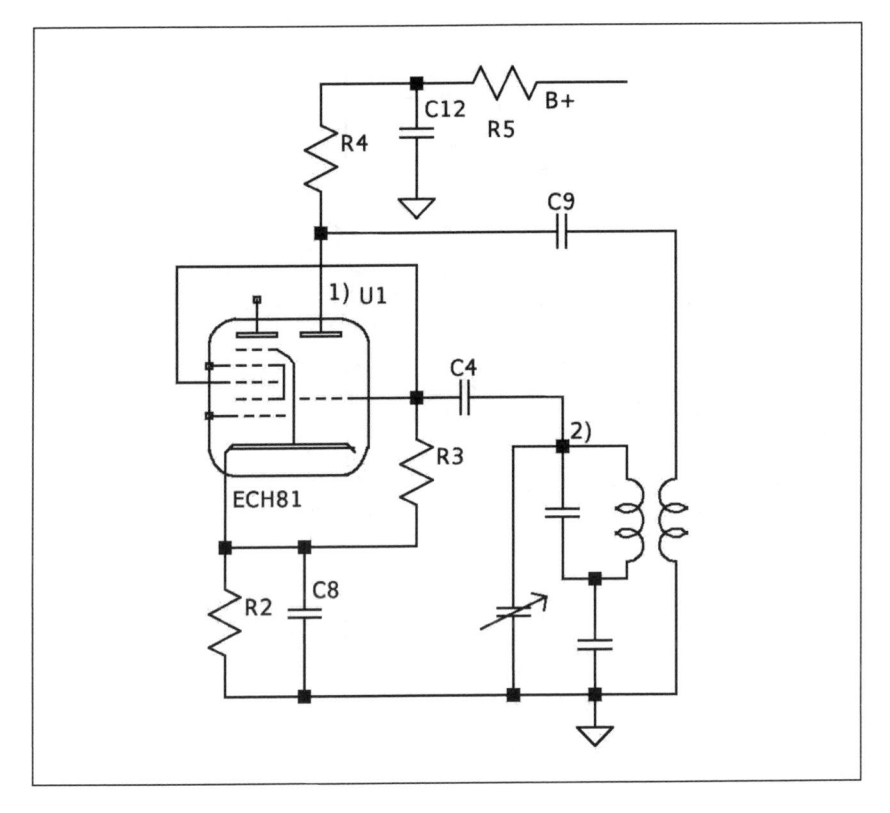

Fig 269: Typical oscillator design

A simple way of checking the oscillator would be to tune a separate receiver (or a wave meter, see the section on grid dip oscillators) to the frequency where the oscillator is expected to oscillate.

In domestic receivers, this frequency would be 455 or 465kHz above the frequency to which the receiver under test has been tuned. No indication means no oscillations.

The IF Stage

A typical IF stage is shown in **Fig 270**. The signal is taken from the secondary coil of the mixer output transformer. The points to check are:

Point 1): The anode voltage should be about the same as the B+ voltage. If there is an R/C-filter in the anode circuit above the coil, the voltage should be lower.

Point 2) is the cathode. A couple of volts.

Point 3) is the screen grid voltage. It could be about the same as the anode voltage, or lower, depending on which pentode is used.

Point 4) is the input to the next stage, and checks the secondary of the IF transformer. The best check here would be measuring the resistance across the coil (with the power off!).

If you attempt to look at the output signal with an oscilloscope, be aware that you might fail. The anode impedance of a pentode stage is very high.

VALVES REVISITED

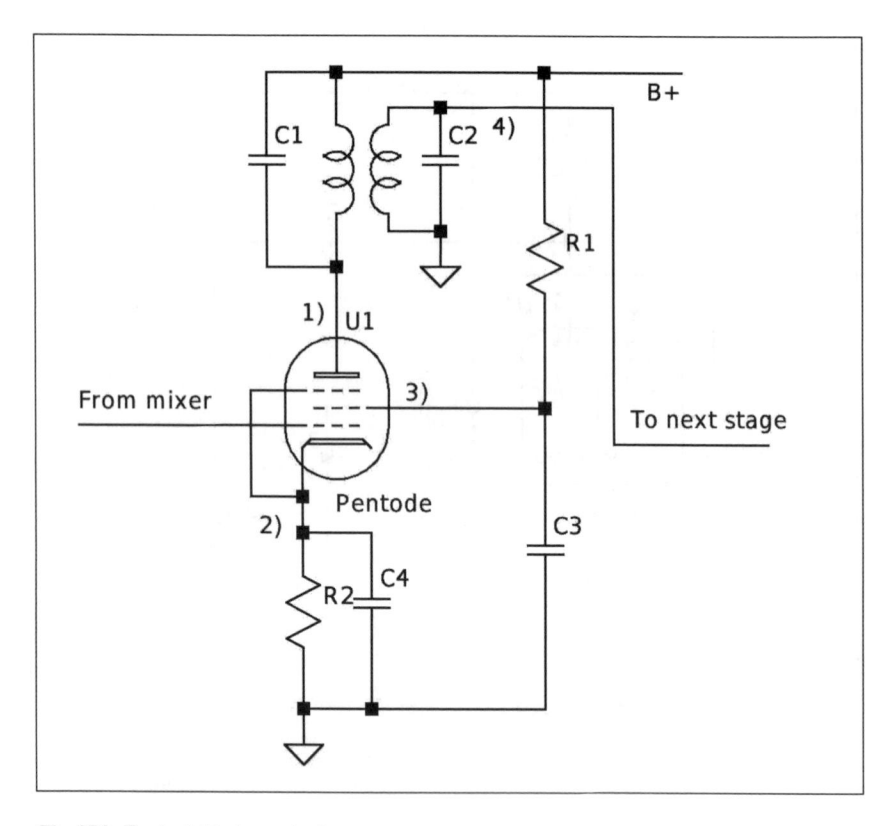

Fig 270: Typical IF stage design

The Detector

Detectors come in many flavours. One that is fairly common is the 6AT6 which is a triode with two built-in diodes - a duodiode-triode, such as the detector shown in **Fig 271**.

Where both diodes are used, one generates the audio signal and the other provides the AVC voltage. This is a simple detector, found in small table top radios.

Only one diode is used here, which provides both AVC voltages and audio signal. Each AVC controlled stage has its own R / C filter to prevent them from interacting. R4 / C5, R5 / C6 and R6 / C7 are shown here for clarity (normally, they would be drawn closer to their respective stages).

The active diode draws current during the positive half period of the IF signal from the tuned circuit L1 / C1 inside the screening can. The primary is connected to the anode of the (last) IF stage.

At point 1) is a smoother copy of the detected IF signal. C2 provides the smoothing of the signal. C2 is also grounding the lower end of the tuned circuit. The signal is then fed to the control grid of the 6AT6 via R2 and P1 (which acts as the volume control), and C3. R3 is the grid leak of the triode. The signal is amplified in the triode and fed to the power amplifier through C4.

It was common to route AVC signals to the mixer, the IF amplifier and the 'magic eye', if any. RF amplifiers were controlled too.

CHAPTER 17: FAULT FINDING

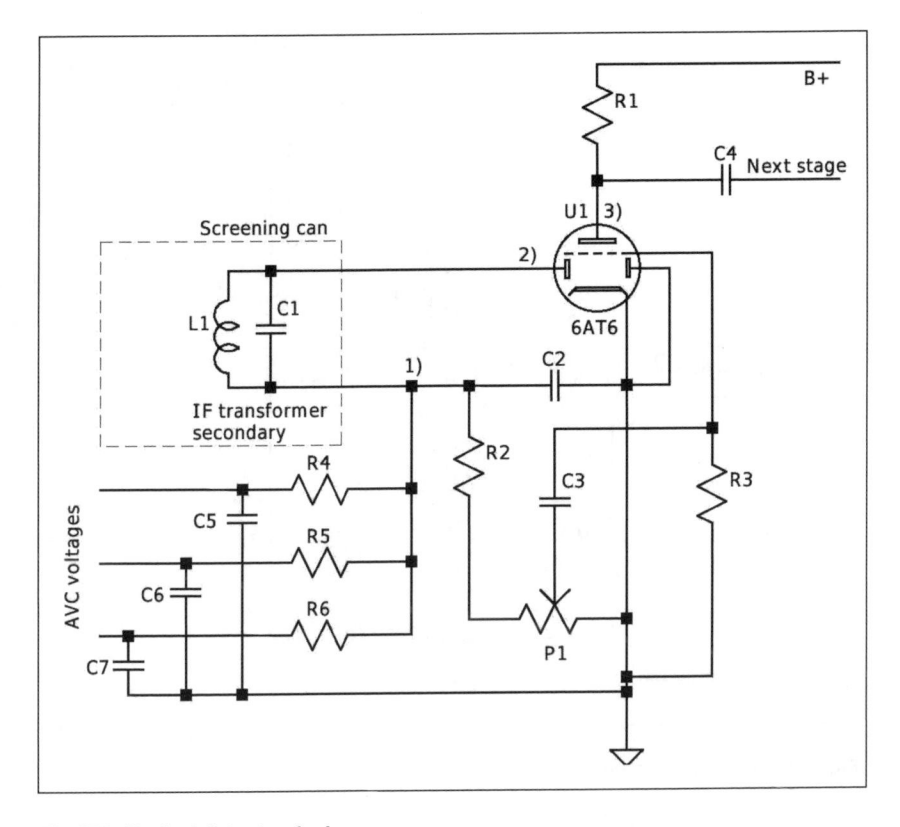

Fig 271: Typical detector design

The voltages at 1) and 2) should be the same, and the signal should vary with the modulation of the input signal. The voltage at 3) may be rather low - check the value of R1 and the voltage at B+.

The AVC voltages should vary with the amplitude of the input signal and go negative when the amplitude increases.

R4 - R6 are big, in the order of a couple of megohms. The corresponding capacitors may be in the order of 50nF. This gives a decent isolation between the controlled stages.

Other radios use other valves for detection. EABC80 (a triode with three diodes) is not uncommon in radios for FM/AM use, where one diode is used for AM detection and AVC generation, and the other two for FM detection. The principle is the same, though - the IF signal is connected to the anode of the diode and the AF signal is taken from the bottom end of the transformer secondary, filtered and then fed to the triode. Instead of a triode, the amplifying section of the valve can be a pentode.

The Output Amplifier

Sometimes the detector / pre-amplifier is followed by a network of resistors and capacitors. This is the bass/treble control.

The final stage is the power amplifier, like the one shown in **Fig 272**. This is frequently a power pentode, and its purpose is to generate enough power to

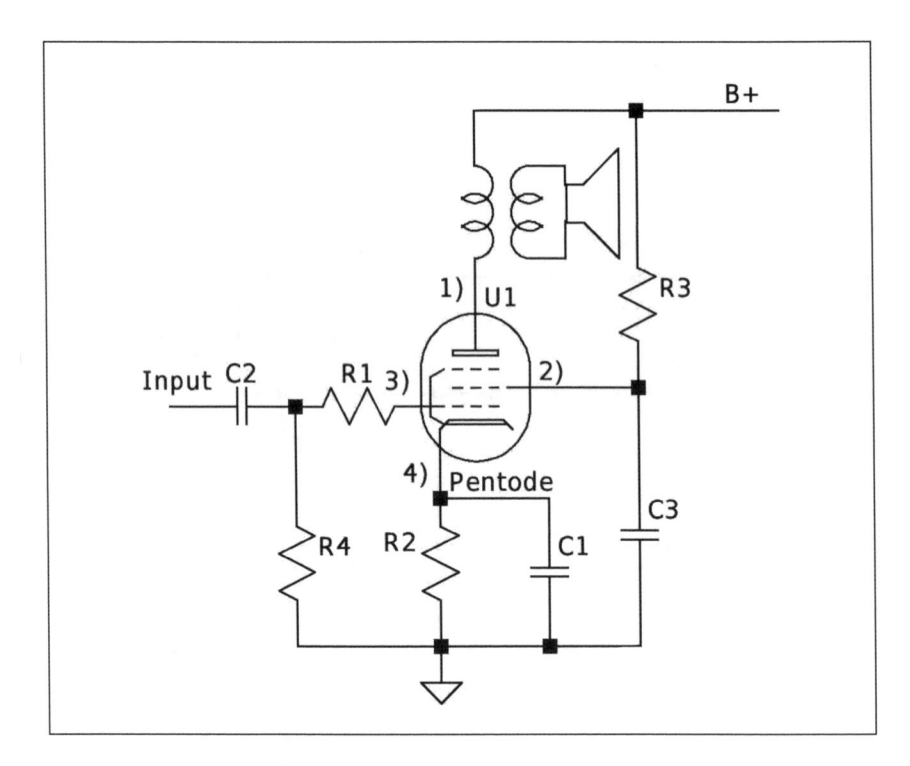

Fig 272: Typical power amplifier design

drive the loudspeaker, which, for matching purposes, is connected to a transformer at the anode.

The voltages are measured at points 1) - 4) and should correspond to the information in the data sheet.

Some pentodes have an internal connection between the cathode and G3; others don't. In some cases there may be a negative feedback for improved sound quality. The feedback chain usually starts at the secondary of the output transformer and ends at the grid or cathode of a preceding stage.

18

Transmitters

SO FAR, THIS BOOK has described valve technology mainly as used in radio receivers, which is a fascinating subject in its own right. However, in order for you to be able to communicate with other people all over the world, through the fascinating medium of radio, you also need a transmitter.

However, a transmitter imposes some additional requirements as compared to receivers:
- If the aerial impedance does not match that of the transmitter output, the transmitter may be destroyed.
- The voltages required to run a transmitter are in most cases much higher than those required for a receiver.
- If the output power of a transmitter is high enough, RF power may be induced into metal objects nearby, and touching those may result in bodily damage.
- The output of the transmitter requires filtering to avoid RF radiation on frequencies other than the intended.
- The output stage needs to be screened from the rest of the transmitter to avoid RF feedback into other circuits.

Stages Needed

The simplest type of transmitter is a CW transmitter for Morse code. It consists of an oscillator, possibly followed by a keying stage, followed by a power amplifier.

The need for a keying stage is motivated by the fact that an oscillator takes some time to reach its nominal frequency, even if warmed-up. During this time, it still emits RF, and when the signal is received, it sounds 'chirpy', a phenomenon which was not unusual during the 1960s and 1970s when home-built simple equipment was commonplace in the ham community.

One way of getting round most of the 'chirp' problem is to use a crystal controlled oscillator, which has very high frequency stability. This is illustrated in

Licensing
Note that the use of radio transmitting equipment of the kind described here requires a licence. Details of amateur radio licensing in the UK can be found at www.rsgb.org

VALVES REVISITED

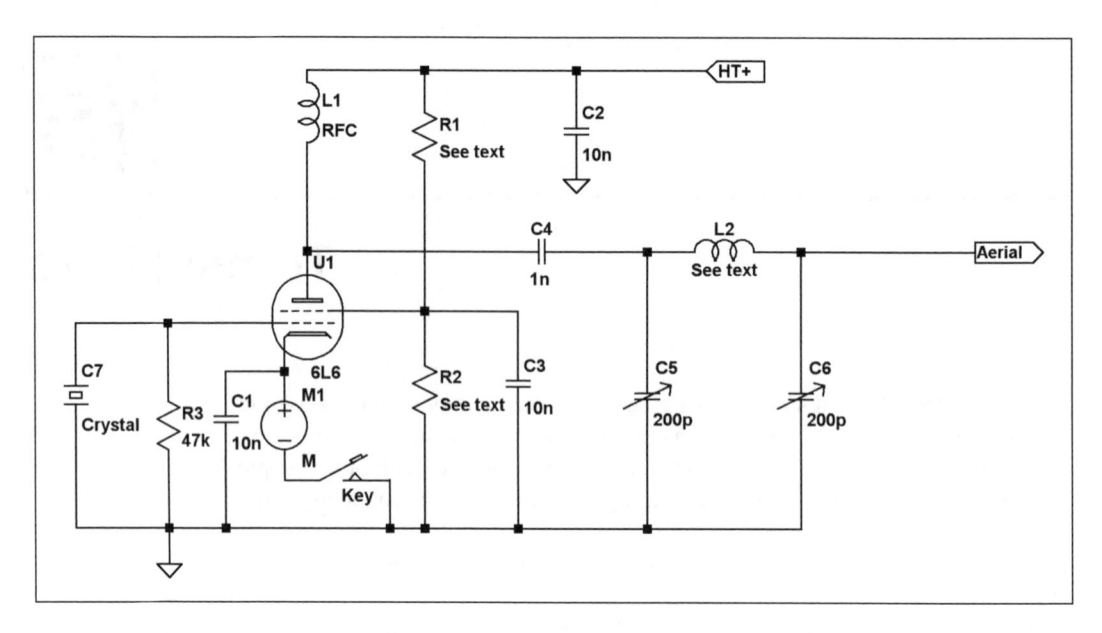

Fig 273: Crystal controlled valve CW transmitter

Fig 273. It eliminates the keyer stage as well as the power stage. The valve illustrated in the figure, the 6L6, is capable of running at 15W input power, which is sufficient for most CW needs.

Other valves may be chosen as well, such as 6V6, 6F6 or 6W6. A 6146 would be able to run at 50W or more. If a 6146 is used, the transmitter may be powered by 400V. 10k for R1 and 47k for R2 would give a 6146 an appropriate screen grid voltage.

A CW transmitter also needs some sort of keying filter. If the carrier is turned on and off too abruptly, the transmitter begins to generate 'clicks' every time the carrier is switched.

A simple filter is shown in Fig 273, and comprises a capacitor between cathode and ground. Since the capacitor is charged initially, it will discharge through the key and meter M1, which takes a little while, enough for the carrier to start smoothly, thereby reducing the 'clicks'. Likewise, when the key is released, the capacitor needs to charge again, with the same result.

The values of the voltage divider R1 / R2 depends on which valve is used and the high voltage. Check the data sheet for information on the valve used.

For the 80 meter band, L2 should be wound on a 30mm diameter former, about 36 turns. For 40 meter operation, 16 turns will be enough.

The combination C5 / L2 / C6 form together a low pass filter which, as mentioned above, will reduce radiated frequencies other than the wanted signal. Additionally, a pi-filter at the output of the transmitter will, when correctly tuned, match the output impedance of the transmitter to the input impedance of the aerial.

C2 may be an air-spaced receiver type capacitor - its vanes do not need to be far apart. This is a low power transmitter and does not generate voltages that are big enough to damage your capacitor.

The crystal and pi-filter can be made switchable to facilitate band switching or frequency switching within the band.

Buffer

Often a low-amplification 'buffer' stage is inserted between the oscillator and the power amplifier, as in the AM transmitter shown later in this chapter. This allows the use of a tunable oscillator which is not affected by the tuning of the PA. It is also a useful place to key the transmitter, thus avoiding chirp.

Tuning the Transmitter

Tuning a transmitter is inevitably more complicated than tuning a receiver. In case of pi-filter output, this method is suggested.

The best indicator would be a dummy load. This could consist of a number of low-inductance (eg carbon) resistors connected in parallel so that the resulting resistance matches the impedance of the aerial. The total power rating of the resistors must exceed the expected output power of the transmitter.

A switch in series with the output could select between the dummy load and aerial. Tuning should initially be carried out into the dummy load, then re-adjusted when connected to the aerial (antenna).

The tuning method is a follows (refer to Fig 273):
1. For high power amplifiers, reduce the high voltage so as not to damage the power valve(s).
2. Set C2 to maximum capacitance (all vanes in).
3. Tune C1 to minimum current as shown by the meter.
4. Decrease C2 by about 10%, thereby increasing output into the aerial.
5. Repeat from point 3 above until satisfied.

A Two-stage Transmitter

Another mini transmitter is shown in **Fig 274**. It is a single valve crystal controlled transmitter, based on a valve which is usually found in TV equipment, the ECL80.

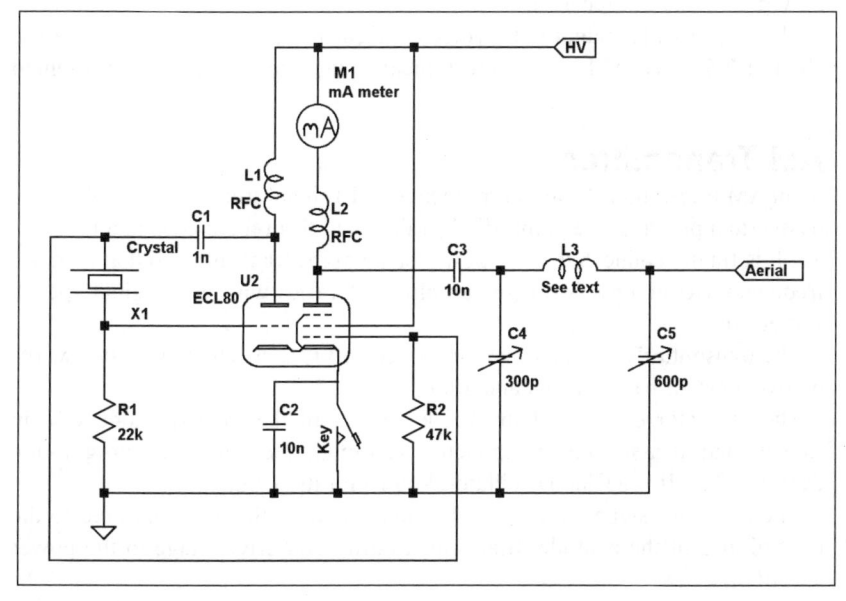

Fig 274: Simple two-stage CW transmitter

The circuit is simple. The triode part of the valve acts as the crystal controlled oscillator, and the pentode section is the power stage. Note that, since the two valve halves have a common cathode, both sections of the valve are keyed at the same time.

For that reason, the ECL82 would probably be a better choice. The triode section of the ECL82 would be grounded directly, whereas the cathode of the pentode section is keyed.

L1 and L2 are both RF chokes. A suitable value for both is about 2.5mH. The resistors are both rated at 0.5 watts. The meter measures the anode current and should have a range of 30mA.

The coil L3 is the pi-filter coil and needs to be wound according to which band you are transmitting. Typical values for a 36mm former are: 80m band, 32 turns; 40m, 20 turns; 20m, 10 turns.

For 20m band operation, a 40m (7MHz) band crystal may be used, and the power stage tuned to the second overtone (harmonic). It is important to use a wavemeter to ensure that the correct harmonic has been selected.

Tuning is performed as described above. The optimum point is reached when the meter reads 20 millamps at the bottom of the dip, with a high voltage supply of 200 - 220V, which is easily obtainable from a receiver type power supply.

Communicating by Voice

Voice communication is a lot faster and easier than using Morse, but the transmitter is more complicated.

There are several types of voice transmission. Amplitude modulation (AM) is one method, rarely in use by amateurs any more, but still used extensively on the LW, MW and SW broadcast bands.

Nowadays single sideband (SSB) is the most popular amateur voice mode. However, SSB requires a BFO at the receiver end, whereas AM can be received on virtually any old receiver.

The majority of amateur radio voice contacts in the heyday of valves (before about 1965) used AM, and some typical transmitter projects are presented here.

AM Transmitter

In an AM transmitter the audio frequency (AF) signal from the microphone is passed to a pre-amplifier, amplified further, and from there goes to a modulator. It is finally connected to the power stage of the transmitter so that its radio frequency (RF) output level (or 'amplitude') varies with the original speech waveform.

The transmitter's RF stages are presented in **Fig 275**. The transmitter works on two amateur bands, 160m and 80m.

The oscillator consists of the triode section of the triode-pentode U1, an ECF80, and its associated components. You might recognise the configuration from Fig 229. It is a Clapp oscillator, known for its stability.

The oscillator signal is taken from the cathode of the triode and fed to the control grid of the pentode. It acts as a buffer and driver stage to the power amplifier.

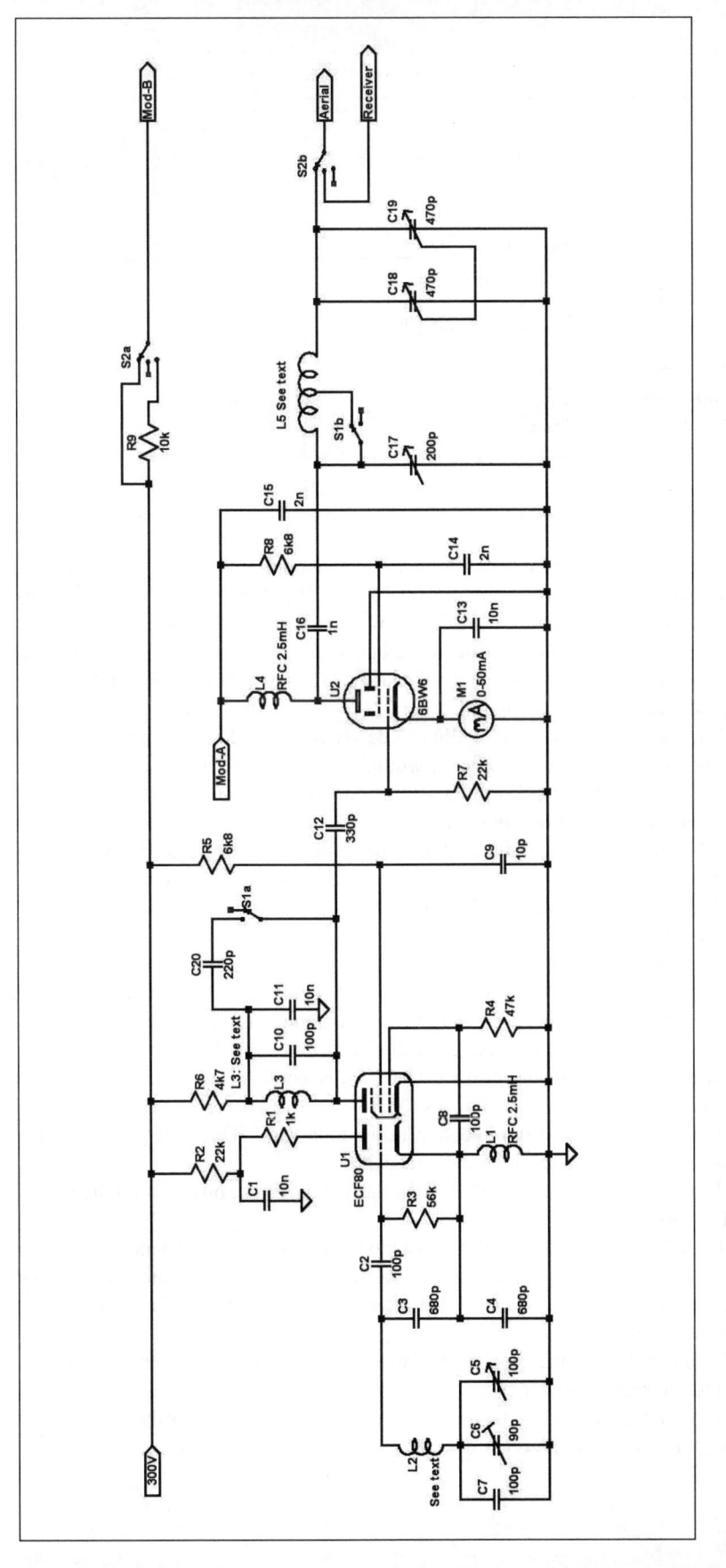

Fig. 275: Transmitter section of the AM transmitter

VALVES REVISITED

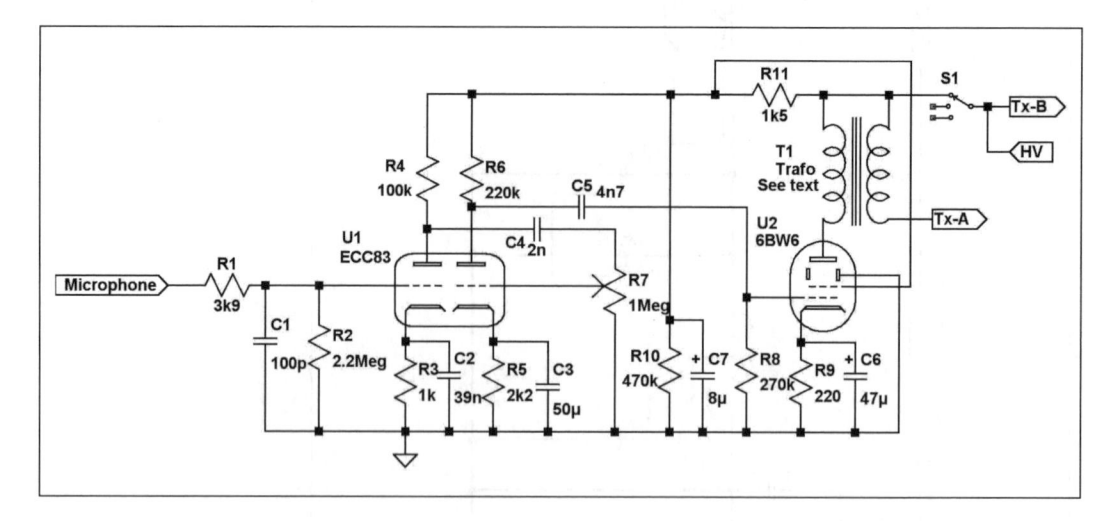

Fig 276: Modulator section of the AM transmitter

The pentode serves as a buffer on 160m and a buffer/doubler on 80m. Its tuning coil, L3, is tuned by C10 and, on 160m, C20. Once tuned, the circuit needs no further tuning. Tuning is achieved by the adjustable core in L3.

The power valve, U3, is a 6BW6, a beam tetrode, capable of 4.5 watts anode dissipation. It may comfortably be loaded to 10 watts, though. A 50mA meter in its cathode lead is a tuning aid.

The output structure is a pi-filter to decrease radiated harmonics.

The coils L2 and L3 should be wound on formers with an iron dust core to allow for adjustments. Formula 10 in Chapter 5 is a good help, but allow for the core to increase the inductance.

L2 is the most critical one and must be solidly constructed so that you end up with a good, stable oscillator. L5, the pi-filter tank coil, is best air wound.

The signal from the modulator (see below) is fed to the anode circuit of the power stage.

Two switches are used to change bands and mode of operation. S1a and S1b are band switches. The switches are shown in position for 80m operation. Section b short-circuits a part of L5, which is the tank coil, and section a adds or removes a capacitor to the resonant circuit in the anode of the pentode.

Instead of tapping the coil, different coils could be used, selected by the switch.

S2a and S2b have a third section in the modulator. This is the mode switch. The sections are shown in 'transmit' position.

The others are 'receive' and 'net'. In the 'net' position, power is applied to the VFO and buffer stages only.

Fig 276 shows the schematic of the modulator. It uses transformer modulation and controls the power stage of the transmitter. At the modulator power stage anode is a transformer with a 1-to-1 ratio. It could be a mains valve transformer, for instance.

This transformer picks up the variations of the anode current and brings them to the transmitter power stage.

The other end of the transformer secondary is connected to the high voltage line, the transmitter receives a varying DC from the modulator.

When connecting the modulator and transmitter together, connect 'Tx-A' to 'Mod-A' and 'Tx-B' to 'Mod-B'. The 'HV' terminal receives 300V from the power supply.

Like the transmitter, the modulator is a simple design. A crystal microphone is connected between the 'microphone' input and ground. U1, an ECC83, serves as a two-stage AF amplifier, simple and straight-forward. The R1 / C1 combination serves as a low pass filter for RF and is intended to prevent parasitic oscillations and RF feedback from the transmitter.

R7 serves as a modulation depth control and should not allow the signal to overdrive the transmitter output. Instead of a panel mounted potentiometer, this could be a trimmer. As a note of interest - sometimes a grid DC connected to a potentiometer like this causes a noisy pot over time. A better solution is to use a capacitor in series with the grid and a grid leak next to it. This solution saves, however, two components.

From the output of the AF amplifier, another 6BW6 receives the microphone signal. It is amplified and fed, as described, to the transmitter's RF section.

The transmitter could easily be fitted for CW operation as well. This requires that the high voltage fed to the ECF80 be stabilised. Best result is achieved if both sections of the valve are stabilised. The purpose of this is to avoid 'chirps' when keying, as described earlier, and this protection is all the more important in a transmitter based on a VFO. The key would be inserted into the cathode line of the transmitter power stage, with the precautions mentioned earlier. Another switch could then be used to select between AM and CW, disconnecting the modulator when running CW.

RF Power Amplifiers

Most RF power amplifiers are built with the same configuration. The RF signal is fed to the control grid of one or more valves and is taken from the anode, from where it feeds the aerial through a pi-filter arrangement. What may differ is the valve(s) involved, and, of course, the power that feeds them. However, the desired power determines the dimensions of variable capacitors and coils, and rating of the components.

Included here is a sample power amplifier to illustrate the techniques used. This book simply gives an introduction to this complex subject. For much more information and higher power amplifiers, see *RF Design* by John Fielding, ZS5JF, available from the RSGB (www.rsgb.org).

The most complicated unit in a power transmitter is the power supply. The power valves need much higher voltages and currents than receiver or small transmitter applications, and bias voltages may need to be introduced. High voltage power supplies are beyond the scope of this book. For further information see the *Power Supply Handbook* by John Fielding, ZS5JF, also available from the RSGB (www.rsgb.org).

Also, stabilising high power amplifiers is not a simple task. In addition, a power stage must be screened from the rest of the transmitter, and RF chokes need to be introduced to prevent parasitic oscillations. On the plus side, a well built and well functioning power amplifier becomes the pride of the station.

Before you set about building a power stage for your station, make sure you know that it doesn't exceed the maximum power rating allowed in your country.

VALVES REVISITED

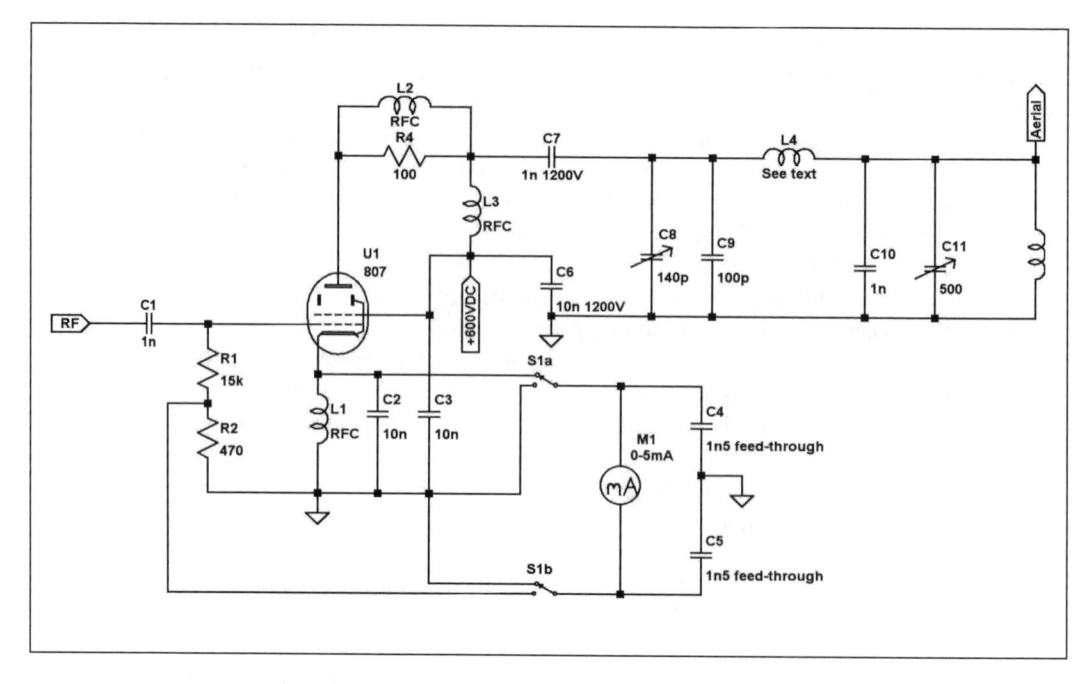

Fig 277: 75W power stage

75W RF Amplifier

It is human nature to always want more. Even if a 10W transmitter does the job well, many people express the opinion that a 100W transmitter would do a better job. Be that as it may, some additions are presented here.

Power amplifiers are frequently based on tetrodes or pentodes. The amplifier shown in **Fig 277** uses the classic beam tetrode 807.

As an alternative, you could use a 1625 which is a power pentode claimed to be capable of some 75W output.

The design does not significantly differ from that of the previously described transmitters. It requires a couple of watts drive, so the AM transmitter shown in Fig 275 could serve as a driver.

The signal is fed to the control grid of the valve and taken from its anode. A pi-filter output reduces harmonics. L4 is the tank coil and could be switchable for different bands, or be of a plug-in type. The heater voltage of an 807 is 6.3V.

The transmitter can cover the amateur bands from 80 to 10 metres, by using different coils.

19

Further Information on the Web

When searching for links on valve equipment, whether it be technical information or suppliers of valve related components, or collectors, you will find that there is a surprisingly large number of people interested in valve technology and valve radios.

The list is by no means complete, and the categorisation by no means perfect. However, if you find valve technology interesting, you are sure to find much to read and enjoy here.

At the time of writing, neither of the links was broken. No guarantee is given, though, for the listed link pages. The Internet is what it is, and that is something we all have to live with and accept.

There is a bigger interest in old radios and the history of radios than one might think. A Google search for 'radio history' yields 965 hits, and searching for 'old radio' returns 766 hits. 519 hits for 'valve radio', and 654 for 'tube radio'. It is, of course, an impossible task to collect all links in one listing.

Here are, however, some links that might be of interest to you. An attempt has been made to categorise them, but it isn't all that easy. Where do you put a link that points to a site where there is technical information as well as a picture gallery of collections, and restoration information? Enjoy!

History

http://earlyradiohistory.us/ *US site*
http://inventors.about.com/od/rstartinventions/a/radio.htm?rd=1 *General*
http://www.radiohistory.org/ *General*
http://didyouknow.org/history/radiohistory/ *General*
http://www.old-time.com/otrhx.html *Links*
http://www.qsl.net/n7jy/radiohst.html *General*
http://www.localhistory.scit.wlv.ac.uk/Museum/Engineering/Electronics/history/radiohistory.htm *General*
http://www.ac6v.com/history.htm *Amateur radio history*
http://www.mds975.co.uk/Content/ukradio.html *UK radio history*
http://www.radiolicence.org.uk/ UK *Radio license history*
http://www.radiolondon.co.uk/kneesflashes/stationprofile/hist.html *UK Radio stations history*
http://www.radiorewind.co.uk/ *BBC history*
http://en.wikipedia.org/wiki/Pirate_radio_in_the_United_Kingdom *UK Pirate radio history*

255

http://www.mds975.co.uk/Content/ukradio2.html *UK radio history*
http://www.dmoz.org/Arts/Radio/History/ *Link list*

Technical Information

http://www.john-a-harper.com/tubes201/ *How valves work*
http://ken-gilbert.com/vacuum-tube-faq *FAQ on valves*
http://www.milbert.com/tstxt.htm *Transistors vs valves*
http://www.crtsite.com/ *Valves in measurement instruments (CRTs)*
http://www.atatan.com/~s-ito/vacuum/vacuum.html *Some valves*
http://grantfidelity.com/site/tubebasics *Valves in audio applications*
http://www.r-type.org/static/basics.htm *The basic function of valves*
http://www.fortunecity.com/rivendell/xentar/1179/theory/vasfda/vasfda.html *Thermionic Valve Analogue Stages for Digital Audio*
http://www.thermionic.org/ *Devoted to electronic valves [also known as tubes], valve radio circuits and equipment*
http://www.archive.org/details/thermionicvalvei00flemrich *Download of a book by Fleming*
http://www.st-andrews.ac.uk/~www_pa/Scots_Guide/audio/part9/page1.html *Article on thermionic emission*
http://science.jrank.org/pages/49568/valve-thermionic.html *Saga of the Vacuum Tube, 70 Years of Radio Tubes and Valves*
http://www.cjseymour.plus.com/elec/valves/valves.htm *Introduction to Thermionic Valves*
http://www.fazano.pro.br/ing/indi05a.html *Early valves*
http://livinginthepast-audioweb.co.uk/vtheory/vtheory.htm *Theory of Operation, Types, and Basic Amplifier Circuits*
http://www.r-type.org/static/valves.htm *Aid to dating sets*
http://www.soundonsound.com/sos/jun04/articles/vickeary.htm *Article on current valve designs*
http://www.valvesntubes.co.uk/How-Valve-Works *Valves / Vacuum Tubes - a quick guide.*
http://www.oldatheart.co.uk/one-valve.html *Story about valve receivers*
http://www.amazon.co.uk/Build-Your-Own-Valve-Amplifiers/dp/0905705394 *Books on valves*
http://www.valveamplifiersound.co.uk/valve-info/valve-vacuum-tube-amplifiers-demystified-a-laymans-guide-part-1/ *On-line magazine*
http://www.physics.uq.edu.au/physics_museum/tour/electronics.html *Australian site about valves*
http://www.museumoftechnology.org.uk/stories/valves.html *A Brief History of the Vacuum Tube Valve*
http://www.marshallamps.com/resources/secret_life_of_valves/secret_life_valves.asp *Article: The Secret Life Of Valves*
http://www.vintage-radio.com/ *Old radios*
http://www.youtube.com/watch?v=ehbqYw1jXd0 *One valve radio (video)*
http://www.oldvalveradio.com/ *Old radio*
http://www.langrex.co.uk/sales01.html *Valve supplier*
http://www.siliconchip.com.au/cms/A_109837/article.html *Sells build instruc-*

tions of three valve radio
http://www.justradios.com/valve.html *Sells components, diagrams etc*
http://www.divdev.fsnet.co.uk/radxref.htm *Valves vs radios listing*
http://www.hanssummers.com/valve.html *Radio info*

Valve Datasheets and Equivalence List

http://www.nj7p.org/Tube.php *Valve database*
http://www.duncanamps.com/tdslpe/ *Valve database downloadable program*
http://www.classiccmp.org/rtellason/tubes.html *Data sheet collection*
http://www.radio-electronics.com/info/data/thermionic-valves/vacuum-tube-equivalents/tube-equivalents.php *Equivalent list*

Valve and Valve Radio Collections

http://www.oneillselectronicmuseum.com/page10.html *For collectors*
http://philsvalveradiosite.co.uk/philsvalveradiosite/default.htm *Collector and builder of valve radios*
http://vintageradio.me.uk/valve/valve1.htm *Collector*
http://www.tubecollector.org/ *Valve collectors*
http://www.tompolk.com/radios/radios.html *Collector*
http://www.gumbopages.com/chuck/radios.html *Collector*
http://www.radiophile.com/ *Collector*
http://www.radiotimeline.com/radios.htm *Collector*
http://www.archive.org/details/oldtimeradio *Radio programmes*
http://www.laud.no/la6nca/staaland/ *Collector WW2 radios*
http://www.stevenjohnson.com/collection.htm *Collector*
http://www.nebraskahistory.org/lib-arch/research/manuscripts/business/neb-radio.htm *Collector*
http://people.cs.uu.nl/gerard/FotoAlbum/RadioCorner/index.htm *Collector (sounds too)*
http://hem.passagen.se/cfn/ *Collector*
http://langaitis.zenonas-old.radios.fotopic.net/ *Collector*
http://www.mindspring.com/~cacutts/radio/radio.html *Collector*
http://www.jllacer.com/radio_collection.htm *Collector*

Boatanchors

In amateur radio and computing, boat anchor is a slang term used to describe something obsolete, useless, and cumbersome - so-called because metaphorically its only productive use is to be thrown into the water as a boat mooring. (Wikipedia)

http://www.boatanchor.com/ *Boatanchors*
http://oak.cats.ohiou.edu/~postr/bapix/ *Boatanchors*
http://www.ac6v.com/antique.htm *Boatanchors*
http://www.dealamerica.com/deal/cgi-bin/ads/bcads.cgi *Boatanchor advertised*
http://www.virhistory.com/ham/rrab.faq.htm *Boatanchor FAQ*

http://www.boatanchors.org/ *Boatanchors*
http://hub.webring.org/hub/ba *Boatanchor links*
http://www.mindspring.com/~cacutts/radio/ba/ba.html *Boatanchors*
http://www.ominous-valve.com/ba-mfrs.html *Boatanchor manufacturers' history*
http://s88932719.onlinehome.us/Boatanchors_Directory/ *Boatanchors directory*

Valve Suppliers

Sometimes it is necessary to look outside your own country for valves.

http://www.business.com/directory/industrial_goods_and_services/industrial_s upplies/pipes_and_valves/valves/valve_circuits/weblistings.asp *Links to suppliers*
http://www.radiotubesupply.com/ *Radio Tube Supplier*
http://www.vintage-radio.net/forum/showthread.php?t=61203 *Forum on valve radios*
http://everything2.com/title/tuning%2520eye%2520tubes *Magic eyes*
http://home.mira.net/~gnb/elec/valveaus.html *Australian site*
http://www.pcsandthings.com/Valves-vacuum-tubes-edinburgh.htm *UK supplier*
http://homepages.tesco.net/~D.G.Hewitt/valves.html *UK Supplier*
http://www.vac-amps.com/tubeprice.htm *US supplier*
http://www.vacuumtubesupply.com/otherresources/browse/30/Vacuum_Tube.html *Supplier links*
http://www.radiotubesupply.com/ *US supplier*
http://www.twenga.co.uk/dir-Garden-DIY,Electric-components,Pentode *UK supplier*
http://www.tubecollector.org/documents/ *UK supplier*
http://uk.rs-online.com/web/6784108.html *RS components and valves*
http://www.rapidonline.com/1/1/272-kt88-tetrode-valve.html *Components and valves*
http://kernowelectron.com/ *Audio valves, shop and info*
http://www.sequoia.co.uk/shop/browse.php?d=40 *Valve supplier*

Radio Restoration

http://www.valveradio.co.uk/ *Restoration of valve radios*
http://www.vintage-radio.com/ *Repair and restoration of valve radios*
http://thebakeliteradio.com/page102/page102.html *Describes restoration*
http://www.radiorestorations.htmlplanet.com/ *Restoration site*
http://www.radio-restoration.com/ *Restoration*
http://www.pasttimesradio.co.uk/resto/repairs_restoration.html *Restoration*
http://homepage.ntlworld.com/neil.fairley/ *Restoration*
http://resurrectionradio.street-directory.com.au/ *Restoration*
http://www.radio-workshop.co.uk/ *Restoration*
http://www.justradios.com/links.html *Link listing*
http://www.thevalvepage.com/index.shtml *Restoration and other stuff*

CHAPTER 19: FURTHER INFORMATION ON THE WEB

Sound Archives

http://www.jhepple.com/freestuff/streaming_media.htm#Radio_Comedy
http://www.oldtimeradiofans.com/
http://www.best-otr.com/
http://www.radiolovers.com/
http://www.downloadradioshows.com/
http://www.archive.org/details/oldtimeradio
http://www.rusc.com/
http://www.freeotrshows.com/
http://www.free-otr.com/
http://www.oldtimeradiofans.com/template.php?show_name=CBS%20Radio%20Mystery%20Theater
http://www.n8elq.com/ShowLists/Free%20Shows.html
http://www.old-time-radio-otr.com/ Some links
http://www.vintageradioshows.com/
http://www.otr.net/
http://www.otr.com/index.shtml
http://www.otrfan.com/

Misc

http://blog.makezine.com/archive/2008/01/make_your_own_vacuum_tube.html
Video of how to make a valve triode in your home

Power Supply Handbook
By John Fielding, ZS5JF

Have you ever wondered how your power supplies work? Have you ever wanted to build or modify a power supply but haven't had the confidence? Do you need a supply that is difficult to find or expensive to buy? The *Power Supply Handbook* answers all of these questions, and more. It provides all that is required to understand and make power supplies of various types. Packed with this and much more, *Power Supply Handbook* teaches the reader how to be confident with building, maintaining and using power supplies of all types. From the new home constructor looking for a straightforward guide through to those seeking a practical reference book, all will find this book a useful and a must for their bookshelf.

Size 240x174mm, 288 pages, ISBN 9781-9050-8621-4

ONLY £15.99

Circuit Overload
The bumper book of circuits for radio amateurs
By John Fielding, ZS5JF

This is the book that all keen home constructors have been waiting for! *Circuit Overload* includes 128 circuit diagrams, complemented by an additional 89 other diagrams, graphs and photographs and is a unique source of ideas for almost any circuit the radio amateur might want. Circuits are provided that cover wide range of topics. Chapters have been devoted to audio, metering & display, power supplies, test equipment, valves and antennas. If you are interested in home construction this book provides simple circuits and advice for the beginner with more complex circuits for the more experienced.

Size 240x174mm, 208 pages, ISBN 9781-9050-8620-7

ONLY £14.99

RF Design Basics
By John Fielding, ZS5JF

RF Design Basics is a practical guide to Radio Frequency (RF) design. Aimed at those who wish to design and build their own RF equipment, this book provides an introduction to the art and science of RF design. The chapters of *RF Design Basics* cover subjects such as tuned circuits, receiver design, oscillators, frequency multipliers, design of RF filters, impedance matching, the pi tank network, making RF measurements, and both solid-state and valve RF power amplifiers. One chapter explains the meaning of S parameters, while another is devoted to understanding the dual gate Mosfet. Much attention is given to the necessity of cooling valve PAs and there is even a practical design for water cooling a large linear amplifier. *RF Design Basics* neatly fills the gap between a beginner's 'introduction to radio' and RF design books.

Size 210x297mm, 192 pages, ISBN 9781-9050-8625-2

ONLY £17.99

RSGB shop

Radio Society of Great Britain
3 Abbey Court, Fraser Road, Priory Business Park, Bedford, MK44 3WH
Tel: 01234 832 700 Fax: 01234 831 496

www.rsgbshop.org

E&OE All prices shown plus p&p

Appendix: European Valve Designations

EUROPEAN VALVE TYPES follow a fairly logical system of designators. A type designator consists of at least two letters, which are followed by some digits. **Table 19** shows the system.

1st letter	Heater rating
A	.4V
B	.180mA AC/DC
C	.200mA AC/DC
D	.0.5 - 1.5V
E	.6.3V AC
F	.12.6V car battery
G	.5V
H	.150mA AC/DC
I	.20V
K	.2V DC
L	.450mA AC/DC
O	.Cold cathode
P	.300mA AC/DC
T	.7.4V; Misc.
U	.100mA AC/DC
V	.50mA AC/DC
X	.600mA AC/DC
Y	.450mA AC/DC

Remaining letters	Type(s) of device(s)
A	.Diode (excluding rectifiers)
AA	.Double diode with separate cathodes (excluding rectifiers)
B	.Double diode with common cathode (excluding rectifiers)
C	.Triode (excluding power triodes)
D	.Power triode
E	.Tetrode (excluding power valves)
F	.Pentode (excluding power valves)
H	.Hexode or heptode (of the hexode type)
K	.Octode or heptode (of the octode type)
L	.Output tetrode, beam tetrode or pentode
M	.Tuning indicator
N	.Gas filled triode or thyratron
P	.Secondary emission
Q	.Nonode
S	.TV sync oscillator
T	.Beam tube
W	.Gas rectifier
X	.Gas full wave rectifier

VALVES REVISITED

Y .... Half wave rectifier
Z .... Full wave rectifier

Note: A nonode is a valve with nine electrodes. One type is known: EQ80, which was used for FM detection purposes.

Numbers	Base type & serial number
1-10	Various side contacts, octals, specials (exceptions are ECH3G, ECH4G, EK2G, EL3G, KK2G which have octal bases)
11-19	8-pin German octal
20-29	Loctal B8G; some octal; some 8-way side contact (Exceptions are DAC21, DBC21, DCH21, DF21, DF22, DL21, DLL21, DM21 which have octal bases)
30-39	International octal
40-49	Rimlok B8A
50-59	B9G; loctal B8G; octal; 3-pin glass; disk-seal; German 10-pin with spigot; min. 4-pin; B26A; Magnoval B9D
60-69	B9G; some submins
70-79	Loctal Lorenz; wire submins
80-89	Noval B9A
90-99	B7G
100-109	B7G; Wermacht base; German PTT base
110-119	8-pin German octal; Rimlok B8A
130-139	Octal
150-159	German 10-pin with spigot; 10-pin glass with one big pin; Octal
160-169	Flat wire submins; 8-pin German octal
170-179	RFT 8-pin; RFT 11-pin all glass with one offset pin
180-189	Noval B9A
190-199	B7G
200-209	Decal B10B
230-239	Octal
270-279	RFT 11-pin all glass with one offset pin
280-289	Noval B9A
300-399	Octal
400-499	Rimlok B8A
500-529	Magnoval B9D; Novar
600-699	Flat wire-ended
700-799	Round wire-ended
800-899	Noval B9A
900-999	B7G
1000-	Round wire-ended; special nuvistor
2000-	Decal B10B
3000-	Octal
5000-	Magnoval B9D
8000-	Noval B9A

Table 19: Designations for European valves

As you can see, the designations were not entirely consistent. The American numbering system is entirely different. However, many American valves have a European replacement.

Several equivalent tables have been made over the years. A very good program that runs under Windows (and using *Wine* under Linux) is freely downloadable from www.duncanamps.com/tdslpe/. It is a database that contains thousands of valves, their replacements and links to data sheets. It is an invaluable source of information if you are interested in valves.

Index

1625 254
1N4148 . . . . .9, 10
300B 161
5687 . . . . .163, 165
5998 161
6146 248
6189 161
6BW6 . . . . .251, 252
6CG7 162
6CW4 5
6F6 248
6L6 248
6V6 248
6W6 248
807 .15, 254

Acorn valve 6
Aerial 142
Aerial tuning unit (ATU) 142
Alexanderson, Frederick Werner . .102
Amplification factor (mu or μ)
 . . . . .11, 12, 15, 17, 19, 20,
 . . . . . .25, 26, 27, 28, 29, 163, 197
 finding .32
Amplifier 67-92
 anode follower 76-77
 audio (AF) (see Audio amplifier)
 buffer 249, 250
 cascode 80-83, 150-151
 cathode coupled . .78, 143, 153-154
 class A 19, 88, 157
 class AB 90, 158
 class B 88, 89, 158
 class C 90-91
 class D .91
 classification of 87-91
 compound 78-87
 earliest 1
 grounded anode (see Cathode follower)
 grounded cathode . .68-69, 152-153
 grounded grid 69-71, 201
 in receivers 2
 IF (see Intermediate amplifier)
 in transmitter (see Transmitter)
 lunar grid (see Audio amplifier)
 operational (op-amps) 86-87
 push-pull 88-89, 158, 160-161
 radio frequency (RF)
  . . . . .2, 3, 13, 16, 17, 19, 20,
  . . .44, 67, 103-104, 133, 175, 240
 SRPP 83-85
 triode 12, 161
Anode 1, 12
 resistance (ra)
  . . . . .12, 15, 17, 20, 25-29, 32
 voltage
  . . . . .7, 8, 11, 12, 13,
  . . . . .14, 15, 16, 29, 191
Antenna (see Aerial)
Armstrong, Edwin Howard
 . . . . .95, 104, 106, 109, 194
Audio amplifier
 . . . . .3, 16, 17, 18, 19, 42, 45,
 . . . .67, 77-78, 102, 125, 135, 245
 assessing qualities of 166-171
 digital signal processing 147
 hi-fi 145-172
 lunar grid 77-78, 162-165
 power 154, 159-165
 preamplifier
  . . . . . .150, 155, 156, 162-163, 174
 RIAA compensation 148-150
 stereo .147
 tone controls 146, 167
 valve vs semiconductor . . .145-146
Audion .1, 10
Automatic Gain Control (AGC/AVC)
 . . . . .20, 23, 44, 45, 49-50,
 . . . . .112, 120, 125, 137, 244
 delayed 50
Automatic Volume Control (AVC)
 (see Automatic Gain Control)

Band switch 2, 173, 193
Bandwidth
 . . . . .12, 102, 118, 121, 135,
 . . . . . .141, 145, 152, 153, 154, 252
Battery power 3, 174, 180
 car radio 3, 12
Baudot, Jean-Maurice-Emile 98
Bias (see Grid)

Capacitance
 inter-electrode
  . . . . . .6, 11, 12-13, 15, 17, 20,30,
  . . .69, 71, 73, 77, 7, 79, 82, 84, 106
 measurement of (see Measurement)
 stray 147
Capacitor
 blocking 44, 222
 decoupling 162
 in oscillators 192, 201
 variable
  . . . . . .3, 44, 63, 174, 192, 201, 248
Cathode 1, 7, 8, 12, 16
 cold 1
 resistor 50
 varying the temperature 3
Cathode follower
 . . . . .67, 72-76, 105, 149,
 . . . . .152, 193, 196, 203, 206
Characteristics 25-40
 dynamic curves 36-40
 making your own curves 30
 using the curves 32-36
Clapp, James Kilton 197
Colpitts, Edwin Henry 195
Coil (see Inductor)
Construction 173-178
 chassis 176-178

of oscillators 191-193, 200
Coupling 41-52
   aerial (antenna) 143
   capacitive 60
   critical 59-60
   DC .45-46
   glow lamp 46-47
   link 60, 61, 193
   reactive 42
   resistor / capacitor 42, 47
   stray .147
   transformer 41, 43

DC70 5
de Forest, Lee 1, 10
Detector 127-134, 244
   anode bend 129, 134
   comparison 134
   diode 9, 127, 134
   frequency discriminator 96-98
   heterodyne 127-128, 131
   infinite impedance 130, 134
   leaky grid 102, 128, 134
   product (see Mixer)
   regenerative 134
   selecting an appropriate 134
   synchronous 106, 131-133, 134
   triode 12, 102, 128-129
Diode (see also Rectifier) 1, 8-10
   applications 9
   double 8, 24
   fact sheet 9
   mixer 116
   pentode as 17, 18
   semiconductor 8
   triple .24
   valve .8
   valve & semiconductor compared
  9-10
Diode-pentode 24
Digital sampling errors 146-147
Distortion 154, 158
   low 19, 152
Double-diode-pentode 24
Drake, Robert L 66

E88CC 38, 39, 130
EAA91 . . . . .8, 9
EB91 8
ECC81 (12AT7)
   . . . . .28, 32, 39, 82, 84,

. . . . .102, 129, 152, 206, 211
ECC82 . . . . .11, 68, 88, 152, 216
ECC83 . . . . .11, 83, 86, 152, 252
ECC86 12
ECF80 . . . . .216, 252
ECF82 214
ECF85 216
ECH35 .22
ECH81 . . . . .195, 216
ECL80 249
ECL82 . . . . . . .16, 104, 135, 156, 250
ECL86 155
Edison, Thomas Alva 3
EF80 . . . . .16, 25, 38, 212
EF83 . . . . .39, 40, 125
EF86 . . . . .39, 40, 219
EK90 .214
EL34 . . . . . . .16, 18, 157-158, 161, 187
EL81 . . . . .90, 91
EL84 . . . . .89, 90, 160
Electron 1, 3, 7, 8, 13, 14, 19, 40
   direction of flow 8

Facsimile (fax) signals 95, 98, 134
Fault finding 237-246
   the detector 244
   the IF stage 243
   the output amplifier 245
   the RF stage 240
   the mixer 241
   the oscillator 242
   using a signal generator 238
   using a signal tracer 238-240
Feedback
   . . . . .67-68, 102, 104, 136,
   . . . . . .152, 156, 159, 191, 194, 247
Filament (see heater)
Filter
   ceramic 125
   crystal 120-124, 135, 141
   for FM 124
   IF (see Intermediate frequency)
   keying 248
   L/ C 118-120, 141
   Mechanical 124
   pi (low pass) 91, 101, 183, 252
   power supply (see Power supply)
   RIAA 148-150
   wobbulator output 216-217
Fleming, Ambrose 1, 3-4
Foster D E 97

Fourier, Jean Baptiste Joseph 166
FP265 .19
FP285 .19
FP400 .19
Franklin, Charles Samuel 198

Further information 255-259

General Electric 2, 19
Gm (see Transconductance)
Goldstein, Eugene 3
Grid
   bias
   . . . . .3, 11, 19, 29, 47,
   . . . . .48-52, 91, 189, 194, 197
   control 11, 12, 16
   invention of 1, 10
   leak .51
   screen 13, 15
   suppressor 15-16

Harmonics 140, 166, 192
Hartley, Ralph V L 196
Hazeltine, Louis Alan 106
Hearing aid valves 5
Heater 1, 2
   direct 1, 7
   indirect 7
   supply 179
   varied as volume control 3
   voltage 29, 162, 206
Henry, Joseph 53
Heptode 20-21
   fact sheet 20
Hexode 22
Hi-fi amplifiers (see Audio amplifiers)
Hittorf, Johann Wilhelm 3
Hum 8, 162, 180, 192

Image rejection 140
Impedance 58-62
   aerial (antenna) 142, 247
   amplifier
   . . . . .69, 71, 73, 77, 7,
   . . . . .79, 82, 84, 148, 221
   and Miller effect 13
   load .29
   matching amplifier 79, 143
   surrounding an oscillator 192
   valve input 29
   valve output 13, 16, 17, 29

Inductance
    measuring (see Measurement)
    mutual .59
Inductor (see also Tuned circuits) 53-58
    for oscillators (see Oscillators)
    honeycomb 58
    plug-in 55, 173, 254
    pot core 55
    skin effect 54, 56
    spider 57
    toroidal 56
    variometer 56-57
    winding 53
    wire 54-55
Intermediate frequency (IF) . . .63, 243
    alignment of 211
    amplifier
         . . . . .17, 20, 59, 67, 107,
         . . . . .119, 125-126, 174, 175
    choice of
         . . . . . .108, 109, 111, 140, 141-142
    filter 118-125
    transformer 43, 175
Instability (see also Feedback)
    in audio amplifiers 156, 162
    in receivers 2, 106
    in transmitters 11, 15

Loudspeaker
    . . . . .3, 16, 17, 102, 104,
    . . . . .146, 154, 155, 156, 167

Magic eye 137
    tuning indicator 137-138
Matched pairs 158
Measurement 3, 221-236
    AC and RF voltages 232-233
    amplifier impedances 221-222
    capacitance 223-224, 233
    current 235
    frequency 227-230
    grid-dip oscillator 227-230
    inductance 224, 235
    Q 224
    resistance 235
    voltages 230-232
    with signal generator (see Signal
         generator)
Miller effect . . .12, 149, 152, 153, 163
Miniature valves 4
Mixer (see also Modulator)
    . . .16, 105, 107, 140, 175, 214, 241

balanced / product detector 115
conversion gain 214
diode .116
heptode 20, 214
hexode 22
in superhets 111-117
octode 22, 112
pentode 17, 20, 113, 214-216
triode 17, 114, 115
Modulation 93-100
    amplitude (AM)
         . . . . .93-95, 105, 127, 135, 250
    digital 98, 145
    double sideband (DSB) 95
    FM vs AM 98
    frequency (FM) 95-99
    frequency shift keying (FSK)
         . . . . .95, 98-99
    independent sideband (ISB)
    phase shift keying (PSK) 99
    pulse 99-100
    single sideband (SSB) 95, 109,
         . . . . .116, 117, 119, 135, 250
Modulator
    . . . .20, 116, 204-205, 213, 252-253
Morse code
    transmissions (CW)
         . . . . . .95, 109, 117, 134, 135, 247
Morse, Samuel Finlay Breese 95
Mu (m) (see Amplification factor)

Neutralisation (see Instability)
Nixie valve 6
Noise
    grid 40
    in receivers 2
    in valves . .11, 12, 15, 16, 17, 20, 40
    shot 40
Nuvistor 5

Octode .22
    curves 23
    oscillator / mixer 112
Oscillator 22, 191-201, 242
    Armstrong 191, 194-195
    audio (AF) 205
    beat frequency (BFO / CIO)
         . . . . .117, 134-135
    bistable multivibrator 198
    building 191-193
    Butler 191
    cathode coupled

 . . . . .191, 199-201, 206, 228
Clapp .197
Colpitts 191, 195, 196, 229
crystal controlled 108, 193
earliest .1
electron coupled (ECO) 23, 191
Franklin 198
grid-dip (see Measurement)
Hartley 191, 196-197, 229
heptode 20
inductors 58, 192
in signal generator 203
in superhet 107
in transmitter (see Transmitter)
octode 112
R / C .205
sawtooth (Miller integrator)
    . . . . .16, 212-213, 219-220
Seiler .196
spectral purity (see also Harmonics)
    . . . . .193-194, 197, 201
stability . . .191, 194, 197, 199, 201
Tesla .197
transitron 212
Vackar 197
Oscilloscope
    . . . . .5, 16, 167, 193-194, 211, 217
Overtones (see Harmonics)

Parameters 29-31, 163
Pentode 15-20, 51
    amplification 16
    connections 16, 17
    fact sheet 17
    in sawtooth generator 212
    miniature 5
    power 15, 16, 17
    UHF 6
Pentode-indicator 24
Phase inverter 88-89, 158-159
Phase shift
    . . .69, 71, 73, 77, 7, 79, 82, 84, 115
Philips 18
Power supply 174-175, 179-189
    battery (see Battery power)
    bleed resistor 181
    dual 47, 162
    fault finding 238
    filter 180, 183
    for lunar grid amplifier 165
    for test equipment 189
    for transmitter 253

lab 179
negative voltage`` 181
stabilised 183-189
unregulated 181
voltage regulation 180-181
Product detector (see Mixer)

Q factor 17, 53, 54, 55, 56, 58-59,
. . . . . . .91, 120, 141, 191, 192, 201
measuring (see Measurement)
Q multiplier 135-137

Radio teletype (RTTY)
. . . . .95, 109, 117, 134
Reactance valve 212, 213
Receiver 101-110
  choosing 139
  crystal set 101, 139
  communication
  . . . . .16, 111, 120, 135, 156
  Collins 51-S1 . . . . .17, 66, 114, 142
  designing 139-144
  direct conversion 109, 140
  Drake R4a 66, 142
  earliest 2
  frequency considerations 140
  Front end 142
  General Electric KL-500 2
  Heathkit .66
  homodyne 104-105, 109
  neutrodyne 106
  pentodes in 16
  Racal RA117 111
  reaction control 2
  reflex 2, 109-110
  selectivity 2, 102, 103, 125
  sensitivity 2, 101, 102, 103-104
  stenode 110
  superheterodyne (see Superhet)
  super-regenerative 109
  synchrodyne 105-106, 109
  tuned radio frequency (TRF)
  . . . . .56, 102-104, 139-136
Rectifier 9, 179-180
  bridge 181, 183
  dual valve 181, 182
  earliest 1
  metal .181
  semiconductor 181, 182
RENS1204 14

Robinson, James 110

Safety 179, 237
  RF power 247
Schottky, Walter Hermann 13
Secondary emission 13, 22
Seeley, Stuart William 97
Seiler, E O 196
Selectivity 17
Signal generator 167, 203-220
  adding functions 209
  clipper stage 208, 209-211
  complete 206-208
  modulating 204
  sawtooth generator (see Oscillator)
  sine generator 205-206
  wobbulator 211-220
Signal tracer 238-240
Space charge 7, 17, 19, 20
Superhet receiver
  . . . . .104, 106-109, 139, 237
  double conversion 111, 132
  frequency converter 111-117
  IF (see Intermediate frequency)
  in detail 111-138
  triple conversion 111, 142
  tunable IF 108

Tellegan, Bernard D H 15
Tesla, Nikola 3, 197
Tetrode 13-15, 51
  double .24
  fact sheet 15
  miniature 5
  pentode as 17, 19
Tetrode-pentode 24
Transconductance (Gm)
  . . . . .25, 26, 27, 28, 29, 163
  finding .31
Transformer
  audio output 156, 162, 163, 175
  coupling (see Coupling)
  IF (see Intermediate Frequency)
  mains supply 179
Transmitter 5, 247-254
  amplitude modulated (AM) 250
  crystal controlled 248, 249
  CW (Morse) 247
  keying filter (see Filter)
  licensing 247

oscillator 247
power amplifier 247, 252-253
single valve 248, 249
spark .101
tuning 249, 250
two-stage 249
voice .250
Transmitting valves 6, 15
Triode 1, 10-12
  double 11, 24, 102, 114
  fact sheet 12
  miniature 5
  mixer .114
  pentode as 17, 18, 19
  UHF 6
Triode-double-diode 24
Triode-hexode 22
Triode-pentode 16, 24
Triode-triple-diode 24
Tuned circuit 53-66, 107
  impedance 58-59
  in oscillators 191
  in receivers 2, 44
  Q factor (see Q factor)
  reading .61
  selectivity (see Selectivity)
  writing 61-62
Tuning 62-66
  capacitive 63
  permeability 64-66
  tracking 63

Valve designations, European .261-262
Vackar .197
VHF/UHF
  valves for 6
Volume control 2, 3, 157
  automatic (AVC) (see Automatic
  Gain Control)
VT51 162
Vacuum tube voltmeter (VTVM)
  . . . . .222, 230-236, 237
  enhancements 232-236
  RF probe 232
  VTVM versus modern or universal
  meter
 230

Wheeler, Harold Alden 49
Wobbulator (see Signal generator)